Environmental Impact Assesssment and Management

–Editiors–

B.B. Hosetti & A. Kumar

2014

Daya Publishing House®

A Division of

Astral International Pvt Ltd

New Delhi-110 002

© 2014 EDITORS, BASALING BASAPPA HOSETTI (b. 1957-)
A. Kumar (b. 1957-)
First Impression, 1998
Reprinted, 2014
ISBN 9789351240266

Published by	:	**Daya Publishing House®**
		A Division of
		Astral International Pvt. Ltd.
		– ISO 9001:2008 Certified Company –
		4760-61/23, Ansari Road, Darya Ganj
		New Delhi-110 002
		Ph. 011-43549197, 23278134
		E-mail: info@astralint.com
		Website: www.astralint.com
Laser Typesetting	:	**Classic Computer Services**
		Delhi - 110 035
Printed at	:	**Thomson Press India Limited**

PRINTED IN INDIA

DEDICATED TO :
GANDHIAN,
PHILOSOPHER,
SCIENTIST,
AND
A
HUMANITARIAN

—Late Prof (Dr.) H. Sanjeeviah

ACKNOWLEDGEMENTS

The idea of editing a book on Environmental Impact Assessment was kindled in my mind while I was participating in a National Symposium at Hyderabad during November 1995. The initial thought was to collate all the important case studies related to EIA of India and write a book, but due to the non-availability of sufficient information regarding the various case studies, the scope of the book was also re-thought to include a few further important aspects of Environmental Management.

I would like to place on record the cooperation, permission encouragement and suggestions offered by various persons including : Dr Afroz Ahmed, Director, Narmada Control Authority, Government of India; Dr. H.S. Patil, Karnatak University, Dharwad ; Dr Ashutosh Gautam, India Glycols Ltd. Kashipur; Dr S. Gopal, Vice Chancellor, Mangalore University, Mangalore; Dr G.S. Puranaik, Professor, Kuvempu University; Dr. Murugesan, Sri Param Kalyani Center for Environmental Sciences, Alwakurichi; Dr Jayaram Hegde, Chairman, Department of Biosciences, Mangalore University; Dr Frost, Environmental Resources Department, University of Salford, Salford, UK; Shri Pattabhirama Somayaji; Shri Ravishankar, Department of English, Mangalore University.

Most important is also the assistance provided by Mr Mahabal Bhatt, Mr Mahesh, Mr Satish and Ms Bhaghya Hosetti during the compilation of the entire manuscript.

EDITORS

FOREWORD

In June 1992 at Rio de generio in Brazil, about 25000 people gathered for the worlds largest Conference. It was also widely called as Earth Summit, because a large number of heads of Nations and Government leaders assembled to discuss the health of the earth. This has become an important milestone in the field of environment and environmental considerations and were incorporated into the political arena. This meet also signifies the relationship between science and politics.

It is in this urgent need for a multidisciplinary approach that the book Environmental Impact Assessment and Management is contributed by the editors. This book has a broad coverage on matters related to the environment and has an integrated style which I hope will find importance in spreading the knowledge and in training Environmental Managers and Scientists. This book will also help in generating interest among the students of Environmental Biology and related subjects.

Dr. M. Abdul Rahiman
Vice Chancellor,
Malabar University,
Kannur-670001,
Kerala State,
India

PREFACE

This book entitled "Environmental Impact Assessment and Management," embodies 15 chapters covering a wide spectrum. Chapter 1 contains details of EIA, Basic Elements and Methodologies. Further, important guidelines to be followed before establishing industries, water related projects and transport related projects are narrated in Chapter 2. Important case studies of EIA viz, second stage expansion of Malpe fishing harbour, Human impact on Himalayan Ecosystem, Konkan Railway Project, Heavy metal pollution in the Coovum river, an impact prediction for an irrigation project and impact of cement making on the air quality of Coimbatore, are dealt in Chapter 3.

Once a project comes into existence, it results in side effects on the immediate environment which are evident on the consecutive 5-10 years of operation. Therefore, its efficiency and pollution abatement systems should be checked and corrected in a manner similar to financial auditing and this aspect has been incorporated in Chapter 4. Now-a -days due to the rapid process of urbanization, all the cities of India have accumulated a lot of filth leading to ghasty smell, loss of property value and the spread of diseases. In order to have a clean environment, we have to have a systematic refuse management package for Indian conditions. As illustrations of this view, the urban solid waste management in Hyderabad, Shimoga and Rourkee cities have been explained in the Chapters 5 and 6 respectively.

Though the Coastal Zone Regulation Act 1992 (CZRA) has been in force in India, its strict application and executions is not to be found in practice. Densly populated coastal areas are suffering from various manmade degredation problems. There management packages through the consideration on Mangalore coast as a model zone are discussed in Chapter 7.

Deforestation is one among the four important causes for environmental degradation. This aspect is discussed critically in Chapter 8. An example to explain the impact of human population on the depletion of forest cover around Bhadravati town is also incorporated as Chapter 9.

BIMARU is the term representing the problems of population explosion and density and the associated societal, environmental problems in Bihar, Madhya Pradesh, Rajasthan and Uttar Pradesh states. The problems cropped up due to coal mining and its impact on the life style of fish farmers in and around Chitra coal mines, Chapter 9 is a case study of this kind.

Wildlife management is another burning problem of the environmental sciences, all over the world. As a demonstration of the nature of the problem the status of Indian wildlife in terms of endangered flora and fauna, and protected species of India, and the termite infestation problems in the foot hills of Western Ghats near Shimoga are discussed in Chapter 10 and 11 respectively.

According to Kudesia (1980) almost all rivers of India are polluted and transformed into industrial and domestic sewers at least near the cities and townships. About 90 per cent of water pollution is caused by domestic sewage and the rest is by Industrial discharges. The conventional abatement methods like activated sludge plants and trickling filters are costly. In many of the places in India liquid waste is directly flooded to low lying areas which in course of time become waste wetland. It is interesting to note that the hydrophytes growing in the wetlands produce some exudates which kill pathogenic bacteria and the hydrophytic plants also absorb nutrients and reduce BOD of the sewage leading to the purification of sewage. Such a case of aquatic plants treating wastewaters of Mangalore city is discussed in Chapter 11.

Chapters 12, 13 and 14 incorporate aspects of potable water budgets of Daksina Kannada, use of remote sensing in the EIA evaluation and primafacie baseline studies for establishment of nuclear power plant at Kaiga respectively. The last important topic of the book is about Food preservation by ionising radiation techniques and the possible economy of food and energy.

CONTENTS

LIST OF CONTRIBUTORS

1. Dr. M.D. Alur, Scientific Officer, Food Technology Division, Bhaba Atomic Research Centre, Trombay, Mumbai-400085.
2. Dr. Arvind Kumar, Reader, Post Graduate Department of Zoology, S.K. University, Dumka-814101, Bihar, India.
3. Dr. Ashutosh Gautam, Assistant Manager, India Glycols Ltd, A-1, Industrial Area, Bazpur Road, Kashipur-244713, UP.
4. Dr. K. Ayyadurai, Associate Professor, Institute of Food and Dairy Technology, Tamil Nadu Veterinary and Animal Sciences University, Koduvalli, Madras-600052.
5. Prof A. S. Chandrashekar, Principal, Sahyadri Science College, Vidyanagar, Shimoga–577201, Karnatak.
6. Dr. N.C.L.N. Charyulu, Professor, Department of Chemical Engineering, Karnatak Regional Engineering College, Surathkal-574157, District Mangalore,.
7. Dr S.N. Doke Scientist, Food Technology Division, BARC, Trombay, Mumbai-40085.
8. Dr B.B.Hosetti, Hydrobiology Division, Post Graduate Department of Biosciences, Mangalore University, Mangalagangothri -574199.
9. Dr. C.V.R. Indira, Professor, Department of Life Sciences, Avinashilingam Institute for Home Sciences and Higher Education (Deemed University) Coimbatore-641043, Tamil Nadu.
10. Dr. K. Jayappa, Post Graduate Department of Marine Geology, Mangalore University, Mangalagangothri-574199, Mangalore.
11. Dr. K. Jeevan Rao, Assistant Professor, College of Agriculture, Andhra Pradesh Agricultural University, Rajendra Nagar, Hyderabad-500 030.

12. Dr. R. Kadirvel, Department of Animal Nutrition, Madras Veterinary College, Madras-600007.
13. Dr. V. Krishnasamy, Department of Chemistry, Anna University, Madras-600025.
14. Mrs. P.Y. Latha, P.G. Department of Environmental Sciences, Kuvempu University, B.R., Project-577115.
15. Dr. B.R. Manjunatha, Post Graduate Department of Marine Geology, Mangalore University, Mangalagangothri-574199.
16. Mr. B.S. Mohan, Research Fellow, Department of Biosciences, Mangalore University, Mangalore-574199.
17. Mr. Naveed A., Department of Environmental Sciences, Kuvempu University, B.R. Project-577115.
18. Dr H.S. Patil, Professor, Water Pollution Research Laboratory, Zoology Department, Karnatak University, Dharwad-580003.
19. Dr. E.T. Puttiah, Professor, Department of Environmental Sciences, Kuvempu University, B.R. Project-577115, District Shimoga.
20. Dr. Rajammal Thirumalnesan, P.G. Department of Life Sciences, Avinashilingam Institute of Home Sciences and Higher Education, Coimbatore-641043.
21. Mr. R.S. Sabhahita, Civil Engineering Department, J.N.N. Engineering College, Navile, Shimoga.
22. Dr. H.M. Somashekarappa, Technical Officer, University Science Instrumentation Centre, Mangalore University, Mangalgangothri-574199.
23. Dr. G. Srinikethan, Department of Chemical Engineering, Karnatak Regional Engineering College, Surathkal-574157.
24. Dr. Swayam Prakash Rout, Department of Chemistry, Utkal University, Bhubaneswar-751004, Orissa.
25. Dr. C. Swaminathan, Department of Chemistry, Anna University, Madras-600025.
26. Ms N.C. Tharavathy, Research Fellow, Department of Biosciences, Mangalore University, Mangalore-574199.
27. Dr. S.G. Torne, Department of Botany, Smt Parwatibai Chowgle College, Margao, Goa.
28. Ms. M. Umamaheshwari Research Fellow, Department of Life Sciences, Avinashilingam Institute of Home Sciences and Higher Education, Coimbatore-641043, Tamil Nadu.

29. Dr. K. Veena Nadig, ICP Division, Central Food Technology and Research Institute, Mysore-570013.
30. Dr. A.V. Veeresh, Department of Botany, Smt. Parawatibai Chowgle College, Margao, Goa.
31. Dr. D. Venkat Reddy, Division of Geology, Department of Civil Engineering, KREC, Surathkal-574157, District Mangalore.

LIST OF TABLES

LIST OF FIGURES

LIST OF SYMBOLS/UNITS

myr^{-1}	Per million year
mgd^{-1}	Million gallons per day
$mgnm^{-3}$	Milligrams per nano cubic meter
$mgkg^{-1}$	Milligrams per kg
mgm^{-3}	Milligrams per cubic meter
kmh^{-1}	Kilometer per hour
gd^{-1}	Grams per day.
K Cal	Kilo calorie
Mha	Million hactare
Km^{-2}	Per square kilometer
$mg\ l^{-1}$	Milligrams/litre
NH_3N	Ammonia Nitrogen.
lm^{-2}	Litre per square meter
Ma	Million annum
Mm^3	Million cubic meter
KGy	Kilogray
MeV	Million electron volts
PPm	Parts per million
mm	Millimeter
TMC	Trillian meter cube

LIST OF ABBREVIATIONS

AAEAM	-	Advanced Approach to Environmental Assessment and Management
APHA	-	American Public Health Association
AWWA	-	American Water Works Association
BDL	-	Below detectable level
BFDS	-	Blast Furance Dust and Sludge
BHC	-	Benzene Hexa Chloride
BIMARU	-	Bihar Madhya Pradesh Rajasthan and Uttar Pradesh
BOD	-	Biochemical/Biological Oxygen Demand
BWFDA	-	Brakish Water Fish Developmental Agency
CZMA	-	Coastal Zone Management Authority
CFTRI	-	Central Food Technology and Research Institute
CITES	-	Convention on International Trade on Endangered Species
COD	-	Chemical Oxygen Demand
CPP	-	Captive Power Plant
CZRA	-	Coastal Zone Regulation Act
CWPCB	-	Central Water Pollution Control Board
CZCS	-	Coastal Zone Colour Scanner
DMG	-	Department of Mines and Geology
DO	-	Dissolved Oxygen
DOEN	-	Department of Environment
EIA	-	Environmental Impact Assessment
EIS	-	Environmental Impact Statement
EMP	-	Environmental Management Plan
EPA	-	Environmental Protection Agency
ESP	-	Electrostatic Precipitator
FAO	-	Food and Agricultural Organization
FCA	-	Forest Conservation Act

FF	-	Fabric Filter
GTF	-	Global Tiger Forum
IBWL	-	Indian Board for Wild Life
ICBP	-	International Council for Bird Preservation
IEE	-	Initial Environmental Examination IEE
IRS	-	Indian Remote Sensing Satellite
IUCN	-	International Union for Conservation of Nature
KIE	-	Kalunga Industrial Estate
KPC	-	Karnatak Power Corporation
KRC	-	Konkan Railway Corporation
KRP	-	Konkan Railway Project
KRRAC	-	Konkan Railway Realignment Committee
KSD	-	Kharif Semi Dry
KSPCB	-	Karnatak State Pollution Control Board
LACIPE	-	Large Area Crop Inventory Programme Experiment
LD	-	Lime Dolomite
MCF	-	Mangalore Chemicals and Fertilizers
MCH	-	Municipal Corporation of Hyderabad
MOR	-	Ministry of Railways
MPM	-	Mysore Paper Mills
MZEE	-	Mandovi Zuari Estuarine Ecosystem
NCAER	-	National Council of Applied Economic Research
NEPA	-	National Environmental Policy Act
NH47	-	National Highway 47
NIUA	-	National Institute of Urban Affairs
NOAA	-	National Ocean Administration Agency
NPDES	-	National Pollutant Discharge Elimination System
NRSA	-	National Remote Sensing Agency
OSHA	-	Occupational Safety and Health Administration
RCC	-	Reinforced Cement Concrete
RCFL	-	Rastriya Chemical and Fertilizers Limited
RITES	-	Rail India Technical and Economic Services
RSP	-	Rourkela Steel Plant
SMS	-	Steel Melting Slag
SPM	-	Suspended Particulate Matter

TDS	-	Total Dissolved Solids
TOC	-	Total Organic Carbon
TPD	-	Tonnes Per Day
UNWCED	-	United Nations World Commission of Environment and Development
USWM	-	Urban Solid Waste Management
UTP	-	Upper Tunga Project
VISL	-	Visweswarayya Iron and Steel Limited
WFP	-	World Food Programme
WHO	-	World Health Organization
WLF	-	Wild Life Fund
WPCF	-	Water Pollution Control Federation
WRDO	-	Water Resource Development Organisation
WWLF	-	World Wild Life Fund

CHAPTER 1

ENVIRONMENTAL IMPACT ASSESSMENT : GENERAL INTRODUCTION, BASIC ELEMENTS AND METHODS

B.B. Hosetti and G. Srinikethan

GENERAL ASPECTS

Introduction

Studies on academic literature on Environmental Impact Assessment (EIA) and a random selection of EIA reveals that an impact is an event which results as the direct consequence of a prior event. Many methods for EIA have been devised to decipher the identification, prediction and assessment of impacts and preparation of Environmental Impact Statements (EIS). All these methods focus on impacts. Impact is the word used in EIA, but little attention has been given in academic and professional literature to the nature of impacts. What is an impact ? How do we know which impact is likely to occur ? These are the questions to be answered in this chapter. Greater attention will be given to socio-economic characters of the impacts and other aspects.

Aim of an EIA

EIA is a planning aid and is concerned with identifying and assuring impacts arising from proposed activities such as policies, programs, plans and developmental projects, which may affect the environment. The main aim of EIA is to improve decisions on development by increasing the quality and scope of the

(1)

information on likely impacts presented to decision makers and members of public groups. Although concerned with different levels of activities EIA are implemented mostly for major developmental projects such as highways, Nuclear power stations, water related projects and rehabitation. Now-a-days the importance of policies and plans are increasingly recognized, but are proving difficult to implement them because of institutional, political and technical problems. Consequently most EIAs are undertaken for major projects in which a number of alternatives are assessed. EIS are produced to a uniform standard in accordance with established mechanisms and procedures.

Environmental Impact Assessment

EIA or EIS is a term which originated in the USA after the Federal Government passed National Environmental Policy Act (NEPA) in 1970. NEPA made it compulsory that all the development projects in the country should undergo environmental scrutiny before execution. Since then the practice of the preparation of EIA/EIS and Environmental Auditing is carried out in several countries. In India EIA programs run under the guidance of the Ministry of Environment. In simple terms the EIA is an exercise to be carried out before any project is undertaken and any major activity or plan is executed. EIA is carried out to ensure that the action will not harm the environment in any way on a short or long term time scale. Scientifically describing, EIA is an activity designed to identify and predict the impacts on biophysical environment, human health and the associated things legislative proposals, policies, programs, projects and operational procedures and the interpretation and communication by information about the impact are included in the EIA (Munn, 1979). It is thus an assessment of the consequences of any decision on the quality of the total environment on which man largely depends for his well being.

The environment means the whole complex of physical, social, cultural, economic and aesthetic factors which affect individuals and communities and ultimately determine their character, forms of relationships and survival. The objective of EIA is to see and check that the environment is not disturbed during executing the projects, like industries and nuclear power projects. It also aims to find out the consequences on the local environment after the completion of the project and finally it aims to study the impacts after few or several years to know the possible occurrence of long term effects.

Types of cases

1. EIA of Air port construction
2. EIA of Railway lines
3. EIA of complex buildings
4. EIA of Industrial plants
5. EIA of Mining activities
6. EIA of construction of dams
7. EIA of new schemes like insect control, flood control, establishment of National parks etc.

Organizations Responsible for EIA

An EIA may be conducted by the project authorities, Environmental engineers, Environmental managers or some professional bodies involved in EIA analysis. In India major Environmental projects are referred to the Department of Environment (DOEN) which functions under the Ministry of Environment, Forestry and Wild Life. However the final responsibility of granting, or okaying the project lies in the hands of the decision makers. For example in the case of Tehri dam in India the EIA was conducted by the Department of Environment and the ultimate decision has to be taken by the Ministry of Energy.

Similarly a Chief Minister of state has the power of sanctioning the project although it has to be referred to the central Government for approval. According to Munn (1979) the decisions are actually shaped and not taken. Thus environmental considerations should be included throughout the planning process. In other words, a person's judgement is as good as his knowledge.

Contents of EIA

An EIA is conducted only when primafacie it appears that the project will affect the environment. It also contains the study of impact during the period after the installation of projects in order to observe whether particular conditions change or not. An EIA document could be outlined as follows.

1. Description of the proposal of the project in details. For example, in case of sugar factories, it includes, sources of raw materials, irrigation sources, availability of water and sugarcane, power requirements, type of processes in the factory, constructions involved, effluents, solid and gasesous discharges produced and their quality and quantity, fuels used, man power required and some alternatives.

2. It includes description about the nature and magnitude of possible environmental changes in water, air and soil and human style qualities, possible spread of diseases and socio-economic impacts etc.

3. It comprises in detail the criteria followed by EIA people in assigning importance in terms of cost benefits ratio.

4. It clearly says whether the project should be accepted or rejected or modified after considering the alternatives.

5. It also includes the provision for future post audit to be conducted.

Prediction of Changes Due to the Project in the Environment

The following paragraph discuss basic steps for the prediction and assessment of impacts on air, water, soil, noise, biological, cultural and socio-economic environment.

(A) Air Quality

(i) Identification of pollutants emission.

(ii) Description of the present air quality level and changes in the recent years. Table 1 presents the typical environmental air quality parameters.

(iii) Meteorological data summarization - windroses representing direction and speed of wind, max. & min. temperature, rain fall and humidity.

(iv) Estimation of ground level concentration of recognized polluting components.

(v) Mitigation plan in the event of adverse effects due to proposed activity.

(B) Water Quality

(i) Identification of water pollutants through effluent during construction and operational stages

(ii) Description of present surface water quality with frequency of variation and historical trends.

(iii) Description of the present ground water quality.

(iv) Meteorological data (acquisition) of temperature, humidity, evaporation and precipitation

(v) Estimation of bacterial waste loads.

(iv) Mitigation methods.

(C) Noise

Undesired sound or sound in the wrong place at the wrong time as decided by the majority at that instance is termed as noise. Sound is an energy form originated from a vibrating surface transmitted in cyclic form by a series of compression and refraction of molecules of the medium (gas or liquid or solid) through which it passes.

(i) Identification of noise levels and noise sources.

(ii) Description of the present level.

(iii) Prediction of the noise level by the proposed activity.

(iv) Mitigation methods like reduction of noise at the source and the installing of noise absorbing systems.

(D) Biological Setting

(i) Description of the flora and fauna-community and species.

(ii) Identification of rare and endangered species and their relevant characteristics.

(iii) Past and present practices adopted for maintaining the flora and fauna with special reference to the protected species, if any.

(iv) Prediction of impact on the biological setting.

(v) Summarization of critical impacts associated with all alternatives considered.

(E) Cultural

(i) Identification of historical and archaeological sites; ecological, scientifically and geographically significant and ethnic areas; potential cultural resources with their significance to local, regional and national context.

(ii) Determination of possible impacts due to the different alternatives on the above resources during different phases of the project (i.e. reconstruction, construction, operation and post operation).

(iii) Development of mitigation methods for preservation of the culture.

(F) Socio-economic

(i) Description of socio-economic setting of the region such as population data base - sex, age, education, employment, profession, family size, language, religion etc.; health and social services; life styles; transportation facility; housing; land holding its usage and value etc.

(ii) Identification of critical social factors which are either marginal or less than the desired average standards of living.

(iii) Prediction of likely changes in the socio-economic patterns with and without the proposed project.

(iv) Identification of factors which may change from satisfactory level to below the standard.

(v) Mitigation methods to alleviate the likely hardships.

Factors to be Selected

The impact of a project can be depicted only through certain parameters. It is the most important aspect of an impact assessment to identify such parameters. It is natural to rely upon the quality of air, water, soil and the total environment for most of the projects. The selection of parameters is also related to the specific nature of the project.

General Parameters

1. Crop productivity, air quality, water quality, aquatic resources, nutrient status of the water, drinking water quality, availability of agricultural land, soil quality, texture, and geological aspects. wild life, endangered species, natural vegetation, grazing natural drainage, recreation, open space, visual disruption, etc. Selection of parameters will be done by the EIA team members comprising of following personnel.

1. Administrators
2. Biologists
3. Local leaders
4. Geologists
5. Journalists
6. Sociologists
7. Economists
8. Engineers.

EIA in India

In India the Department of Environment (DOEN) which was started in 1980 has initiated such procedures. The Department is now converted into a full fledged Ministry of Environment, Forests and Wildlife and is responsible for decision making on environmental projects. Whenever a project is proposed and has unfavourable effects then the Ministry constitutes an adhoc committee or

Table 1 : Typical list of Environmental Parameters of Air and Water, normally monitored with the method/instructions used

S.No.	Parameter	Method/Instruments used
I.	Air Analysis	
1.	Suspended particulate matter	High volume sampler
		Gravimetric method
2.	Oxides of Nitrogen	High volume sampler
3.	Sulphur dioxide	High volume sampler
4.	Hydrocarbon	G.C.
5.	Methane	G.C.

Contd...

S.No.	Parameter	Method/Instruments used
II.	**Water Analysis**	
1.	Temperature	Thermometer
2.	pH	pH meter
3.	Turbidity	Nephelometer
4.	Conductivity	Conductometer
5.	Total Dissolved solids	Gravimetric
6.	Hardness as $CaCO_3$	Chemical
7.	Calcium	Flame photometer
8.	Magnesium	A.A.S.
9.	Dissolved oxygen	Chemical
10.	Permanganate value	Chemical
11.	C.O.D.	Chemical
12.	CO_2	Chemical
13.	Acidity as $CaCO_3$	Chemical
14.	Phenolptaline alkalinity as $CaCO_3$	Chemical
15.	Total Alkalinity	Chemical
16.	Chloride	Chemical
17.	Sulfate	Gravimetric
18.	Silica	Chemical
19.	Iron	Chemical
20.	Copper	A.A.S.
21.	Manganese	A.AS.
22.	Oils & Grease	Gravimetric
23.	Sodium	F.P
24.	Potassium	F.P.
25.	Nitrate	Chemical
26.	Carbonate	Chemical
27.	Bicarbonate	Chemical
28.	Coliform count	Microbial method
29.	B.O.D. at 20° for 5 days	Chemical
30.	Hydroxide	Chemical
31.	Total cations as $CaCO_3$	Chemical
32.	Total anions as $CaCO_3$	Chemical
33.	Suspended solids	Filtration
34.	Sulfite	Chemical
35.	Zinc	A.A.S.
36.	Colloidal silica	Chemical
37.	Phosphate	Chemical
38.	Free chlorine	Chemical
39.	Fluoride	Chemical

expert group of a multidisciplinary nature or a task force. The recommendations of the committee are not however, binding on the government. Some of the major EIA projects undertaken by the Ministry of DOEN are :

1. Thal-Vaishet fertilizer project near Bombay and the committee recommended pollution control measures.
2. Hydroelectric power project near Madumalai Wildlife sanctuary in Tamil Nadu.
3. Effects of air pollutants of Mathura oil mills on Taj Mahal
4. Rail link between old and New Bombay
5. Silent Valley project in Kerala
6. Konkan Railway Project in three states, Goa, Karnataka and Maharashtra.
7. Tehri dam project on Narmada
8. Heron Reservoir
9. Lalpur dam project in Gujarat
10. Kaiga project

COMPONENTS OF EIA

Purpose

It is a planning and decision making tool. It identifies beneficial and adverse impacts and examines the significance of environmental impacts as to whether they are easily mitigable or not. If not, the purpose of EIA is to state clearly and concise information about the project to the decision makers.

Screening and IEE

The screening and initial environmental examination (IEE) is carried out to select the projects which have potentially significant effects/impacts and some projects require only an IEE. This screening criteria includes the evaluation of threshold project type and sensitive areas.

Scoping

The scope is to determine the extent of EIA study and identify issues to be considered. Mentioning the priorities for making environmental assessment and also to make a plan for public involvement.

Preparation of Terms of Reference

    (a) Outline the specific activities to be conducted.

    (b) Describe professionals required for the study area.

    (c) Establish a schedule and budget for activities.

    (d) Place the EIA study into the content of existing policies, legislation and regulations.

    (e) Give importance to most important element of study.

    (f) Provide technical guidance.

EIA Report

It should be concise and limited to addressing significant environmental issues which include :

    (a) Determination of baseline conditions like environmental base map and analysis of existing data.

    (b) Types of impacts including the physical resources, resource depletion, surface water quality, hydrology, ground water, soils, geology and air quality.

    (c) Ecological resources include fisheries, and aquatic ecology, wildlife, forests and environmentally sensitive areas.

    (d) Human use and values include water supply and land use patterns.

    (e) Quality of life values include socio-economic, occupational, public health and aesthetics.

Assessment of Methodologies

It includes adhoc, checklists, matrices, overlays, cost benefit ratio and simulation modelling methods etc.

Mitigation Measures

    (a) Action should be taken to maximize project benefits and minimize undesirable effects.

    (b) Suitable number of alternative proposals should be given.

    (c) Possible preventive and compensatory or corrective measures.

    (d) Project plan for implementation.

Review of Draft of EIA

    (a) Plan for public review and comments

    (b) Ensure technical and scientific validity of the report

    (c) Presents a quality control assurance check

Impact Monitoring

    (a) Study of base line data

    (b) Seeing that impacts do not exceed the legal and other specified standards and guidelines

    (c) Providing early warning of potential environmental damage.

    (d) Check compliance with mitigation measures specified in the report.

    (e) It sects impacts and indicators.

Evaluation of EIA Process

    (a) Assesses performance of environmental protection programs.

    (b) Complements the environmental monitoring process.

    (c) Draws conclusions from monitoring the data.

Overall Assessment

After the evaluation of effects, during the planning stage assessment can be formulated.

    (a) List of each significant effects with quantification

    (b) Comparing their relationship with each other

    (c) Deriving conclusion on over all environmental integrity of the project and on feasible modifications to the project plan.

(d) Carrying out the present recommended environmental monitoring program as a part of project operation and maintenance.

Numerous techniques and methods have been used in evaluating and presenting the impacts of proposed and on going developmental activities on the environment. These methodologies display a variety of frame work, data, format, man power monitoring and time resource requirements. Every EIA study require various methodologies.

Selection of Methods

The EIA team or the person will face a vast amount of raw data to be organised. Hence each technique and method should be tried for the evaluation of the quality and characters of impacts.

1. It should be systematic in approach and able to organize a large mass of heterogeneous data.
2. It should be able to quantify the data and impacts more relatively and accurately.
3. It should be capable of summarizing the entire data.
4. It should be able to aggregate data into least loss of information, because of aggregation.
5. It should have good predictive ability.
6. It should extract salient features.
7. Finally It should be able to display raw data and derive information in meaningful fashion.

All the EIA methods should be flexible enough to allow modifications and changes through the course of study and they should be simple.

Impacts

The impact identification should be scientifically comprehensive to contain all possible and alternatives. There should be specific impacts identified accurately as to the location and extent of the impact. The significance of the methodology to assess the explicit nature of the impacts on local, regional and national scale, what would be the condition with and without the project are to be explained. Try to aggregate vast amount of data, in-depth and knowledge, alternatives, composition and public involvement etc. Impact communication is made to the following groups.

1. Affected parties, affects and social groups
2. Summary format should be such that it should be understood by laymen as well as experts.
3. Highlight the key issues.
4. Compliance with the reference

GENERAL METHODS

A. Adhoc Method

There are several methods and techniques available to evaluate the impacts. Adhoc method is the most simple method. It can be performed without any training. It merely presents impacts without any compression. This will help in case of IEE impacts. This method has the following drawbacks.

1. It gives no assurance about compliance of all impacts.
2. It lacks consistency in analysis.
3. It is inherently inefficient as it does not study the impact in detail. Because of these impact drawbacks it is not recommended for detailed EIA. This can be used when no expertise and resources are available.

B. Checklists

These are strong in impact analysis and identification. These are consistent in nature. Impact assessment is an important and fundamental function of an EIA. The simple and descriptive check lists merely identify the possible potential impacts without any rating. Hence this method is more applicable at IEE stage.

The Oregon checklist goes a step further and provides an idea of the nature of impacts by means of the rating of impacts as long term directive. This is also not much useful in detailed EIAs. Such checklists help in identifying impact and also interpretation and evaluation. The checklist is not useful in identifying diverse impacts. Again there exist different checklist methods. The EIA system considers secondary effects which could not be considered by checklist methods. Scaling and weighing of checklists are capable of quantifying the impacts.

C. Matrices

Matrices provide the cause effect relationship between various project activities and their impact on environmental components. Matrices provide graphic display of impacts weights assigned to environmental components and consequently impacts with actual guarantee of facts during operation. Aggregation of numerical impacts through suitable transformation function is possible through matrices. Matrices are also strong in identifying impacts unlike checklists since they can also inform higher order effects and interactions. These help in communicating results in an easily understandable format to audience. But these cannot compare the alternatives in a single format.

D. Networks

These identify direct and indirect and higher order impacts. These are able to identify and incorporate mitigation and management strategies into planning stages of a project. These are suitable for expressing ecological impacts, but are of lesser utility in considering social, human and aesthetic aspects. This is because weighing and rating are not features of network analysis. Networks usually consider only adverse impacts of the environment and hence decision making in terms of cost benefit of a development project to a region is not amenable to network analysis. Since networks can incorporate several alternatives into their format, the display becomes very large. Networks are capable of presenting scientific and factual information but provide no evaluation for public participation.

Overlays

This system is useful while addressing questions at site and route selection. They provide suitable and effective modes of presentation and display to the audience. But overlays analysis cannot be the sole criteria for EIA. There is no provision for quantification and measurement of the impacts. It is not assured that all impacts will be covered. The considerations in Overlay analysis are purely spatial. Social, human and economic aspects are not covered as considerations. Further, higher order impacts cannot be identified. Overlays are very subjective and they rely on the judgement of analysts to evaluate and assess questions on the capability relating to existing land use patterns and prospects of developmental activities. Overlay methods are useful for industrial EIAs.

Cost Benefit Analysis

This provides the nature, expenses and benefits accruable from a project in monitory terms and helps in decision making in all EIAs. The difficulty encountered here is that the impacts have to be transformed and stated in explicit monetary terms which is not always possible. The cost-benefit is concerned with the effects on environmental quality. This system is not suited for small scale developmental projects.

Simulation and Modelling Workshops

This is a combination of several models and methods. For example, the analysts have developed an advanced approach to environmental impact assessment and management (AAEAM), also referred to as adaptive environmental assessment and management. AEAM broadens the potential scope of various simulation models to predict impacts. This is beneficial during project planning. This method evaluates the impacts in three phases : (1) The initial workshop; (2) The secondary phase workshop, and (3) transfer of workshop. The AEAM technique cannot handle higher impacts and interactions between impacts. This does not involve public participation.

Additional Reading

1. Royston, M.G. 1981. Environmental impact assessment as a tool for environmental quality management, cited in Environmental Management, Eds. Deshbandhu, India Environmental Society, New Delhi. 15-28.

2. Munn, R.E. 1979. Environmental impact assessment: Principles and Procedures, Scope-5, Toronto, Canada.

3. Pandey, R.A., P.R. Choudhary and A.S. Bal 1975. Impact assessment studies with reference to aquatic environment cited in Recent researches in aquatic environment, Eds. Ashutosh, G. and N.K. Agarwal. 1-24.

CHAPTER 2

GUIDELINES FOR EIA OF INDUSTRIAL ESTABLISHMENTS, WATER RELATED PROJECTS AND TRANSPORT RELATED ACTIVITIES

B.B. Hosetti

I. INDUSTRIES AND ENVIRONMENTAL GUIDELINES

1. Introduction

On the basis of environmental impacts, industries are categorized into two types.

(i) Site dependent

(ii) Site independent.

Site dependent impacts are those which depend on the accessibility of the environment or the quality of prevailing environment.

Site independent ones are those which are related to raw materials, process technology, intermediate and finished products, infrastructure and equipments.

A brochure (Booklet) for the purpose is prepared by a team comprising Dr. Nilaya Chaudhari, Chairman, Central Board for the prevention and control of water pollution, Shri Brijendra Sahay, Joint secretary, Dept. of Industry and Development, Dr. D.K. Biswas, Director, Dept. of Industry and development.

Industrial development significantly contributes towards economic growth. However, industrial progress brings along with it a huge amount of environmental problems. Many of these problems

(16)

could be avoided if industries are set on the basis of environmental considerations. Pollution abatement steps are to be considered at the time of starting the industry rather than going for setting abatement measures at a later stage.

The Industrial policy statement 1980 recognized the need for preserving ecological balance and improving living conditions in the urban centers of the country. On the basis of this policy indiscriminate expansion of existing industry and the setting up of new industries within the limits of metropolitan cities and larger towns are not permitted. However, the policy has not touched upon the implication of setting up an industry in sensitive areas both ecological or otherwise which would have effects on the environment.

Industries are being located on the basis of raw material availability, access to the market, transport facility and other techno-economic conditions. Very little attention being paid to environmental considerations in recognising the criteria for setting up of industries.

To prevent pollution of water, air and soil due to industries, Industrial licensing procedure requires that the entrepreneurs, before setting up the industry, should obtain clearance from the Central/State pollution control boards. The Central/State pollution control Board stipulates that gases and licensed effluents emanating from industry should adhere to certain quality standards. However, these stipulations do not prevent the industry from affecting the total environment by wrong siting. Also cumulative effect of a number of industries at a particular place is not being studied, with the result that environmental pollution is increasing.

In the case of certain industrial activities it is not only necessary to install suitable pollution control equipments but also to identify appropriate sites for their location.

For the select group of industries, industrial license should be issued only after the following conditions have been fulfilled:

(1) The state Director of Industries confirms that the site of the project has been approved from environmental point of view by the competent authority.

(2) The entrepreneur commits both to the state and center that he will install appropriate equipments and implements

which are the prescribed measures for the prevention of pollution.

(3) The concerned state pollution control board has to certify that the proposal meets the environmental requirements and that the equipments installed or proposed are adequate to meet the requirement.

The state department of environment is the competent authority to approve the project, though it has to be passed by the center. In case of state's approval, such department, a nodal agency appointed or the center will approve.

(4) The entrepreneur has to submit half yearly progress reports on the installation of pollution control devices to the respective state boards.

(5) Depending on the nature and location of the project, the entrepreneur has to submit a comprehensive EIA and EMP reports.

Environment Guidelines

In order that the concerned authorities could help the entrepreneurs, it is necessary to frame broad guidelines for siting an industry. The parameters should also be identified. All this is done to ensure the use of natural and manmade resources in a sustainable manner with minimum depletion, degradation and/or destruction of environment, according to Industries Development and Regulation Act.

Areas to be avoided

While siting industries, care should be taken to minimize adverse impacts of industries on immediate neighbourhood and on distant places.

The industries

(1) Should be about 25 km about away from ecologically sensitive areas depending on geoclimatic conditions.

(2) About half km away from high tide in coastal areas.

(3) At least half km from flood plain or modified flood plain or upstreams.

(4) At least half km from highway and railway.

(5) Should be away from major settlements (3,00,000 popula-
    tion). In this case, specifying the distance is difficult be-
    cause of urban sprawl. The preferred distance is 25-50 km.

Ecologically Sensitive Areas

(a) Religious and historic places
(b) Archaeological monuments
(c) Beach resorts
(d) Health resorts
(e) Coastal areas rich in all manmade breeding grounds of
    specific species
(f) Gulf areas
(g) Biosphere reserves
(h) National parks and sanctuaries
(i) Natural lakes and swamps
(j) Sessile zones
(k) Tribal settlements
(l) Area of scientific and geological interest
(m) Defense installations
(n) Border areas
(o) Air ports
(p) Scenery areas
(q) Hill resorts, etc.

Siting Criteria

In case of siting an industry economic and social factors are
recognized and assessed. Also environmental factors should be
taken into consideration. Proximity of water source, height, major
settlements, market for products, raw material resources etc.

The following must be away from industries for environmental
protection; the important factors include

1. No forest land should be converted into non forest activity
   (Forest Conservation Act 1980).

2. No prime agricultural land should be converted into
   industrial site.

3. Acquired land should be sufficiently large to accommodate appropriate treatment plants and also to store and recycle effluents. Reclaimed wastewater may be used to raise green belt and create water body for aesthetics, recreation and aquaculture. The green belt should be half km wide around the industry.

4. The green belt distance between two adjoining large scale industries should be one km.

5. Enough space should be provided for storage of solid waste.

6. Layout and farm of industry should be attractive (scenic beauty).

7. Industrial township should be created at a proper distance.

8. Each industry should install their ambient air quality measuring station within 120 degree angle between each station.

Environmental Impact Assessment

The purpose of EIA is to identify and evaluate the potential impacts (Beneficial and adverse) of development projects on environment. It is useful aid for decision making based on understanding the environmental implications including social, cultural, and aesthetic conditions/concerns which could be integrated with the analysis of the project cost and benefits. This exercise should be undertaken early enough in the planning stage of projects through the selection of environmentally compatible sites, process technologies and such other safe guards.

All Industrial projects have similar environmental impacts but, all may not be significant enough to warrant elaborate assessment procedures. The road for such activity can be decided after initial evaluation. The projects for detailed EIA include the following :

1. Those which can significantly alter the land scape and land use pattern.

2. Those which need upstream development activity like assured mineral and forest products supply or downstream industrial process development.

3. Those involving manufacture, handling and use of hazardous materials.

4. Those which are sited near ecologically sensitive area, urban places, hill resorts, places of scientific and religious importance.

5. Industrial estates with constant emissions of various types that could cause environmental damage.

EIA Preparation

EIA is prepared on the basis of existing background pollution levels vis-a-vis contributions of pollutants from the proposed plant. The EIA should address itself to some of the basic factors listed below :

(a) Meteorology and air quality : Ambient levels of pollutants such as CO, NOx, SOx, SPM should be determined at the center and at other three locations on a radius of 10 km with 120° angle between stations. Additional contribution of pollutants of the locations are to be predicted by taking into account the emission rates of the pollutants from stacks of the proposed plant under different meteorological conditions.

(b) Hydrology and water quality

(c) Site and its surrounding

(d) Occupational safety and health.

(e) Details of treatment and disposal of effect and alternate measures and uses.

(f) Transport of raw materials and handling

(g) Impact on sensitive targets

(h) Control equipment and measures proposed to be adopted.

Environmental Master Plan

Preparation of EMP is required for formulation, implementation and monitoring of environment protection measures during and after siting the projects. The Camp should include cost of measure for environmental safeguards as an integral component of the project cost and environmental aspects should be taken into account at various stages of the project, like :

(a) Conceptualization - preliminary EIA

(b) Planning - detail studies of EIA and design of safe-guards.

(c) Execution - implementation of environment safety measures.

(d) Operation - monitoring of effectiveness of built-in safeguards.

The management plan should be necessarily based on considerations of resource conservation and pollution abatement, these are as under:

The EMP cell addresses the following:

Liquid Effluents

1. Effluents from industries should be treated according to the standards prescribed by central/state water pollution control boards (CWPCB/SWPCB).

2. Soil permeability studies should be made prior to effluents being discharged into holding tanks or impoundments and steps taken to prevent percolation and ground water contaminations.

3. Special precautions should be taken regarding height patterns of birds of the coast. Effluents containing toxic compounds are known to cause extensive deaths of migratory birds. Location of industrial plants is prohibited in such sensitive areas.

4. Deep well burial of toxic materials should not be resorted as it can result in resurfacing and ground water contaminations; and, resurfacing causes damage to crop and live stock.

5. All efforts should be made for rescue of water and its conservation.

Air Pollutants

1. The emission level of pollutants from different stacks should comply to the pollution control standards prescribed by Central/State Boards.

2. Adequate control equipment should be installed for minimizing the emission of pollutants from stacks.

3. Inplant control measures should be adopted.
4. Infrastructural facilities should be provided for monitoring the stack emissions and measuring ambient air quality in the area.
5. Proper stack height as prescribed by board for better dispersion of pollutants over a wide area to minimize the effects should be ensured.
6. Community buildings and townships should be built up 1 to 1/2 km away from the industry and green belt may be developed as a physiographical barrier.

Solid Waste

1. The site for solidwaste disposal should be checked for permeability so that no contaminants percolate into ground water or river or lake .
2. Waste disposal areas should be selected down wind at village or township.
3. Reactive materials should be disposed after immobilization with suitable additives.
4. The patterns of filling disposal site should be planned to create better landscape and be approved by appropriate authority.
5. Intensive programs for tree planting on disposal areas should be undertaken.

Noise and Vibration

Adequate measures should be taken to control noise and vibration in the industries.

Occupation Safety and Health

Proper precautionary measures for adopting occupational safety and health standards must be followed with abatement supports.

House keeping : Proper house keeping and cleanliness should be maintained both inside and outside the industry.

Human Settlements

1. Residential colonies should be located away from the solid and liquid waste dumping areas. Meteorological and environmental conditions should be studied properly before selecting the site for residential areas to avoid pollution problems.
2. Persons who are displaced or have lost agricultural lands as a result of locating the industry in the area should be properly rehabilitated.

Transport System

1. Proper parking facility and avoiding congestion.
2. Siting industry should be away from highways.
3. Spillage of chemical on roads to be avoided. Proper road safety sign should be displayed to avoid accidents.

Recovery and Rescue of Waste Products

Efforts should be made to recycle or recover the waste materials to the extent possible. The treated liquid effluents can be conveniently and safely used for irrigation of lands, plants, and fields for growing nonedible crops.

Vegetation Cover

Industries should plant trees and ensure vegetative covers in their premises. This is particularly advisable for those industries having more than 10 acres of land.

Disaster planning : Proper disaster planning should be done to meet any emergency situation arising due to fire, explosion, sudden leakage of gas etc. Fire fighting equipments and other safety appliances should be kept ready even during floods and earthquake.

B.   Environmental Management Planning Cell

Each industry should set up a department with trained personnel to take up the model responsibility of environment management as required for planning and implementation of the projects.

Table 2 : List of polluting industries

1. Ferrous metallurgical	12. Drug and Pharmaceutical
2. Nonferrous metallurgical	13. Fermentation
3. Mining	14. Rubber
4. Mineral processing	15. Paint
5. Coal	16. Milk
6. Power generation	17. Leather
7. Paper and pulp	18. Electro plating
8. Fertilizer	19. Chemicals
9. Cement	20. Cider
10. Petroleum	21. Synthetic rexin and plastic
11. Petrochemicals	22. Manmade fibre.

II. CRITERIA FOR EIA OF WATER RELATED PROJECTS

Introduction

While conducting a water quality impact assessment, all the applicable water quality analysis and various standards must be known. Water quality criteria is defined as the evaluation of levels of specific concentrations of constituents which are expected, to ensure the suitability of water for specific uses. "Water quality standards" are legal regulations established by the states, limiting the levels of various constituents of water. Stream quality standards apply to ambient waterways while effluent standards are applicable to discharges of liquid effluents into those waterways. The activities that must be performed by the water quality assessment team are basically those that must be followed in any impact assessment process. The general procedure included the following steps :

1. Perform a preliminary review of the existing environment and proposed project.

2. Select aquatic environmental indicators to be used for evaluating the environment and assessing the effects of the project (Table 3 and 4).

3. Describe the existing environment by providing quantitative descriptions to each indicator, using an existing data source.

4. Arrange field sampling programs to complete the description of the environmental setting.

5. Make predictions of the effects of the proposed project on aquatic sources (impact assessment)

6. Propose modifications which could minimize all the adverse impacts resulting from the project.

Table 3 : Environmental indicator-parameters

Category	Sub-category	Indicators
Geophysical	Geology	Bedrock type
		Bedrock characteristics
		Depth to bedrock
	Soils	Soil type
		Soil characteristics
		Depth to water lable
	Topography	Watershed description
		Watershed map
		Drainage areas
		Slope
		Relief
	Erosion/	Locate erosion problems
	sedimentation	Erodability of soils
		Locate sedimentation problems
		Stream bed loads
Hydrology	Surface water	Inventory water sources
		Inventory water withdrawals
		Water budget
		Lake water surface elevations
		Surface area
		Lake stratification
		Depth of flow
		Flow velocity
		Discharge (average flow,
		Peak and seasonal variation)
		Flood and drought records
		(include flood frequency analysis)
		Describe flood control facilities
		Stream order
		Reservoirs (Purposes, operating schedule)
	Ground water	Salt water intrusion
		Permeability of aquifers
		Porosity of aquifers
		Depth to groundwater

Contd......

Category	Sub-category	Indicators
		Yields
		Seasonal variations
		Long-term trends
		Recharge areas
		Recharge rates
		Inventory withdrawals
		Inventory deep-well discharges
	Meteorology	Temperature (daily and seasonal variation, high, low, mean) wind (speed, direction, windroses)
		Precipitation (seasonal variations, extremes storm frequency analysis)
		Snow (monthly distribution. extremes) frost (earliest, latest)
		Humidity (daily and seasonal variations)
		Dew point (daily and seasonal variations)
		Solar radiation (daily and seasonal variations)
		Water quality Surface water
		Classification of stream
		Stream standards
		Temperature
		pH
		Conductivity
		Turbidity
		Total dissolved solids
		Total suspended solids, color
		BOD (5-day, 20°C)
		BOD–ultimate
		COD
		TOC
		Dissolved oxygen
		Hardness
		Alkalinity
		Acidity
		Nitrate
		Ammonia
		Total Kjeldahl nitrogen
		Organic nitrogen
		Phosphate
		Ortho-phosphate
		Organic phosphorus
		Sulfate
		Chloride
		Flouride

Contd......

Category	Sub-category	Indicators
Water systems		Iron
		Manganese
		Magnesium
		Potassium
		Sodium
		Calcium
		Silica
		Mercury
		Phenol
		Total coliforms
		Sodium absorption ratio
		Pesticides
		Radioactivity
		Surfactants
		Heavy metals
		Trace organics
		Carcinogens
	Ground water	(Same indicators as for Surface water)
	Water use	Flow (daily and seasonal variation)
		Residential water use
		Industrial water use
		Agricultural water use
		Commercial water use
		Municipal water use
		Metering systems
		Water importation
		Water diversion
	Water treatment	Intake water quality (See water quality indicators above)
		Describe intake
		Describe plant
		Design capacity
		Current demand (Time variation)
		Chemical additions
		Energy requirements
		Sludge type and quantity
		Sludge disposition
		Product water quality (See water quality indicators above)
		Operational difficulties

Contd......

Category	Sub-category	Indicators
	Distribution System	Size of lines
		Age and condition of lines
		Capacity of lines
		Current flow (Daily and seasonal variation)
		Pressure
		Storage requirements and capacity.
Wastewater systems	Collection Systems	Sewer sizes
		Sewerage and condition
		Capacity
		Current flows (daily and seasonal variations)
		Problems (odor, sludge etc.)
		Infiltration/inflow analysis
		Storm water collection
		(Separate and combined sewers)
	Treatment systems	Describe systems
		Locate facilities
		Age and condition of plants
		Design capacity
		NPDES effluent limitations
		Raw waste characteristics
		(See water quality indicators above)
		Effluent characteristics
		(See water quality indicators above)
		Flows and loads (average and time variations)
		Describe sludge handling systems
		Sludge (type, quantity, moisture content, disposition)
		Outfalls
		Operational difficulties (odour, insects, poor effluent etc.)

NPDES—National Pollutant Discharge Elimination System, administered by the US. Environmental Protection Agency.

Table 4 : Preservation techniques for various samples

Parameter	Sample volume (ml)	Preservative	Allowable holding time
Acidity	100	Refrigerate, 4°C	24 hr
Alkalinity	100	Refrigerate, 4°C	24 hr
BOD	1,000	Refrigerate, 4°C	6 hr
COD	50	H_2SO_4 to pH<2	7 days
Chloride	50	None	7 days
Color	50	Refrigerate, 4°C	24 hr
Cyanides	500	Refrigerate, 4°C	24 hr
Dissolved Oxygen			
Probe	300	Determine on site	No holding
Winkler	300	Fix on site	408 hr
Hardness	100	Refrigerate, 4°C	24 hr
Metals Nitrogen	200	HNO_3 to pH <2	7 days
Ammonia	500	Refrigerate, 4°C H_2SO_4 to pH<2	24 hr
Kjeldahl	500	Refrigerate, 4°C H_2SO_4 to pH<2	7 days
Nitrate	100	Refrigerate, 4°C H_2SO_4 to pH<2	24 hr
Nitrite	100	Refrigerate, 4°C	24 hr
Organic carbon	50	Refrigerate, 4°C H_2SO_4 to pH<2	24 hr
pH	100	Determine on site	No holding
Phenol	500	Refrigerate, 4°C H_2PO_4 to pH<4 1.0g $CaSO_4$/liter	24 hr
Phosphate	50	Refrigerate, 4°C	24 hr
Residue	100	Refrigerate, 4°C	7 days
Specific conductance	100	Refrigerate, 4°C	24 hr
Sulfate	50	Refrigerate, 4°C	7 days
Temperature	1000	Determine on site	No holding
Turbidity	100	Refrigerate, 4°C	24 hr
Coliform bacteria	500	Refrigerate, 4°C	36 hr
Pesticides	1000	Refrigerate, 4°C	24 hr
Radioactivity	1000	None required	

If the environmental setting is not judged to be sufficiently complete, then the other sources of information should be investigated, or field measurements should be made so as to complete the setting. If the impact analysis reveals that additional environmental indicators are required to fully describe the character and magnitude of the impacts, then those indicators should also be developed and described. The evaluation of the impacts may reveal a means of minimizing adverse impacts. As a result of modification of the proposed project, it may necessitate re-evaluation of impacts. This highlights the fact that the impact assessment should be an integral part of the planning process and not simply an evaluation step to be added into the project to satisfy the requirements of the NEPA. Incorporation of the impact assessment into the planning process can minimize adverse impacts as well as public objection, while increasing the compatibility of a project with the environment.

Environmental Setting

The first phase of the process is the description of the environmental setting: an inventory of existing water quality, hydrologic resources, and conditions influencing water quality, and water resources to be developed. This phase of the project involves the selection of the most useful environmental indicators and description of the environment with the help of literature reviews and field investigations. This phase also describes other aspects of the proposed project and alternative course of action, which could influence the water quality or quantity.

At the onset of the assessment, the water quality team must carry out survey in the area of the proposed project and review the characteristics of the project. The object of this survey is to determine possible water quality impacts and to identify sensitive environmental areas. The next step is detailed study in environmental setting is the selection of environmental indicators. The environmental indicators selected should be described as quantitatively.

Environmental Indicators of Water Quality

1. Selection of reasonable number of indicators which should not be too many or too less. The project team consisting of

experts from different fields must take the responsibility for selecting appropriate indicators for each individual project and geographical location.

2. The selection of environmental indicators should be based on the scope or regional environment. A Preliminary review of the proposed project will be valuable to this selection.

3. Another factor influencing indicators selection is the environment itself. Indicators should be chosen in such a way those will enable the assessment team to adequately describe the environment.

Biological Factors

Most of the environmental impact studies primarily deal with the toxicological aspects of water quality only, rather than from the viewpoint of integrated aquatic ecosystem functions. Although in many cases a restricted approach to the detailed investigation is justifiable, precedence of ecologically oriented study to detailed studies is beneficial in two ways, first a general ecosystem level study. Secondly the general study can serve to evaluate the importance of the specific effects within the context of the ecosystem function; e.g. the use of DDT in the 1940's and 1950's was accompanied by toxicological studies indicating relatively low toxicities of the compound for most vertebrate species except some sp. of fish (Pimentel, 1971). In the 1960's however, several studies, with more ecological orientations, revealed very high values at higher trophic levels (Woodwell et al. 1967). Further DDT concentrations may continue to rise in the long lived organisms for years after cessation of DDT use (Harrison et al., 1970). Thus some species of birds which in simple toxicological studies appeared not to be sensitive to damage by DDT, were severely affected in long term field exposures because of accumulation of DDT in the food web and resulting in indirect mortality and leading to sublethal reproductive failures (Stickel and Rhodes, 1970).

The second important reason is that isolated examination of specific effects may lead to conclusions and predictions that may not be valid in the more complex context of the ecosystem. As an example to this fact, Whester (1968) reported that DDT at about 0.01 ppm levels, inhibited photosynthetic activity in the species of marine phytoplankton in the laboratory. However, a number of latter studies indicated that extrapolation of these laboratory results to field conditions is not justified. From these two examples of DDT ecosystem dynamics, which produced unexpected results. It is clear that a limited nonecological approach to impact studies can lead to underestimation. Ward (1978) has rightly introduced biological component in EIA studies.

A. Ecological Indicators

I.   Ecosystem structure.
   (a)   Species composition and abundance
   (b)   Feeding relationships
   (c)   Ecological dominance and key species
   (d)   Indicator species and ecological indicators.
   (e)   Species diversity.

II. Ecosystem functions
   (a)   Productivity
         Primary productivity
         Secondary productivity
   (b)   Trophic structure and energy flow
   (c)   Nutrient cycles
   (d)   Decomposition and detrivores
   (e)   Succession or development of communities
   (f)   Individual species characteristics.

Table 5 : Biotic Indicators of aquatic environment

Sl. No.	Indicator	Method of preservation
1.	Bacterial population	Water sample to be collected in
	- Total count	sterilized bottles and preserved
	- E. coli	in ice-box/refrigerator
	- Fecal coliforms	
	- Fecal streptococci	
2.	Plankton	
	- Phytoplankton	1% Lugol's Iodine solution
	- Zooplankton	4% Formalin
3.	Benthos	4% Formalin
4.	Macrophytes	4% Formalin
5.	Periphyton	4% Formalin

B. Laboratory Studies

   I. Extrapolation from laboratory studies.

      (a) Changed nature of the stressing factor when applied under field conditions.

      (b) Interaction of stress effects with variability in physico-chemical factors

      (c) Cumulative effects of stress with the time

      (d) Individual characteristics of the tested species under field conditions

      (e) Intra & inter-specific interactions.

   II. Physical model systems

   III. Measurement of toxicity

Field Survey

Water quality surveys can generate valuable baseline data, which can be compared to conditions after the proposed project is operational. A well conceived, carefully planned sampling program has a high probability of success in assessing the water quality. The overall plan for the water quality survey must contain the following elements.

— Detailed plan for sample collection
— Provision for laboratory analysis
— Description of the methods to be used for data reduction and manipulation including statistical analysis.

The plan should also address

— Location of sampling stations
— Parameters to be analyzed
— Time schedules, including time of day, month, year, and frequency.
— Method of data collection
— Sampling and handling prior to analysis.

a. Sampling stations.

Sampling points should be located to provide an accurate description of existing water quality. In addition, the sampling points should be selected to maximize the ease of sampling. This is facilitated by locating sampling points on good base maps. The actual location of sampling points is primarily dependent on the physical situation, for example number of sampling points at each cross section of river depends on its width and depth. In determining the location of sampling points, it is important to recognize conservative and nonconservative water quality parameters. Selection of water quality parameters depends primarily on the type of project proposed and the anticipated impacts of the project, e.g. if the project involves thermal discharges, the temperature must be measured along with parameters which are basically temperature dependent such as dissolved oxygen and BOD.

The nonconservative BOD & dissolved oxygen system are frequently of concern in impact assessment because, organic wastes exert an oxygen demand leading to depletion of dissolved oxygen levels in the stream. Usually only one sample is taken at each sampling point on any one day. This is valid only as long as large diurnal variations do not occur. In streams with significant algal activity, diurnal variations in dissolved oxygen of several milligrams per liter may be observed. If this is the situation, it is suggested that at least one day of around the clock sampling should be conducted in order to obtain average values.

Sampling over one year period is desirable to seek good baseline data. This is normally done for large projects, such as proposed power plants. Sampling is conducted on a routine basis, such as once in fifteen days, if one year sampling program is conducted. An alternative to this is to conduct four intensive sampling programs, one during each season, five to ten days in length. For smaller projects budgetary constraints may rule out year round sampling. In this case, a five to ten day sampling program could be implemented. Routine baseline monitoring should be suspended during periods of rainfall except in special studies.

Data Collection

Grab and composite samples are generally collected. Grab samples represent conditions existing at the particular location, while composite samples are a series of samples over a period of time. Composites have the advantage of being more reflective of average conditions, while grab samples require less time to obtain.

In water bodies where flows and concentrations do not change rapidly, grab samples are preferred. In a stream that is subjected to rapid variation, composite samples are better to define average conditions. Several methods of measuring velocity and discharge in flowing streams are available. A simplistic approach is the use of surface floats. In this method, the time required for a float to travel a known distance is observed, and the average velocity (v) is obtained by the relation,

$$v = (d/1.2t)$$

The factor 1.2 account for the fact that surface velocities are normally about 1.2 times the mean velocity. If the cross-sectional area A is measured, the discharged Q is given by

$$Q = (V_A)$$

This method is useful in small shallow streams.

In deep water bodies the most popular method to use current meter is a propeller-type or rotating-cup meter. Samples are analyzed as per Standard Methods for the Examination of Water and Wastewater and Standard Chemical Analysis of Water and Wastes.

The site visit, literature review and interviews with agency representatives are used to complete the description of the environmental setting by describing each selected environmental indicator as quantitatively as possible. The setting can be illustrated through a series of tables or matrices presenting the values or descriptions assigned to each indicator. When the description of the existing environmental setting is completed, the water resources/water quality assessment team must assess its overall quality.

Recent Approaches in the EIA Studies

The emergence of the concept of sustainable development in recent years has brought in the general realization that societal perceptions must shift towards ecological determinism so as to achieve qualitative growth within the limits of ecosystem carrying capacity. The carrying capacity based on planning process, innovative technologies for enhanced material and energy efficiency of production and consumption, structural changes towards less resource-intensive sectors, and preventive environmental management are some of the strategies for developmental goals with ecological capabilities.

Sustainable development is a process in which the exploitation of resources, the direction of investments, and institutional changes are all made consistent with the future as well as present needs. The concept of sustainable development is based on two underlying premises, viz., symbiotic relationship between consumer human race and producer natural systems, and compatibility between ecology and economy.

Carrying Capacity

The concept of sustainable development is closely linked to the carrying capacity of ecosystems. Ecosystem carrying capacity provides the physical limits to economic development and may be defined as the maximum rate of resource consumption and waste discharge that can be sustained indefinitely in a defined planning region without progressively impairing the bio-productivity and ecological integrity. Carrying capacity is determined by the vital resource which are in latest supply. The concept of carrying capacity implies that improvement in the quality of life is possible only when the pattern and levels of production-consumption activities are compatible with the capacities of natural environment as well as social preferences. The carrying capacity-based planning process thus involves the integration of social expectations and ecological capabilities by minimizing demand/supply patterns. Recently comprehensive EIAs based on carrying capacity and sustainable development are being undertaken to create a bank of information on regional basis. This data can be utilized for specific EIA studies for certain activity in the area.

Future Research

The experience gained with EIA has proved that a thorough research associated with the environment in project evaluation and decision making are felt necessary.

1. Compilation of a complete and accurate environmental base line data of unique areas should be completely characterized and data stored so that they are readily available. This type of information are often concerned with soil types, ground water, natural plant communities, climate, current census and school data for low-density and rapidly growing areas, property records, and realistic land-use classification etc.

2. Refinement of techniques for measuring impacts on biological communities may be carried out using sophisticated ecological computer models for describing ecosystems should be developed in order to predict future as well
as present conditions accurately.

3. Refinement of techniques for measuring impacts on social
well-being-techniques already existing in sociology and
psychology should 'be modified and new techniques
researched and developed to assess comprehensively the
impacts of specific projects on the surrounding community.

4. Better methods for predicting cumulative impacts : Quantification techniques are needed for predicting long-term
changes induced by a group of project and for assessing the
cumulative impacts of new land uses that may result.

5. To avoid the controversy often generated by EISs, methods
are needed for objectively determining whether particular
impacts are beneficial or detrimental to a community by an
explicit understanding and consideration of community's
value structure.

6. Refinement of material balance and input-output techniques : These and other sophisticated economic models
should be refined to be more applicable to the impacts of
small scale single projects also.

7. Refinement of environmental impact matrices: Further
research should be carried out to assign impact weighting
factors, to identify areas of impact and relevant project
components, and to separate secondary impacts from primary impacts.

Conclusion

Exploitation of resources and various developmental activities
have deteriorated quality of biotic environment. Lessons from such
activity in developed countries, many developing countries have
become concerned to restore the environment. The experience of
industrialized countries which was often viewed in the past as a
cleaning up process of various sectors of polluted environment. In
case of developing countries including India, are marked by a

determination to achieve sustainable development in an eco-friendly and sound manner. In other words, it is both desirable and essential to pursue the short and long term developmental goals while simultaneously ensuring a better environmental management.

References and Further Reading

Afroz Ahmad, P.S. Ramakrishnan and K.S. Rao. 1989. Environmental Impact Assessment of Water Resources Projects in Himalayas: An Urgent Need for Sustainable Development. In *Environmental Conservation and Development* (Eds. P.R.Singh, O.P. Verma, Ram Boojh), Directorate of Environment, U.P.

Biswas, A.K. and Agarwal, S.B.C. 1992. *Environmental Impact Assessment for Developing Countries.* (Butterworth-Heinemann,Oxford).

Center, L.W. 1977. *Environmental Impact Assessment.* McGraw-Hill, New York.

Harrison, H.L. Loucks, O.L., Mitchell, J.W., Parkurst, D.F., Tracy, C.R., Watts, D.G. and Yanacone, V.J. Jr. 1970. *Science 170*: 505-508.

Jain, R.K. and Webster, R.D. 1977. Computer aided Environmental Impact Analysis. *J. Wat. Resour. November*: 257-271.

Khanna, P. 1994. Role of EIA in sustainable development. In: Technical Papers for presentation. *Indo-British workshop on Environmental Impact and Risk Assessment of Petrochemical Industry and Environmental Audit.* Jan. 8-10, NEERI, Nagpur.

Pimentel, D. 1971. *Ecological Effects of Pesticides on Non target species.* Exec. Off Pres. Off. Sci. Technol., Washington, D.C.

Rau., J. 1980. *Summarization* of *Environmental impact analysis Handbook.* (Eds. John G. Rau and David Wooten) McGraw-Hill.

Stickel L.F.and Rhodes L.I. 1970. *In the Biological Impact of Pesticides in the environment.* (J.W. Gillet ed.) Environmental Health Sci. Ser., No.1. pp. 31-35. Oregon State Univ., Corvallis.

United Nations Environmental Program 1979. *Draft guidelines for Assessing Industrial Environmental Impact and the sitting of Industry.* A report prepared for UNEP by Atkins Research and Development, Epson, Surrey.

United Nations Economic Commission for Europe 1992. *Application of Environmental Impact Assessment Principles to Policies. Plans and Programs.* Environmental Series 5 (UNECE, Geneva).

Ward, D.V. 1978. *Biological Environmental Impact Studies: Theory and Methods.* Academic press.

Woodwell G.M., Wurster, C.F. and Isaacson, P.A. 1967. Science. *156*: 821-824.

Wurster, C.F, Jr. 1968. *Science.* 159: 1474-1475.

III. CRITERIA FOR TRANSPORT RELATED EIA

Introduction

Transportation is an important element for all developmental activities. In turn, the developments generate/intensify transport activities. Many people think environmental degradation and losses in the social fabric are the prices to be paid for the development. This is a wrong notion, as the harmful effects of transport activities can be minimized by thoughtful planning, coordination and management of transport system. The environmental consequences or impact range from noise created by transport vehicles to those complicated issues of social burden of those killed and maimed in accidents; in addition to disturbances caused to natural environment and ecology. Generally, an environmental impact assessment process can be described by the activities described below:

An exposition of transportation impacts should include a long list of items as follows :

1. Visual
   (a) Temporary
   (b) Permanent
2. Social and economic
   (a) Life-style changes.
   (b) Travel patterns.
   (c) School districts.
   (d) Temples/churches etc.
   (e) Recreations

(f) Business

(g) Minorities and ethnic groups

(h) Urban quality

(i) Secondary impacts

3. Relocation

    (a) Number of households displaced

    (b) Neighbourhood disruption

    (c) Available housing

    (d) Number of businesses displaced or impacted

    (e) Documentation of public participation activities

    (f) Unusual circumstances

4. Air Quality

    (a) Microscale impacts

    (b) Mesoscale impacts

    (c) Analysis of methodology

    (d) Documentation of early consultation

    (e) SIP consistency

5. Noise

    (a) Identification of sensitive receptors

    (b) Comparison of future noise levels criteria

    (c) Comparison of future noise levels with existing

    (d) Noise abatement measures

    (e) Noise problem with no reasonable solution

6. Water quality

    (a) Erosion

    (b) Sedimentation

    (c) De-icing and weed control products

    (d) Chemical spills

    (e) Ground waste contamination

    (f) Stream modifications

    (g) Impoundment

    (h) Fish and Wildlife

    (i) Documentation

7. Wetlands and coastal zones.

   (a) Analysis summarized.

   (b) Consultations undertaken

   (c) Practical measures to minimize harm

   (d) Documentation that there are no practical alternatives.

8. Flood hazard.

   (a) Impacts on beneficial floodplain values

   (b) Incompatible flood plain development

   (c) Measures to minimize flood risks

   (d) Evaluation of alternatives

9. Natural resources

   (a) Prime and unique farmlands

   (b) Threatened and endangered species

   (c) Natural land forms

   (d) Ground water resources

   (e) Energy requirements

10. Land use.

    (a) Growth inducement

    (b) Factors which may influence development

    (c) State and/or local government plans or policies

    (d) Planned versus unplanned growth

    (e) Social, economic and environmental impacts likely from induced growth of development.

11. Historic sites.

    (a) Historic, cultural, architectural, archaeological significance

    (b) Documentation of consultation

12. Construction

    (a) Air

    (b) Noise

    (c) Water

    (d) Detours

    (e) Safety

(f) Spoil and borrow

(g) Mitigation measures

Scope of EIA

U.S. Department of Transport (DOT) had developed a checklist of EIA, which is basically a description of the impacts associated with transportation projects organised by a category of impact as well as temporal phases. These impacts as detailed in Table 6 can be either beneficial or detrimental depending on the specific project. Those are :

1. Planning and design phase
   - (a) Impact on land use through speculation in anticipation of development.
   - (b) Impact of uncertainity on account of economic and social attributes of nearby areas.
   - (c) Impact on other planning and provision of public services.
   - (d) Acquisition and condemnation of property for project, with subsequent dislocation of families and businesses.
2. Construction phases
   - (a) Displacement of people
   - (b) Noise
   - (c) Soil erosion and disturbance of natural drainage
   - (d) Interference with water table
   - (e) Water pollution
   - (f) Air pollution (including dust, and burning of debris)
   - (g) Destruction of or damage to wildlife habitat
   - (h) Destruction of parks, recreation areas, historical sites
   - (i) Aesthetic impact of construction activity and destruction or interference with scenic areas
   - (j) Impact of ancillary activities (e.g. disposal of earth, acquisition of gravel, and fill)
   - (k) Commitment of resources to construction.
3. Operation of facility
   - (a) Direct
     1. Noise

2. Air pollution

3. Water pollution

4. Socioeconomic

5. Aesthetic

6. Effects on animal and plant life (ecology)

7. Demand for energy resources

(b) Indirect

1. Contiguous land use

2. Regional development patterns

3. Demand for housing and public facilities

4. Impact on use of nearby environmental amenities (parks, woodlands, recreational areas)

5. Impact of additional and / or improved transportation into congested areas.

6. Differential usefulness for different economic and ethnic groups and resulting problems and solutions

7. Impacts on life styles of increased mobility and other impacts .

8. Impact of improved facility on transportation and related technological development (and consequent impacts)

The following is the information relative to potential impacts of transportation projects organised according to spatial boundaries.

Table 6 : Potential environmental impacts of transportation project

Category	Planning & design	Construction	Operation
I. Noise impacts		+	+
A Public health			
B Land use			
II Air quality impacts		+	+
A Public health			
B Land use			
III Water quality impacts		+	+
A Groundwater			
1. flow and water table alteration			

Contd......

Category	Planning & design	Construction	Operation
2. Interaction with surface drainage			
B. Surface water			
1. Shoreline and bottom alteration			
2. Effects of filling and dredging			
3. Drainage and flood characteristics			
C. Quality aspects			
1. Effect of effluent loadings			
2. Implication of other actions such as			
a. Disturbance of benthic layers			
b. Alteration of currents			
c. Changes in flow regime			
d. Saline intrusion in ground water			
3. Land use			
4. Public health			
IV Soil erosion impacts	+		+
A. Economic and land use			
B. Pollution and siltation			
V. Ecological impacts	+		+
A. Flora			
B. Fauna (Other than humans)			
VI Economic impacts	+	+	+
A Land use			
1. In immediate area of project			
2. In local jurisdiction served or traversed			
3. In region			
B Tax base	+	+	+
1. Loss through displacements			
2. Gain through increased values			
C. Employment			
1. Access to existing opportunities			
2. Creation of new jobs			
3. Displacement from jobs			
D. Housing and public services			
1. Demand for new services			

Contd...

Category	Planning & design	Construction	Operation
2. Alteration in existing services			
E. Income	+	+	+
F. Damage to economically valuable natural resources		+	+
VII. Sociopolitical impacts			
A Damage to or use of		+	+
1. . Cultural resources			
2. Scientific resources			
3. Historical resources			
4. Recreation areas			
B Life style and activities	+	+	+
1. Increased mobility			
2. Disruption of community			
C. Perception of cost benefit by different coesive groups	+	+	+
1. Racial			
2. Ethnic			
3. Income class			
D. Personal safety		+	+
VIII. Aesthetic and visual impacts		+	+
A Scenic resources			
B Urban design			
C Noise			
D Air quality			
E Water quality			

1. In immediate area of the project.
   (a) Displacement by project itself.
      1. Residential
      2. Commercial
      3. Industrial
      4. Recreational
      5. Natural resources
      6. Cultural resources
      7. Scenic resources

(b) Land use choice affected by project

1. Attracted by increased access

   (i) Ancillary uses (Facility or user service)

   (ii) Users benefiting from access (certain industrial, commercial, residential, and public uses)

2. Disrupted by project

   (i) Incompatibility with noise pollution, aesthetic, safety and other effects of facility

   (ii) Incompatibility with access oriented uses

   (iii) In compatibility resulting from increased access of nonresident users (e.g. natural or wild areas)

(c) Neighbourhood (or area) services, facilities, and living patterns affected by facility.

1. Disruption of service districts

   (i) Public facilities

   (ii) Private nonprofit facilities

   (iii) Retail establishments

2. Effects on neighbourhood cohesiveness and stability

2. In land jurisdiction served or traversed by facility

   (a) Effect on land use planning and controls

   (b) Effect on planning and development of public facilities resulting from the land use patterns generated or influenced by the project (including effect on tax base, cost of services)

   (c) Effect on areas not directly contiguous with project of actions by those displaced, disrupted, attracted, or otherwise affected by the project.

3. In region where project is located

   (a) Effect on regional development planning, inducement, and controls

   (b) Revenue effects, influencing other public projects.

   (c) Economic effects, influencing private development in the region as a whole and differentially within it.

Transport being the catalyst for all social and economic activities, such factors need a careful consideration.The factors are listed in Table 7.

Table 7 : Socio-economic factors for transportation projects

Factor	Comment
I **Sociological**	Social relationships
A Community (Local area)	
1. Neighbourhood severance	Violation of neighbourhood boundaries
2. Cultural patterns	Ethnic cohesion, stability, life style
3. Crime	Assault, robbery, breaking etc.
a. Rate	Change in opportunity
b. Police protection	Availability and speed
4. Fire hazard	Type and density of land uses
a Hazards	Dwellings trash etc.
b. Fire protection	Available equipment and time
5. Health	
a. Health factors	Sanitation, dangerous sports etc.
b. Medical services	Time to reach health facilities
6. Religious services	Opportunity to attend
a. Loss of places	Removal of churches/temples
b. Access to	Isolation of members
7. Educational	Loss of or effect on access to :
a. Elementary	
b. Junior high School	
c. High school	
d. Trade and college	
8. Recreational facilities	Other than parks and playgrounds
9. Social services	Gathering places other than previously considered
10. Public utilities	
11. Neighbourhood availability	Pleasantness of surroundings
a. Construction period	Disruption
b. Long run	Cleanliness, repairs, etc.
B Metropolitan area	Loss of and effects on access to
1. Police protection	
2. Fire protection	
3. Medical services	
4. Educational services	
5. Parks	
6. Recreation	
7. Historical sites	
8. National defense	
a. Evacuation	As a link in system
b. Military movements	As a link in system
c. Hazards to critical industry	

Contd......

Factor	*Comment*
II. **Economic Impact**	
A Community (local area)	
1. Employment	
a. Construction period	Change in place or access
b. Long run	Change in place or access
2. Shopping facilities	
a. Construction period	Change in place or access
b. Long run	and loss of customers
3. Residential values	
4. Other property values	
5. Property tax base	
a. Construction period	Loss of taxable values
b. Long run	Potential for changes
6. Displaced residents	
a. Owners	
(1) DSS housing	DSS decent, safe and sanitary
(2) Non DSS housing	
b. Renters	
(1) DSS housing	
(2) Non DSS housing	
c. Ease of replacement	
7. Displaced business	
a. Small business	
(1) Number	
(2) Number of jobs	
(3) Ease of relocation	
b. Other business	
(1) Number	
(2) Number of jobs	
(3) Ease of relocation	
8. Remaining business	Effects on jobs
a. Small business	and solvency
(1) Construction period	
(2) Long run	
b. Other businesses	
(1) Construction period	
(2) Long run	
9. New business	Potential for business
10. Multiple use of right of way	
B Metropolitan area	Outside of local area
1. Access to employment	
2. Access to shopping	
3. Commercial activity	
4. Property values and tax base	

A Case of Airport Development in the USA

A case of eight airports in USA was studied and EIS relative to impacts on the artificial and natural environments were included, as follows :

1. Impact upon human environment
    (a) Real estate impacts
        1. Any public lands involved within boundaries of the proposed projects
        2. Number of acres involved in the project
        3. Number of residences within the boundaries of the project
        4. Number of relocated people as a result of the project
        5. Numbers of housing replacement as a result of the proposed project
        6. Any land use studies of the areas adjacent to the proposed project included
        7. Any roads/highways to be closed due to the project
        8. Any need for roads/highways relocation due to the project
        9. Vicinity maps included in the statement
        10. Maps showing proposed facility layout
        11. Soil description in statement
        12. Any mention to future expandability of the proposed project
        13. Any mention of methods of disposing of debris for clearing (i.e. burning, burial, solid waste treatment)
    (b) Asthetics and visual impacts
        1. Positive or negative effects disclosed from statement regarding land cleaning
        2. What physical features will be retained or removed by the project
        3. Any unique interest areas in the boundaries of the project mentioned
        4. Any scenic beauty areas affected by the project

5. Any recreational areas affected by the project
6. Any forest areas directly affected by the project (i.e within the boundaries)
7. Any land scaping required
8. Any residential developments directly within the project boundaries
9. Any industrial developments directly within the project boundaries
10. Any hospitals within the project boundaries
11. Any churches/Temples within the project boundaries

(c) Community impacts
1. Any division/disruption of an existing community
2. Surface transportation due to the airport discussed in relation to traffic congestion

(d) Public services impact
1. Water supply discussed for project facility
2. Sewage treatment for facility arranged
3. Solid waste treatment for facility outlined
4. Liquid waste and accidental spill precautions discussed
5. Storm drainage (run-off) precautions discussed

(e) Displaced persons discussed

(f) Noise impact
1. Any plans or considerations given to compensate for vehicular noise of the facility
2. Overlay map showing composite Noise Rating (CNR) zones of the project included
3. Detailed noise pollution analysis included
4. Survey opinions of similar situations in other locations included in the statement
5. Air traffic patterns in the area changed due to the project
6. Areas under the new air traffic patterns described thoroughly

    7. Projected air traffic volumes described for affected areas

    8. Argument of noise due to takeoff and landing will be decreased due to runway extension

    9. Any indication given of the land use compatibility studies for adjacent to the project

(g) Airport configuration air space patterns, and safety problems discussed in the statement

(h) Projected level of use for the project included

    1. Expected future operation increase anticipated

    2. Expected future use of facility by larger aircraft discussed in statement

    3. Number of passengers expected to increase as the population of area increases

(i) Employment impact due to jobs created by project discussed

(j) Stimulated population growth in a particular direction discussed

(k) Social, psychological impact discussed

2. Natural Environmental impacts

(a) Wildlife in the projected area

    1. Any rare/endangered species indicated in the project area or boundaries

    2. Bird sanctuary or wildlife reserve involved in or near the project

    3. Existing inhabitants seek more suitable habitats in adjacent areas

    4. Any wildlife relocation necessary

    5. Clearing procedures coincide with periods of least biological activity to cause hardships of relocating

    6. Detailed listing of residing wildlife in the area given

    7. Expected change in the wildlife population in the area

    8. Fewer hunting opportunities due to the project in the area

9. Fewer fishing opportunities due to the project in the area

(b) Water pollution impacts

1. Any streams/rivers within the project boundaries
2. Any ponds/lakes within the project boundaries
3. Any marine life or fish in the project area
4. Any water relocation necessary due to the project
5. Any relocation of marine life or fish necessary
6. Mention made of existing water table characteristics
7. Mention made of adherence to FAA circ. 150/5370-Airport construction controls to prevent Air and Water pollution
8. Water pollutant statement qualified with specific data
9. Methods discussed for handling run-off associated with runways etc.
10. Mention made of the possibility of water table pollution
11. Mention made of associated turbidity due to runoff

(c) Timberland impacts

1. Any Timber within the project boundaries
2. Detailed listing in statement
3. Any timber marketable
4. Timber utilized in landscaping
5. Mention made of disposal means of cleared timber and brush

(d) Any existing physical features (Buildings, cemeteries, etc.) in the project boundaries? Any requiring relocation

(e) Air pollution impacts.

1. Any mention of the effects of the facility upon the oxygen-carbon dioxide exchange process
2. Mention made of dust generation during construction
3. Means for dust alleviation discussed
4. Predicted pollutant emmissions a significant additive over existing pollutant levels

5. Air pollutant statements qualified with data
6. Any assessment gives to visual impact of particulate smoke trials from operations of aircraft in the area
7. Meteorological effects upon air pollutants discussed (i.e. wind currents relative to adjacent cities)
8. Meteorological effects upon neighbouring cities discussed (carrying added pollutants)

(f) Erosion control impact
1. Grading of erosion resistant slopes discussed
2. Turfing discussed
3. Area topography given

(g) General ecological implications (suggestions)
1. Effect (increase/decrease on the full productivity of the area)
2. Influence on the migratory and wintering waterfowl and local birds in the area

Individual Impact Assessment

As discussed in the foregoing sections EIA for transportation activities involve a very wide ranging considerations. Only a few aspects are covered in a bit more detail regarding the assessment and prediction procedures.

1. Noise

Noise cannot be defined in a wholly satisfactory manner. When the sound is unwanted by the recipient or loud enough to be the cause of annoyance, it may be described as noise. Common noise levels, sources and typical reaction are given in Table 8.

Noise is one of the bye product of the mechanized modern developments. It is mainly caused by machinery of one kind or the other, particularly transportation vehicles, industrial processes and constructional works. Table 8 shows the major sources of noise annoyance to people as surveyed in the USA.

Table 8 : Common noise levels and typical reactions

Sound source	Noise level (dBA)	Typical Reaction
Military Jet	130	Painfully Loud
Jet takeoff at 60 m	120	
	110	Maximum vocal
Jet takeoff at 600 m	100	Effort
Freight train at 15 m	95	
Heavy truck at 15 m	90	Very annoying hearing damage (in 8 hours)
Highway traffic at 15 m	80	Annoying
	70	Telephone use difficult
	60	Instructive
Light car at 15m ·	50	Speech noisy office interference
Public library	40	Quiet
Soft whisper at 4.7 m	30	Very Quiet
Threshold of hearing	0	

Table 9 : Percentage of Various noise sources

Sources	Percentage
Motor vehicles	55
Aircraft	15
Voices	10
Radio and TV	2
Home maintenance	2
Construction work	1
Industry	1
Other	14

Human ear can hear over a frequency range of 16 to 20000 hertz (Hz). i.e. 16-20000 oscillations per second. This is the range of infrasonic to ultrasonic. Of course, the threshold of audibility varies with physiological factors and physical factors. These are normally related to the sensitivity of the individual. Age, and efficiency of any protection used. Now a days, the weighted noise level (sound level) is used to express the subjective loudness. This is a measure of overall intensity after filtering the extreme notes. Thus, there are three types of weighting devised values in dBA, dBB and dBC.

The one which is most commonly used is the A- weighted scale. These measures sound levels which are matched to human hearing (in dBA) are normally referred as noise levels.

Excessive noise affects human body, machines as well as structures. The normal work efficiency also gets seriously impaired by continued exposure to high noise levels. The detrimental effects of noise on human being can be considered under the following headings:

1. Subjective effects
2. Behavioural effects
3. Physiological effects.

The disturbance that is evoked by noise depends on one most important factor that is the subjective reaction to noise by the sufferers and individual sensitivity.

Disturbance, annoyance are some of the terms that describe the subjective affect of noise below 40 dBA. The behavioural effects include communication interference, sleep disturbance and hindrance in simple talk performance. Performance in class room and in workshops are affected by irregular bursts of noise which are more disruptive than steady noise up to 90 dBA. ISO, 1999 has fixed the upper limit of tolerable noise level for 8 hours exposure at 85 dBA. Harmful effects on the body is included in the physiological effects of noise. Exposure to loud noise for a long duration may result in a threshold shift of hearing - permanent or temporary. Subjects also experience tinnitus vertigo, headache and fatigue. Exposure to high noise levels for longer duration also cause palpitation, increase the cholesterol level in blood (Blood pressure), dilate the blood vessels of brain and upsets the chemical balance of the body. In addition to headache, nausea and giddiness, it increases secretion of adrenalin and decreases urine formation.

Combination of sound level, duration and repetition that are considered acceptable for human exposures were prepared by the National Research Council Committee on Hearing, Bioaccoustics and Biomechanics, generally referred as CHABA. It is clear that harmful noise begins at level of 90 dBA for occupational situation and it is of interest to find what level of noise fall in the range. The parameters of noise which are important for any assessment are:

1. Temporal distribution of the sound

2. Magnitude of the sound
3. Cyclic frequency of sound
4. Coherence or time-variance of the sound.

Thus, noise varying over time must be analysed statistically. The equivalent continuous noise level is the level of steady sound in a given location which has the same A weighted sound energy. The time varying sound is the basic parameter in evaluating the risk of hearing impairment in noisy environment. Peak measurements are also necessary to check the impulsive noise standards.

The occupational safety and health administration (OSHA) of USA regulations for protecting the workers in noisy environment is probably the most rigorously implemented regulations. The permissible noise exposure limit as per OSHA (1971) are given in Table 10. While the allowable limit as 90 dBA for 8 hour shifts of work, it may be noted that every 5 dBA increase from this limit of noise level in the working environment requires the allowable exposure period to be halved. Controlling noise within tolerable limits is needed. Thus, it seems that all activities of modern life must be guided by noise standards. Only a couple of examples are taken here.

Table 10 : Permissible noise exposure limits per day (OSHA, 1971)

Sound level (dBA)	Time (hours)	Sound level(dBA)	Time (hours)
85	16.00	100	2.00
86	13.93	101	1.73
87	12.13	102	1.52
88	10.57	103	1.32
89	9.18	104	1.15
90	6.97	105	1.00
91	6.97	106	0.87
92	6.07	107	0.77
93	5.28	108	0.67
94	4.60	109	0.57
95	4.00	110	0.50
96	3.48	111	0.43
97	3.03	112	0.38
98	2.83	113	0.33
99	2.25	114	0.28

Source : Occupational safety and health administration (OSHA, 1971)

Aircraft Noise

The population around airports are disturbed by the aircraft movements in the day. The swedish committee for aircraft noise has recommended the critical noise as shown in Table 11.

Road Traffic Noise

Road traffic is considered most noisy activity of modern life. The road traffic noise result from

— road surface

— speed of traffic

— proportion of heavy vehicles and traffic volume

— gradient and intersections

— reflection of noise from buildings

— obstacles to passage of sound

— absence of vegetation and plantations

— meteorological effects

Concrete roads are more noisy (5 dBA higher) than bituminous roads, while wet roads are less noisier. Normally doubling the speed from 50 to 100 kmph^{-1} will increase noise level by 9 dBA, while lowering speed from 98 to 86 reduces noise by 1 dBA. For each 5 per cent noise was found to increase by 4 dBA. Reducing heavy vehicles from 20 to zero reduced noise by 5dBA. The change of noise with change of speed is shown in Table 12.

Table 11 : Critical noise level near airport

Permitted number of Aircraft movements per year	Critical noise in dBA
500	95
1500	90
5000	85
15000	80
50000	75

Table 12 : Mean noise level (L90) and changes with the speed

Original speed	Changes in noise level in dB at other speed				
kmph	50	70	90	110	130
130	-	-	-3	-1	0
110	-	- 5	- 2	0	+1
90	- 5	- 3	0	+ 2	+ 3
70	- 2	0	+3	+ 5	-
50	0	+ 2	+ 5	-	-

Recommended limits of noise for residential areas near main roads for some advanced countries are shown in Table 13.

Table 13 : Limits of noise levels in residential areas

Country	Noise level (dBA)	
	Day	Night
Sweden	55	45
West Germany	55	40
Austria	46-51	36-43
Switzerland	60	50

B. Air Pollution

In developed countries 30 per cent of the fossil fuel is burnt for operating the road vehicles and aircrafts which pollute the environment. Poor maintenance, large proportion of overaged vehicles, traffic jams due to improper planning, extensive overloading, poor upkeep of the road, urban development, all result in about 33 per cent more emmissions of carbon monoxide by a road vehicle in India in comparison to that in USA. The vehicle emission will increase three fold in the next decade due to increased vehicle population and added congestion thereoff. The average emmission from petrol and diesel in kg/100 litres of fuel burnt is given in Table 14.

In developed countries it is found that 63 per cent of the carbon monoxide presents in the air in towns is from the roads. CO affects blood and results in impairment of vision and judgment, and cardiovascular changes. Hydrocarbons and oxides are relatively inoffensive at ambient concentrations. Nitrogen oxides can form acidic compounds and kill plant and animal tissues. Lead is well known metabolic poison which can cause toxicity in infants, Brain

Table 14 : Components of exhaust gas (kg/1000 litres)

Components	Petrol Engines	Diesel Engines
Carbon monoxide	274	7.1
Hydrocarbons	24	16.4
Nitrogen dioxide	13.5	26.4
Sulfur dioxide	1.1	4.8
Organic acids	0.5	3.7
Aldehydes	0.5	1.2
Solid particles	1.4	13.2
Lead (mg/m3 of exhaust gas)	5-30	-

disease and metabolic disorder. Bad driving habits, the rate of emmission varies for different types of engines (types of combustion) and the age of the vehicles. Thus it has an implication on the national policy on automobile manufacture and enforcement on maintenance of vehicles. The dispersion of the emmission also helps in alleviating the situation. But, the meteorological conditions play a big role in dispersion and dilution of the pollutants. The wind velocity, topography and ambient temperature controls the dispersion.

Land Abuse

The modes requiring dedicated right of way eventually proliferate in different forms. Water way does not suffer from this disadvantage, and therefore all new road or rail project must be evaluated for alternatives by which it can be avoided. Construction of roads or railway disturbs the drainage of land unduly by interfering with free flow in most cases (though sufficient flood openings are provided). New construction approach vegetations causing erosion, biological and chemical solution, increasing turbidity or temperature in streams etc. New techniques should be evolved to avoid burrow pits for the embarkments. Railways use of wooden sleeper should be totally stopped and concrete sleepers may be used. Most of the new road or a rail line disturbs the ecological setup. This is even more severe in hilly areas. Blasting, rock cutting and excavation for new construction leads to geological and geophysical disturbances which affects the stability and performance of the road itself. Other associated environmental issues are flash floods, leaching of nutrients and aestheic degradation. Like the new construction, the maintenance techniques result into lot of environmental problems. In many developed countries demolition plan for multistoreyed building is approved along with its construction plan. Such a strategy will have to be adopted for maintenance of roads and railways as much of the waste materials of contruction and maintenance are generally left dumped near the site.

Conclusion

Large scale impact of any transport development or activity cannot be denied. The awareness of this is only a recent one for developing countries. Due to other important priorities, adequate concern is not shown even now. Moreover, the resources and other constraints for implementation of any stringent standard has pushed us to a stage of high tolerance. But, it is a fact that a general concern about environment is a must to protect it from any further deterioration. National policy and action plans can only help in this regard.

References and Further Reading

Carter, L.W. 1977. *Environmental Impact Assessment*, McGraw Hill Pub. New York.

Chaupnik, J.D. 1977. (Ed.) Transportation noises, *Ann Arbor Science*, Ann Arbor,

CHHS, 1971. Road Rail Noise-Effects on Housing, Central Mortagage Housing Corporation, Canada.

Cohn L.F. and Hevoy, G.R., 1982.Environmental Analysis of Transportation systems, John Wiley, New York.

Crocker, H.J. and Price, A.J. 1975. Noise and Noise Control, CRC Press Cleveland.

ISD, 1982. Assessment of Occupational Noise for Hearing Conservation Purposes, ISD 1999 . International Organisation for Standardisation, Switzerland.

Lohani, B.N., 1984, *Environmental Quality Management*, South Asian Pub., New Delhi.

Siksar, P.K. 1990. Impact of Transportation activities on Environment. Q.I.P. course on Environmental Impact Assessment, University of Roorkee, Roorkee.

CHAPTER 3

ENVIRONMENTAL WASTE AUDITING

G. Ashutosh

Importance

Sustainable development will only become a reality if we adopt methods of production that generate less waste and less emissions than traditional industrial processes. Some time the change involves new technologies, however improvement in operation can often dramatically reduce the level of emissions. A reduced level of emissions and wastes of other types frequently means saving cost in production and also to economize accurate information about the origin and source of environmental releases is important for effective waste discharges. Once the sources are identified, then the most effective options for avoiding, reducing and recovering the wastes can be evaluated. This topic deals with a brief procedure for waste auditing useful for factory managers as a management tool that lead to cleaner industrial production (Fig.3.1).

Waste

Waste in broad sense defined as anything that is non-product discharge from a process which may be in the form of gas, liquid and solid phases.

Fig. 3.1 : Objectives of Environmental Audit

(62)

Importance

The concern for the environmental protection has increased manifold in the recent years and all of us are worried about health hazards that are posed by deteriorated environment. In order to reduce environmental deterioration, we require certain good practices to utilize our natural resources more effectively. The environmental audit is an efficient tool to achieve this goal. International chamber of commerce and industry defines environmental audit as : A management tool comprising systematic, documented, periodic and objective evaluation of how well environmental organisation, management and equipment are performing with the aim of helping to safeguard the environment through, facilitating management control of environmental practices, assessing compliance with company policies, which would include meetings of regulatory environment. It is also defined as follows.

The environmental audit is,

(a)   An activity of verification of records.

(b)   A comparison of outputs with expectations.

(c)   An assessment of avoidable errors and wastes.

(d)   An assessment of risk.

(e)   An investigation into the field conditions.

(f)   Evaluation of managerial effectiveness.

(g)   Validation of the environmental data, records and reports.

(h)   A basis to provide recommendations to improve the environmental management systems.

Concept

The concept of environmental audit has emerged out from the behaviour of a progressive person who often puts himself the question am I on right path, can I choose better, how has the other choose? The question of hazards of various industrial/developmental activities has initiated the introspection, analysis and the title of environmental audit has been given to these activities in combination.

Components

Environmental audit has basically two components.

(1) Assessment and (2) Verification.

Assessment

This provides expert judgement in environmental hazards, risks, their control and management. It also provides insight into the prevailing activities with associated risks and recommendations to improve the organisations' approach to control pollution and better environmental management.

Verification

To ascertain and record the performance by appraising the application of, and adherence to, environmental policies and procedures. It also certify the validity of the data and reports related to environmental management and pollution control, and evaluates the managerial effectiveness to the environmental management.

Objectives

The main objective of the environmental audit is sustainable development. Which can easily be achieved by controlling environmental pollution. As depicted in Fig.1. In a cyclic order control of environmental pollution leads to the conservation of natural resources, which ensures good health, better safety (minimization of risk of health and safety) and improved productivity. All these in combination ensures sustainable development.

Enterprising

Most of us are calling driving force behind the environmental audit, but it is the compulsion which is enterprising our activity.

Environmental Consciousness

(a) Sustainable development
(b) Strong economy and a healthy environment are compatible goals.
(c) Our own future.
(d) Changing employees and customer attitude.
(e) Co-operation between environmental groups and industrial units.

Regulatory

(a) Laws and Standards.

(b) Regulations

(c) Fines

(d) Risk of civil and criminal prosecution to the organisations and pollution control personnel.

Methods

Environmental Audit is a well organised systematic study of how well we are performing with the environmental management and environmental quality practices.The major activities, which are involved in environmental audit, can be carried out in three different phases.

(a) Pre Auditing.

(b) On Site Activities.

(c) Post Audit Activities.

Pre Audit Activities

(1) Audit Scheduling

(2) Nomination of the Audit Team

(3) Setting out Tasks and Priorities

(4) Education on Environmental Audit

(5) Review of Background Information

(6) Preparation of Background Paper.

On Site Activities

(1) Interaction with Local Staff

(2) Inspection of Field Conditions

(3) Sampling and Test

(4) Review of Desirable Records and Documents

(5) Tentative Findings.

Post Audit Activities

(1) Preparation of Environmental Audit Report

(2) Solutions and Recommendations

(3) Formulation of Action Plan

(4)   Implementation of Action Plan

(5)   Reviewing on the Progress of Action Plan.

Benefits

The Environmental Audit has several merits, it provides us

(1)   Assurance to Compliance with Regulations and Standards.

(2)   Development in Environmental Management.

(3)   Improvement in Environmental Performances.

(4)   Awareness to Environmental Issues.

(5)   Reduction in Potential Liabilities.

(6)   Improvement in Sharing of Informations.

(7)   Reduction in Potential Pollutants and health hazards.

(8)   Reduction in the Consumption of Water and Other Raw Material.

(9)   Improvement in Productivity.

(10) Abatement of Environmental Pollution.

Environmental Audit in India

In India, the procedure for environmental audit was notified by the Ministry of Environment and Forests in the year 1992 (vide notification No. GSR 329 (E) dated 13th March, 1992) under the Environment (Protection) Act, 1986. In the year 1993, by an amendment (vide notification no. GSR 386E, dated 22nd April, 1993) the term environmental audit has been revised to environmental statement.

Environmental Statement has to be submitted by every concerned person carrying on an industrial, operation or process requiring consent under the section 25 of the Water (Prevention and Control of Pollution) Act, 1974, or under the section 21 of the Air (Prevention and Control of Pollution) Act, 1981, or both or authorization under the Hazardous Waste (Management and Handling) Rules, 1989, issued under the Environment (Protection) Act, 1986. The environmental statement has to be submitted to the concerned State Pollution Control Board in a prescribed form (Form V), for a period ending on 31st March, latest by 30th September every year. Environmental Statement is not merely a document or filling up of a form, it has several good objectives behind it.

FORM - V

Environmental audit report for the financial year ending 31st March

PART - A

(i) Name and address of the owner/
     occupier of the industry,
     operation or process

(ii) Production capacity

(iii) Year of establishment

(iv) Type of industry

(v) Date of the last environmental
     audit report submitted

PART - B

(i) Water consumption mgd^{-1}

Process

Cooling

Domestic

Name of the Product	Water Consumption per unit	
	During the Previous Financial Year	During the Current Financial Year
	(1)	(2)
(1)		
(2)		

(2) Raw Material Consumption :

Name of the Raw Material	Name of Products	Consumption of Raw Material per Unit of Output	
		During the Previous Financial Year	During the Current Financial Year
(1)			
(2)			

PART - C
POLLUTION GENERATED
(Parameters as specified in the consent issued)

(a) Water

(b) Air

PART - D
HAZARDOUS WASTES
(As specified under hazardous Waste management and handling Rules, 1989)

Hazardous Wastes	Total Quantity (Kg.)	
	During the Previous Financial Year	During the Current Financial Year
(a) From Process		
(b) From Pollution Control Facilities		

PART - E
SOLID WASTE

	Total Quantity	
	During the Previous Financial Year	During the Current Financial Year
(a) From Process		
(b) From Pollution Control Facilities		
(c) Quantity Recycled or Re-Utilized		

PART - F

Please specify the characteristics (in terms of concentration and quantum) of hazardous as well as solid waste and indicate disposal practice adopted for both of these categories of wastes.

PART - G

Impact of the pollution control measures on conservation of natural resources and consequence on the cost of production.

PART - H

Additional investment proposal for environmental protection including abatement of pollution.

PART - I
MISCELLANEOUS

Any other particulars in respect of environment protection and abatement of pollution.

CHAPTER 4

CASE STUDIES ON EIA

I. A CASE STUDY ON SECOND STAGE EXPANSION OF MALPE FISHING HARBOUR

N.C.L.N. Charyulu

Introduction

Each generation is a trustee of Environment for succeeding generation. As such every person should strive to maintain the environment for sustained growth in the standard of living. Every development activity will necessarily have an impact on the environment. An impact is defined as any change in the visual, physical, chemical, biological, cultural, social, economical structure and/or quality of the region.

The main purpose of Environment Impact Assessment (EIA) is to identify and evaluate the potential impacts, (both beneficial and adverse), of any proposed developmental activity/project on the environmental system consisting of air, water, soil with all living and non living matter and their interaction and interdependency. Thus the goals of EIA are to provide (i) sufficient clean air, water and soil free of any toxic substance that is injurious to all living beings on the earth, (ii) recycling of wastes and (iii) minimizing noise.

In the present case study, EIA process comprising of main components, major steps and activities (with personnel) to be involved and methodologies adopted for arriving at a proper EIA statement are discussed with reference to expansion of a minor fishing harbour project, Govt. of Karnataka, Dept. of Fisheries, Bangalore, 1994.

The process of EIA is an evolving process with a history of barely two and a half decades (Larry 1977). So much so there is a proliferation of literature with various case studies (Asit and Agarwal 1992), yet there are very many ambiguities about the efficacy and the effectiveness. However, there is a common acceptance on the basic concept that EIA is a management activity and hence a dynamic one (Wood 1988). This is also becoming an integral part of the planning process. In the first instance before undertaking an action plan for preparing EIA it is necessary to determine whether the project falls under relevant act necessitating the exercise. Further a decision is to be arrived whether (a) the project is accepted at all or not, (b) the project is accepted with an amendment or modification, (c) does an alternative which result in the same good for the population, exist or not. Besides the degree of toxicity generated by polluting nature of the project, the threshold value of capital investment has been fixed as the criterion which makes promoter of the project to get the EIA prepared for the purpose. Government of India also categorized the industries depending on the degree of toxicity produced through the effluent generated from the project.

The EIA presumes that thorough analysis of reliable data and the explicit statements derived thereof, will help in changing the policy decision, if necessary, for a better standard. The main components, hence, in an EIA report are: (i) Clear definition of the proposed project, (ii) Scoping or identification of impacts, prediction and assessment, (iii) Report presentation, preliminary draft, public participation and feed back, review/revision, final report, (iv) Clear qualitative and quantitative statements on irreversible and irretrievable commitments of resources, if any, necessary for the project should it be implemented, (v) Impact monitoring and mitigation methods and (vi) Provision for policing and environment auditing after the project implementation.

In this section an attempt is made to describe briefly the EIA process undertaken by the author for second stage expansion of Malpe Fishing harbour project for Government of Karnataka (1994).

Project Description

Scope and definition: Karnataka State which has a maritime coast line of about 300 km can contribute 20 per cent of country's fishing industry by expanding and providing infrastructural facilities in their present fishing harbours. Government of Karnataka have drawn-up plans to improvising harbour facilities at different centers like Malpe, Gangolli, Karwar etc. Accordingly a proposal for 2nd stage expansion of existing Malpe Fishing Harbour at a cost of about 11 crore was prepared in 1989. Their expansion project activity comes under the schedule of activities requiring the pollution control board clearance. Hence an EIA and EMP report for the project was prepared.

Objectives

The main objectives are to (i) improve and modernize the facilities, (ii) increase direct and indirect employment, (iii) increase the productivity in fishing, (iv) improve directly and indirectly the trade, (v) decongest, (vi) provide hygiene in the working environment.

Activity and Location

The proposed project is an expansion of the existing fishing harbour at Malpe which is a village situated very near to historically prominent town, Udipi (13 N -17 E) in Karnataka state along Arabian coast.

The activity envisages different construction works viz. (i) construction of new and extension of existing quays, (ii) construction of jetties, auction halls, (iii) dredging, (iv) dry-dock facility creation and (v) providing other facilities : lighting, water supply for cleaning and drinking, roads, sanitary blocks, fire service station, fuel supply, hygienic vending services.

Present Activity Setting and Likely Changes

Malpe is not an all time weather harbour with small basin of 2000 square meters, a quay of 118 meters and 2 jetties of 100 meters long. 60 meter wide built to serve only, 256 small mechanized vessels and 23 deep sea fishing vessels. However, as against the above now 529 small vessels are used. The present use pattern is given in Fig. 4.1. Actual annual average landing of fish during 1986-91 is presented in Fig. 4.2.

Important varieties of fish that normally land are oil sardine, mackerel, catfishes, croakers, seer fish, panacid prawn etc. The

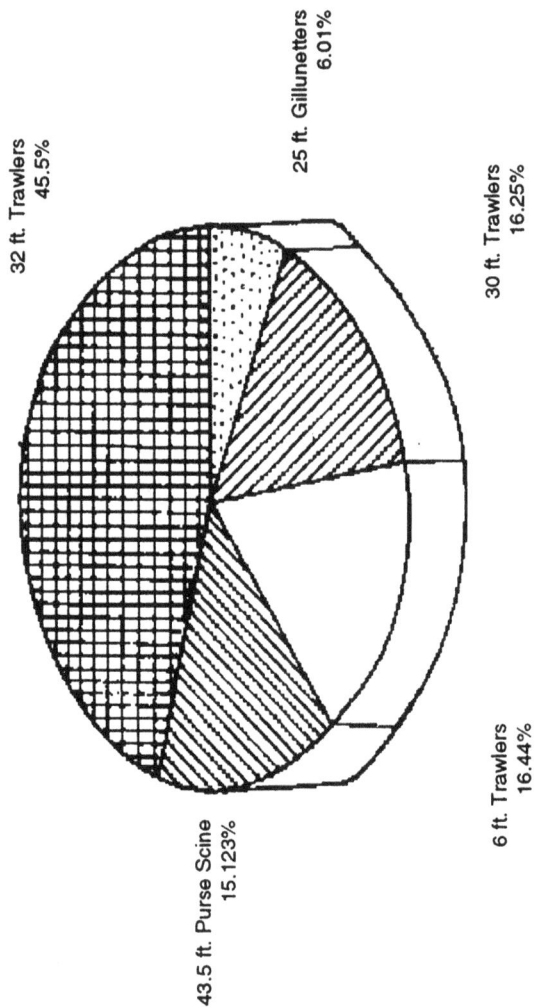

Fig. 4.1 : Types of mechanised vessels used in the Malpe harbour

32 ft. Trawlers
45.5%

25 ft. Gillunetters
6.01%

30 ft. Trawlers
16.25%

6 ft. Trawlers
16.44%

43.5 ft. Purse Scine
15.123%

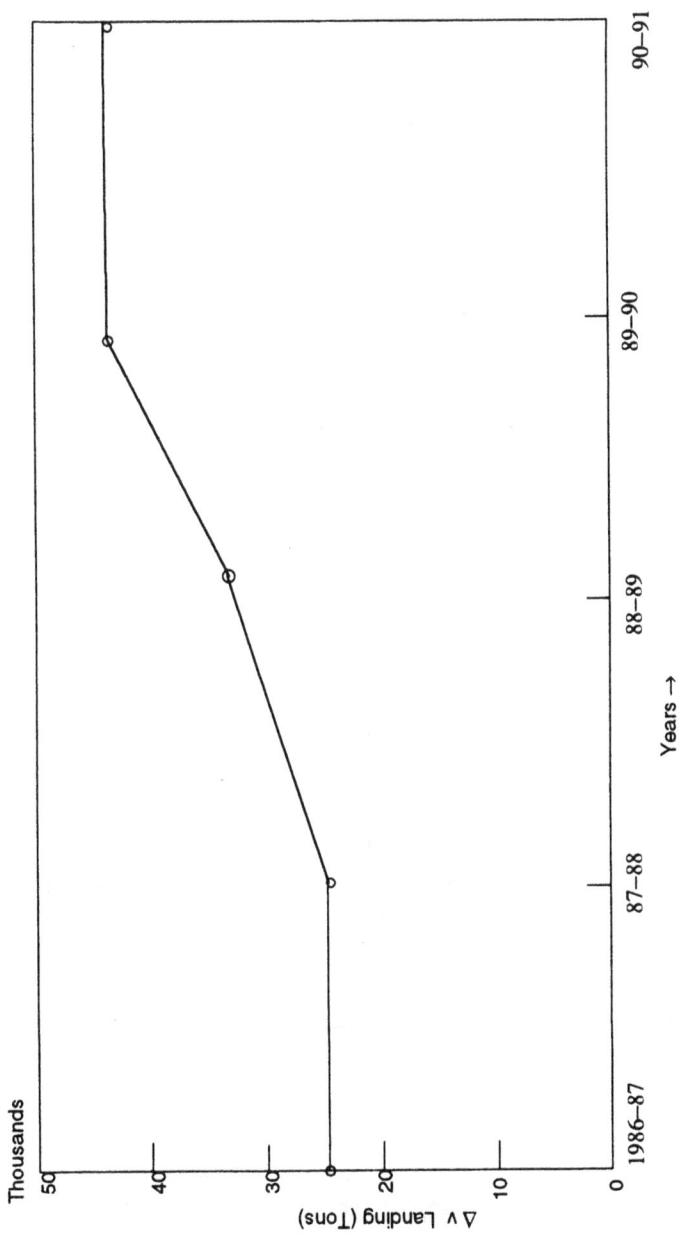

Fig 4.2 : Actual average annual landing of fish at Malpe harbour

expansion will help in doubling the landing fish, increasing productivity, expansion of fish trade to interior markets of state and also export.

Alternatives : Three options were discussed

1. No project
2. Diversion of vessels to other nearby port
3. Development of other ports in preference to Malpe.

Present Environmental Setting Within 5 km Radius

Socio-economic: The base line data on population, community structure, housing, literacy, occupational structure, health, green belt and other amenities were collected with the help of authorities of Malpe Fishing Harbour and D.K. district statistical department. The population with females more than males showed an increase from 17,985 in 1981 to 19,179 in 1991. The fisherman population is 6,933. The literacy rate is about 70 per cent with 6 primary, 2 junior secondary and 1 higher secondary school. The main occupation is in house hold industries. People in the region have no major occupational diseases and health, environment is satisfactory.

Air, Water, Noise

Environmental quality of air, water and noise levels were generated for 14 months with 6 air sampling stations and one sample per month per location, 17 locations for water with one sample per month per location and for noise, 8 locations 2 samples per location. The preproject air quality level was within the prescribed limits. Since no major air pollutant emissions, the quality can be expected to be maintained. Similarly the noise intensity created by the towing boats is likely to be decreased due to distribution of boats. So also the water quality is likely to be maintained.

Impact Assessment

Battle Environment Evaluation System (BEES) is applied to quantitatively assess the impact. Though in this system there is significant arbitrariness but gives a good pointer and thus can be used for relative evaluation between alternatives. Two phases viz., construction and operation, were considered and the over all weighted points found to be between -1000 to 2000 which indicates non-injurious but requires management plan.

Environmental Management and Suggestions

The sources of pollution during construction or operation are from trade, handling of cargo, spills/leaks/failures and solid waste. The assessed impact on environment during operation is more than the construction. However all waste waters generated from auction hall, cargo vessel cleaning, vehicle cleaning, canteen and bath rooms etc. are suggested to be treated by common effluent treatment plant. Further safety audit and risk management was suggested to be carried out periodically. The post construction and operation phase environmental control interaction are suggested for protecting the environment and ecology.

Conclusion

Environmental Impact Assessment due to any proposed developmental activity should be made as a statutory requirement. This is an evolving process. EIA is a management activity and is to be incorporated in the planning process.

References

Larry, W. Canter, 1977. *Environmental Impact Assessment*, McGraw Hill, New York,

Wood, C., 1988. Environmental Impact Assessment in Plan Making in *Environmental Impact Assessment Theory and Practice*, ed. Peter, W. Win Hyman Ltd. London.

Asit, K. Biswas and Agarwal, S.B.C., 1992. *Environmental Impact Assessment for Developing Countries*, Butterworth-Heinemann Ltd.

II. HUMAN IMPACT ON HIMALAYAN ECOSYSTEM

B.B. Hosetti

1. Scope of EIA

The impact of human activities on the Himalayan biogeophysical, socio-economic and cultural environments has been analysed.

The main man induced activities which have affected the equilibrium of Himalayan mountain ecosystem are:

1. unplanned land use
2. cultivation on steep slopes
3. over grazing
4. major engineering activities
5. over exploitation of village community or community forests.
6. shifting cultivation
7. unplanned tourism and urbanization
8. cold desert conditions prevailing at about 41,692 sq. km of the northwest Himalayas.

The geomorphological conditions, arrested succession, and checking the climax formation are major causes for land slides. Sedimentation, changes in surface and ground water hydrology and leaf fall etc. have caused eutrophication. In addition the drying of natural streams and receding glaciers were observed. Wild fauna like musk deer *(Moschus moschiferous)* and snow leopard, *(Panthera unica)* are now under threat due to changes in their habitat. Population pressure, migration and settlement are major causes of poverty and agglomeration.

2. Introduction

High mountain are major subdivisions of the biosphere and cover almost 1/4 of the land surface. About 10 per cent of world population live in mountain areas, while around 40 per cent of the world population are depending upon mountain resource for fuel, fodder, timber, energy, water, agriculture, minerals and recreation.

Humans are the important factor to change most of world's mountain ecosystems. The question is how great is man's impact on the world's mountains. Mountain ecosystems are sensitive to quite small disturbances. This is more true in case of tropical forest ecosystems. These disturbances were explained in UNESCO's Man and the Biosphere (MAB) program in its Project No: 6.

Impact of Human Activities on Mountain and Tundra Ecosystems : This project covers the management and sustainable development of mountain ecosystems.

3. Salient Features of the Himalayas

Indian Himalayan portion covers an area of 523,000 sq. km representing 16 per cent of the total area of India. The Himalayan catchment area is directly fed by Indogangetic and Brahmaputra plains, which cover 1,249, 250 s.q. km. or 38 per cent of total country's land surface. The himalayan hills are geologically young, created about 60-70 million years ago. About 300 million people live in himalayan region and its foothills. Forests have spread in about 38 per cent of himalayan surface which starts from 300 m to 8000 m height. Most of the northern rivers originate from himalayas. The impact of growing human population and the intervention of man in himalayas is illustrated as below.

1. Analysis of environmental components which are influenced by man.
2. Ecosystem approach for resources conservation and sustainable development.

4. Environmental Components

Several human activities on himalaya such as land use, cultivation on steep slopes, over grazing of natural grasslands, major engineering activities such as roads, mining, dams, reservoirs, irrigation canals, hydroelectric projects, over exploitation of village or community forests, cutting broad leaved trees, shifting cultivation etc, have degraded and threatened the equilibrium of mountain ecosystems. These problems have been analysed as physical, biological and socio-economic components.

Physical Components

Soil degradation and loss : 1/3 of the total himalayan land is derelict due to poor land maintenance, altered agricultural practices like cultivation on slopes (50 per cent) and intensive land use which lead to erosion and land slides. Traditional pastoralism i.e. herds are mainly privately owned. The private livestock holders have little interest to adapt to the carrying capacity of the pasture land. The traditional grazing of herds has reduced the regeneration of plants and caused deterioration of soils.

Atmospheric Pollution: It is not a serious problem in himalayas. It was noticed that electronic industries using chloroflurocarbons (CFC's) located in the central himalayas have made this area vulnerable to air pollution. Cement industries and extraction of nonmetallic ores like magnesite and lime have reduced the vegetation near point sources of pollution as well as degradation of land and water. Microclimatic changes include increase in evapotranspiration and the mean temperature induced by construction of dams and reservoirs. Eg. Surya canal and Gandak canal projects in the outer Himalayas.

Land Scape and Land Slide

Beautiful scenic places have been affected as a result of a large engineering activities. The formation of climax vegetation is also very rare due to negative impacts on plant succession. It was found that construction of roads and highways are responsible for the landslides.

Receding Glaciers and Impact on Water Resources

The abandonmen of grasslands of mountain areas and changes in vegetation cover have changed all the major avalanches. The climate change has lead to continued recession of snowline. Glaciers like Pindari in the central himalaya used for drinking and irrigation, and the rivers in Kumoan himalaya are now fast receding due to climate changes. The average rate of retreat at Pindari glacier is 23.74 myr^{-1}.

Water Resources : Decreasing waterbodies and eutrophication have been observed in the entire himalayas. Changes in water courses, fresh floods due to poor vegetation cover, low infiltration, diminishing perennial water resources are the common problems. Natural aquifers are drying up due to these disturbances. Reduced water and soil storage in mountains is also became cause for the problems.

Components Impact on Vegetation: These with high biological diversity and rich genetic resources are particularly fragile. Human activities have resulted in arrested succession leading to impoverishment of the germplasms of larger shrubs and trees and invasion of exotic weeds. Large scale transformation of land for chir pine - *Pinus roxburghii* plantation has resulted in changes in soil physico-chemical and biological properties. Over grazing also lead to poor recycling of nutrients. Forest ecosystems are failing to recover original states leading to desertification. In the entire west Khasi, Meghalaya, Ladakh, Western himalaya and in several parts of Kumaon and Garhwal, the replacement of original forest by fast growing exotic species like Eucalyptus, pine, Japanese Cedar has affected the native forest cover. The various important factors responsible for the decline in forest cover are shifting cultivation, expansion of agricultural land, forest exploitation for fuel, fodder, timber and major engineering activities like dams, hydroelectric projects and human settlement. Excessive deforestation and over use of mountain slopes is increasing the poverty that is closely related to the deteriorating biological, physical and socio-economic environments.

Impact on Fauna: Himalaya was once very rich in wild fauna, when the population pressure was low in the region, important fauna like musk deer - *Moschus moshciferous,* snow leopard - *Panthera unica,* Himalayan thar - *Hemitragus jamalhicus,* Brown bear - *Ursas arctos,* Black himalayan bear - *Selenactos himalayans* are now under serious threat due to changes in their habitats which include, introduction of exotic species, construction of roads, dams, reservoirs, mining, hydroelectric projects and illegal poaching. Fishes in high altitudes, the brown trout *Salmo trutta* farie, mahseer, - *Tor putitora* and in the lakes and streams of Jammu and Kashmir, Himachal Pradesh, have been seriously affected by obstruction in their migratory routes and changing in their breeding grounds. Bird migration has also been seriously affected.

5. Socio-economic Components

Population and poverty are closely linked. Human and livestock population are increasing in himalayas. Natural resources on which population depends are limited. The life supporting capacity of the mountain lands and the productivity of low lands are under extreme pressure. The increase in population have now exceeded the carrying capacity of the available habitat land. It has been resulted into overstocking of grazing animals and increase in need for new agricultural and pastoral land. Consequently the once, thickly forested upper land slopes get disturbed due to to erosion. These are being cleared for cultivation, fodder, firewood, grazing and timber.

Low agricultural productivity, low income, low standard of

living, lack of proper infrastructural facility are serious environmental problems in the himalaya resulted into migration. Out migration of hill people include temporary, seasonal, juvenile, periodic, and permanent migrations. This act of migration of hill people
to urban areas due to decline in agricultural production has increased the environmental imbalances.

Tourism and Recreation

Tourism has disrupted the normal ecosystem equilibrium of
the area. The main causes are (1) Tourist influx (2) hazardous
growth of tourism and adhoc planning in environmentally sensitive areas. Tourists have been known to spoil the environment. For
example they have spread and thrown containers or glass bottles
and tins everywhere they visited.

Health problems: Malnutrition, poor hygienic conditions,
migration from mountain to low land urban areas have resulted
into health problems. Tuberculosis has become a serious concern in
central himalayas due to malnutrition and poor working conditions. Migration is also posing serious pressure on housing, health
service and amenities in urban area. Mountain people living in
humid low altitude areas often develop bacterial, parasitic, viral
and microbial infections against which they have no natural immu-
nity.

Women : The act of migration of male population has increased
the work load on woman. Besides house hold duties, collecting fuel
wood, drinking water, fodder, childcare, cooking, cleaning, washing, the women also work in agriculture. Fuel, fodder and famines
compel the women to trek for hours to fetch water, timber for fuel
wood and fodder for animals. More than 90 per cent of the work
force in Kumaon central himalayas is conducted by women.

6. Resources and Conservation

The existing problems in himalayas are serious which need
management of natural resources in order to check further degradation and to meet the growing demands of the people. In view of
this, environment management should include natural, cultural
and landscapes, interacting with hydrological, climatological,
geological, biological, cultural and technical systems. Detailed

strategies are formulated for the management of physical, biological and socioeconomic environments where special emphasis on application of EIA of developmental activities in fragile area are needed.

In order to solve this, involvement of local people and use of locally designed and suitable strategies have been recommended for resolving the problems ensuring ecologically sustainable development.

7. Recommendations for Himalayan Ecosystem Management

Water shed and water resource :

1. Detailed knowledge about site, soil characteristics and land use is needed before accepting watershed management programmes.

2. Peoples committee should be formed impressing representatives from every village in the command area.

3. Attention should be given to decrease the run off, increase in infiltration, deep percolation by deep ploughing, mulching, well managed terraces on slopes, construction of check dams and subsurface drains.

4. Agroforestry programmes should be launched for watershed management and to meet the growing demand for fuel and fodder. The sustained yield capacity of the plantation (10-25 m3 ha/yr.) should also be maintained.

5. Rain water harvested from roofs may be used as safe drinking water.

Land

1. Land capacity with emphasis on degree of degradation.

2. Land with steep slopes (45%) should be utilised for revegetation only. However land with slopes (30%) should be utilized for agriculture.

3. Gulley erosion should be controlled by check dams and plantation on gulley banks.

4. Land deterioration due to shifting cultivation can be controlled by cultivation of perennial red grass (Cajanus cajan). or some leguminous creepers which fix nitrogen to improve soil fertility.

5. Erosion is a serious problem in himalayas. It should be controlled by improving arable farming, deep ploughing, contour cultivation, wing and strip cultivation, crop rotation, mulching, and well prepared terraces etc.

Forest

1. Massive scale afforestation programmes are needed with emphasis on area seeding in different topography.
2. Human interference should be avoided to maintain natural regeneration of vegetation.
3. Shelterbelts in industrial areas should be developed by selecting different tree species.
4. All the marginal land and common lands available in Himalayas should be brought into plantation programmes giving priority to broad leaved species for good covers.
5. Forest panchayat (Peoples committee at village level) should be placed in the centre to development efforts so as to provide protection to common property resources.
6. The right to utilize forest resources i.e., fodder, fuel, timber also should be given to villagers who started the plantation.
7. Villagers and school children should be encouraged to take part in plantation activities through "shram dan".
8. The practice of uncontrolled forest fire should be checked to avoid erosion.

Wild life

1. Careful analysis of conservation status of species concerned.
2. Evaluation of habitat (Carrying capacity)
3. Control of exotic species
4. Adequate planning for pest control
5. Alternative habitats for rare and endangered animals
6. Establishment of game breeding farms with active involvement of tribals captive breeding for rare and endangered animals.

7. Establishment of biosphere reserves/protected areas for regular monitoring and management.

8. Cottage industries should be developed based on wild life produces through tribal agencies.

9. Species from over stocked area should be translocated to understocked areas. Census of more uncommon species such as leopard, bear, wolf, thar, hyena, rated gibbon, pheasants, falcons and eagle should be under taken.

Poverty elevation

1. Fertility rates should be controlled by improving status of women. Special emphasis should be given to establishment and extension of small scale industry to check outflow of male workers.

2. Population/resource ratio and its use to be studied, carrying capacity of human population in himalayas should be considered.

Health

1. Primary health centres, aiming to improve health education, immunization campaigns and family planning services should be provided with involvement of government and voluntary agencies like Red Cross Society of India, village health workers, physicians etc.

2. Local health guidance and medical stores should be provided to each village to avoid long trekking.

Tourism

Tourism is considered together with agriculture, live stock production and forest management as well as health facilities, communication and transport systems. Tourism is to be taken into account together with management of natural resources of mountain areas. Detailed EIA is need to be carried out before opening for tourism.

Energy

1. Small scale hydroelectric projects should replace large scale projects,

2. Planting fast growing tree species
3. Utilization of wind and solar energy in rural and urban areas
4. Biogas installation
5. Use of smokeless storey should be encouraged
6. Use of LPG should also be encouraged

Pasture

1. Planting and protection of suitable shrub, trees and grasses to solve fodder and fuel problems
2. Implementation of soil and water conservation programmes
3. Pasture farming should be combined with crop farming
4. Storing fodder for dry season
5. Pasture is intensified by irrigation
6. Pastures should be maintained by biological technical and administrative and social measures.

Women : Women are sustainable developers and could be trained for a number of specific tasks, eg. Mushroom cultivation, handicrafts, plantations, bee keeping, cultivation of herbs, medicinal plants, sewing, sericulture, rabbit breeding, water harvesting for irrigation and drinking and the supply of fuel wood and fodder. Women should be consulted for all planning and extension of more Mahila mandal dal in villages to improve skills in women and to increase their income. Dissemination of environmental awareness and environmental education at all levels of learning is needed. Environmental education should be considered as a part of formal education system at all levels.

III. KONKAN RAILWAY PROJECT

B.B. Hosetti

1. Project Plan and Present Status

The konkan railway project, consists of a rail line of 760 km along the west coast of India. It runs from Roha of Maharashtra near Pune to Mangalore of Karnataka.(Fig 4.3). It passes through the coast of three states, Karnataka, Goa and Maharashtra. It has an

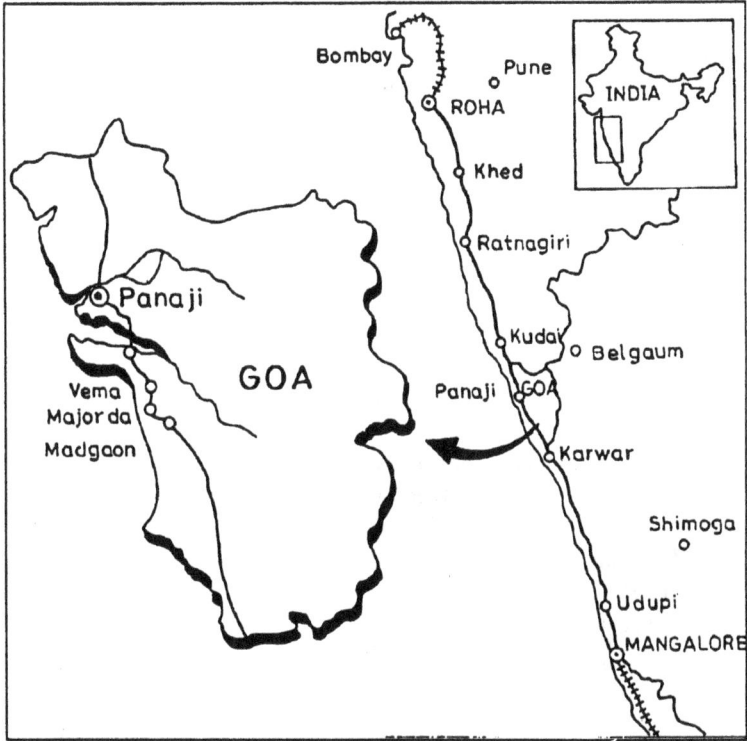

Fig. 4.3 : The Konkan Railway Route

estimated budget cost of 1400 crores. The Mangalore Udipi route has started functioning from the year 1993. It has completed about 105 km in Goa.

The then main dispute was over the decision on a route track between Mayem and Bali in Goa, a distance of 54 km, which constitutes only 7 per cent of the total proposed line. Till the issues raised by the localities of Goa are effectively attended to, this project cannot be completed before the deadline.

2. Opposition Groups

The chief opposition to this (KRP) Konkan Railway Project is being lead by (KRRAC) Konkan Railway Realignment Committee, a coalition of environmentalists. As the name of the group suggests, it wants the present alignment of KRP through Goa shifted from the coastal area to the midland in order to minimize its impact and safeguard Goan interests.

The rail line is to pass through the Mondovi Zuari estuarine ecosystems (MZEE), Khazan lands, and the carambolin lake, which is a notified wetland and the site of archaeological heritage of old Goa.

3. Alternatives

The alternative alignment suggested by KRRAC would increase the cost by 55 crores and length of the rail line by15 km with an addition of 189 km of tunnel. The KRP claimed that this would delay the project by one year and that it cannot afford the additional costs. Further, the Konkan railway corporation said that the alternative alignment suggested will destroy 350 ha of forest and defended that without realignment only 54 ha of forests may be affected.

To examine the issues raised by this controversy the Ministry of Environment and Forests set up an expert committee headed by Ms Kamla Choudhry. A majority of the committee members agreed on the historical and cultural issues raised. The members could not come to a unanimous decision on the destruction of Khazan wetlands and on other environmental issues. Therefore a sub-committee was set up to evaluate the issues raised by the controversy in relation to possible alignments.

1. The present coastal route.

2. A midland route

3. An eastern route

This sub-committee has stated in its report that every development project, especially large infrastructural projects like Narmada or Tehri or Konkan railway, is bound at large to have environmental and social, economic and cultural implications.

After assessing the three possible alignments the sub-committee concluded that the presently proposed line by KRP between Mayem and Bali is undesirable since it violated the crucial environmental, social, cultural and economic ecology of Goa. Therefore the subcommittee suggested to the KRC to opt for the midland route in the context of the ground situation 1992. The members of this sub-committee were —

Dr. V.A. Pai Panadikar, Prof E.F.N. Riberio, Mr Shyam Chainani, Ms Usha Albuquerque, and Dr. Kesavan Nair.

4. Impact Assessment

A team of experts like Madhav Gadgil, M.D. Subashchandran and scientists from Remote Sensing Institute and others were asked to undertake EIA work of KRP. EIA projects were first introduced in India by the planning commission in 1977 as desirable and not mandatory but now it has become a routine procedure. Gadgil and Subashchandran stated that they did not know exactly which was the best route, since no technical experts had studied, the tidal circulation, ecology of Mosquitos in coastal swamps which cause Japanese encephalitis, but they were forced to give their decisions.

These two ecologists started working with Rail India Technical and Economic Services Ltd., (RITES), a Government of India undertaking, on the EIA. The EIA was insisted on because the west coast highway cuts through the area and breaks tidal circulation, and the consequent water stagnation would cause the out break of Japanese encephalitis. The impact assessment was converted into a closed door process. The detailed report is still not available. It is the secret material of the KRC.

The biologists were asked to list the biological communities along that route and do little else. These persons were told not to consider any alternative routes and the construction along the route for instance Dasgoan-Panjim, was in full swing before start of EIA. Some suggestions, by the team like doubling the current line between Madgaon-Majorda, were incorporated but others were not. They diverted the main route plan near Kumta in Karnataka and near Madgaon - Panjim in Goa. The rout passes through Carambolin Lake which is the abode of thousands of water birds. The work on the KRP goes on with the contractors bulldozing hills, burrowing the earth, filling up wetlands recklessly and cutting mangroves.

The proposed fast rail linking Mangalore to Roha is known as Konkan Railway. When completed, it would offer high speed service on broad gauge tracks. Its design is based on the technology conductive to the use of diesel and supportive electric traction to enable the speed upto 160 kmh with a provision for separate lines in each direction. It requires a series of underbridges and over-bridges. Also emerging stoppage at high speeds would make level crossings difficult along the total stretch. The KRC line has to pass through the littoral districts of states Maharastra (382 km), Goa (106 km) and Karnataka (273 km). According to KRC, by Oct 1994 one track should start working and by 2004 both the tracks should start to work.

Moreover, it was a corporate scheme in terms of finance, and the upgrading of present gauge to broad gauge, addition of new spars, roads and other forms of development was not part of KRC budget. It is to be borne by the respective states. It is not clear now how these will be provided for and at whose cost. To meet the deadline of 1994 imposed by the MOR - Ministry of Railways, KRC searched for speedy solutions. Accordingly environmental, heritage and cultural issues were not given adequate weightage.

To partially integrate such doubts the EIA report was obtained. This has been prepared by RITES, a railway subsidiary, only to justify a predetermined alignment. This report was commissioned when the work was going on near Mayem and Panjim. The report was prepared and submitted in support of the present KRC alignment. There was no mention of Khazan lands or the damage to Carambolin lake, or the damage to archaeological and heritage sites.

This place had a rich civilization, for centuries including, Kingdoms, primarily Bhoja's of the 3rd century and the great Chalukyas, Kadambas, and finally the Yadavas of Devagiri until the Portuguese period. All of them, in the long history of Goa, have left their indelible mark in this fragile belt.

The KRC gave importance to engineering sustainability but not to the rest of the issues involved. In midland taluks there are no natural tropical forests. There is very little dense forest cover.

5. Khazan Lands

Khazan lands of Goa spread in 18,000 ha and about 2000 ha which are used for one crop of paddy and pisciculture (mostly illegal) and about 4000 ha are inundated currently which are used for pisiculture. The average production of salt resistant variety of rice was around 40,000 tonnes per year in the Khazan. In order to enhance Khazan land production the Government of Goa set up a Brackish Water Fish Farmers Development Agency, BWFFDA, in April 1992 and the state government started giving a subsidy of 50 per cent on the costs of repair of bunds and slice gates.

The point behind the agitation with regard to the same Khazan lands is to safeguard the important production of Khazanlands which are the life line for thousands of people. The agricultural land developmental panel of the government of Goa in its report of March 1992 stated that Khazan lands have to be studied as integrated systems specific to the estuarine region. Any bund like raised structure with a broad platform for constructing a road or railway track will cause divisions and a loss of drainage pattern leading to loss of huge fertile Khazan lands.

Estuarine pisciculture is offering new avenues for employment and for earning foreign exchange. The committee is convinced that a high 18 km long railway mud embarkment will cause irreversible ecological as well as economic damage since the income potential of Khazan lands is large.

Recommendations

The Mayem to Bali route is feasible from the point of view of engineering as the shortest route from Roha to Mangalore and therefore a careful environmental audit is proposed.

6. Present Status

This Konkan Railway controversy cropped us in March 1993 over the possible negative impacts of rail line on the coastal Goa and has delayed the completion for two year. The KRC has now practically gone away from its original proposal and has accepted the minimally modified route suggested by the Justice Oza committee. This change and delay has burdened the KRP in terms of financial budget and also has delayed the target date of completion from middle 1994 to 1996 . Now the KRRAC is silent , and the work is going on and would complete shortly. On July 26, 1996 the Central

Railway Minister, Mr Ram Vilas Paswan has stated that the rail line work of KRC will be completed by the end of December 1997. Presently 95 per cent of the work is over and the remaining 5 per cent will be completed shortly and the rail line may start functioning from February 1998 onwards.

IV. IMPACT PREDICTION OF AN IRRIGATION PROJECT : A CASE STUDY OF UPPER TUNGA PROJECT (UTP) AT SHIMOGA

Veena Nadig NK

1. Introduction to EIA

EIA is potentially one of the most significant valuable, interdisciplinary, pragmatic decision-making tool with respect to alternate routes for development, technologies and project sites. It is an ideal anticipatory mechanism which establishes quantitative values for parameters indicating the quality of the environment before, during and after the proposed developmental activity, thus allowing measures that ensure environmental compatibility.

The main objective of environmental impact assessment concerns the evaluation of the environmental implications of a development so that decision makers are informed of the likely environmental effects before any decision is taken. Evaluation is the determination of the magnitude of the impacts against the standards. These standards can be legislative policies or the wishes of the public in general or more appropriately the local community affected by the proposal.

Public participation is a key element in EIA. EIA is a part of the democratic process whereby key issues are highlighted through public participation. One advantage of good public participation is that the public can act as an information source for those carrying out the assessment. A good EIA should allow the developer to identify potential environmental impacts and consider the design and operational requirements to minimize or mitigate any negative environmental impacts. The scope of the EIA should concentrate on key impacts and include social, economic and environmental implications of the project. In its most comprehensive form,

EIA is an interactive process starting at the feasibility/conceptual phase and carries right through the life time of the project. The effectiveness of the EIA is dependent on the links with project planning from concept to abandonment.

EIA has often been static linear exercise characterized by a tendency to re-invent the wheel rather than a dynamic interactive process of continuous learning and improvement. Monitoring and auditing of impacts once project commences is an essential element of EIA. The central philosophy of this is that unforeseen impacts can be highlighted and changes in others may come to light so that actions to mitigate them can be taken. Monitoring and auditing are important for the irrigational projects, assessment linked to monitoring which leads to audits should be strongly encouraged. However, its effectiveness is dependent on the scope and scientific vigour of the monitoring program. EIA should also suggest alternatives whether of site or technology if applied in a comprehensive manner.

The benefits of environmental impact assessment (EIA) are now widely accepted and there is a growing belief that environmental assessment of policies, plans and programs (strategic EA i.e. SEA) may also be necessary to ensure that alternatives and impacts which cannot be fully considered at project level are adequately evaluated.

The policy of environment for management of land-use in India has been quite muddled. The beneficiaries of various water harvesting structures had not assumed responsibility for maintenance after the works were completed even when the benefits were substantial. None of the water shed have been used as an onfarm research site with experiments designed and implemented jointly by scientists and farmers for developing location specific technology. Participating in people's plan requires respecting their knowledge and experimental ethic. There are very few examples illustrating documentation of indigenous innovations and on-farm and station research on the validation of a value addition in the same.

2. Background Information About Irrigation

No branch of engineering has contributed more to the development of civilization than the art and science of controlling the flow of water. Irrigation projects can contribute greatly to increased income of agricultural production compared with rainfed agriculture. Irrigation contribution to food security in India is 55 per cent which is widely recognized. Irrigation is a key component of the technical package needed to achieve productivity gains.

Irrigation is an essential component of sustainable agricultural development but it is not a unique sector, since it faces challenges similar to those confronting other public and private sector economic activities. Irrigation allows for better and more diversified choices in cropping patterns and the cultivation of high value crops. Successful irrigation is a crucial determinant of the worlds future development because of its influence on the supply price of food.

Irrigated area claimed to an estimated 244 million hectares in 1991, some 4 million hectares were added to the irrigation base, enough to lift the per capita irrigated area slightly to 45.2 hectares per thousand people. Since about a third of the global harvest comes from 16 per cent of world cropland that is artificially watered, irrigated area is an important agricultural indicator. By far the largest demand for the world's water comes from agriculture. More than 2/3 of the water withdrawn from the earth's rivers, lakes and aquifers is used for irrigation. In India about 30 per cent of all public investment has gone into irrigation. Globally around 70 per cent of water withdrawals are used for agriculture.

In India, it is estimated that the submergence area of reservoirs roughly forms 20 per cent of the irrigated area. During last 20 years it is estimated that about 2 million ha of fertile land have gone out of cultivation when only 3.5 million ha have been provided canal irrigation between 1970-71 & 1989-90.

Irrigation can help to make yield increasing innovations a more attractive investment proposition but it doesn't guarantee crop yield increase. The overall performance of many irrigation projects has been disappointing because of poor scheme conception, inadequate construction and implementation or ineffective management. The mediocre performance of the irrigation sector is also contributing to many socioeconomic and environmental problems.

Water is not an easy sector in which to promote cooperation, but the potential gains are high, which makes renewed efforts worthwhile. Resolution of many water allocation and development problems requires a common willingness to forego personal benefits for the social welfare. Yes, irrigation is a service with customers and users, it's not a production industry. The salination, which is costly to remedy, while an additional 60-80 million hectares is believed to be moderately affected. Moreover, salinization may be spreading by as much as 1-1.5 million hectares per year. A recent world bank report finds the problem to be widspread in many important agricultural regions, affecting an estimated 11 per cent of the India's irrigated land area. Another sign of unsustainable irrigation is the overpumping of ground water. This problem is perbasic in important crop growing regions.

Water is certainly essential for human survival and is the staff life for all species but water related illness are the most common health threat in the developing world. An estimated 25000 people die every day as a result of water related sickness.

3. Project Proposal

3.1 Objective of UTP and technical detail of present Tunga anicut

The river Tunga is a major tributary of river Tunga Bhadra which in turn is a major tributary of river Krishna. River Tunga and Bhadra both take their origin at Gangamula in the Agumbe range of western ghats. River Tunga runs through a hilly terrain under the influence of south west monsoon and receives a copious supply of water from the hilly and wooded catchment of the ghats, having the mean annual rainfall of catchment area varying from 10795 mm (425 inches) at Agumbe and 762 mm (30 inches) at Gajanur. As per the gazed data of the river at the Gajanur site, the 50 per cent and 75 per cent dependable annual yield works out 6.398 M cum (225.96 TMC) and 5.22 M Cum (184.184 TMC) respectively.

River Tunga commands a total drainage area of 2240.35 sq. kms. (865 sq. miles) upto the existing Tunga anicut site and lies in the districts of Chickmagalur and Shimoga. In order to utilize 11.5 TMC of water from river Tunga, in 1946 an anicut was proposed across the river at Sakarebylu near Gajanur, lying between latitudes 13"-45'. 30" and longitude 75"-31'-0" which is 11.2 km from Shimoga. The area irrigated from this anicut is about 8669 hectares

(21500 acres). This project was commenced in 1946 and completed in 1956 at an expenditure of Rs 3.31 crores.

Table 15 : Technical details

Reservoir		
Average river bed level		RL 571.50 M (1875')
Lowest foundation		RL 566.00 M (1857')
Dead storage level	LBC	RL 582.14 M (1910 ft)
	RBC	RL 582.75 M (1916.25 ft)
Minimum draw down level		RL 584.04 M (1916.25 ft)
FRL		RL 584.04 M (1916.25 Ft)
MWL		RL 588.24 M (1930 Ft)
TBL		RL 590.06 M (1936 ft)
Rigid crest		RL 584.04 M (1916.25 Ft)
Gross storage capacity		37.89 M. cum (1338 mc. ft)
Dead storage		23.47 M.cum (829 mc. ft)
Live storage		14.42 M.cum (509 mc. ft)
Dam		
Type of Dam		Earthen dam with central puddle core
Maximum height of Dam	-	13.10 M (43 ft)
Length of Dam	-	176.77 M (580 ft)
Top level of Dam	-	RL 590.06 M (1936 ft)
Top width of Dam	-	5.48 M (18')
Free board	-	1.82 M (6 feet)
Side slope of earthern dam	-	Upstream 1:1 Downstream 2:1

Contd......

Spillway

Type of spillway	-	High off weir
Length of spillway	-	365.74 M (1200 ft)
Max flood discharge	-	7362.30 cusecs
		260000 cusecs

Submergence

Total submergence/acres	-	1632 hectare (4032 acres) at RI 1930 ft.
Villages affected	-	14 viz.
Sakrebail	-	Shimoga taluk
Tinkapura	-	Shimoga taluk
Naidile	-	Shimoga taluk
Hale lakkavalli	-	N R pura taluk
Bilagal	-	Shimoga taluk
Chatnahalli	-	Shimoga taluk
Bommenahalli	-	Shimoga taluk
Kanagalakoppa	-	Shimoga taluk
Kanabur	-	Shimoga taluk
Lingapur	-	Shimoga taluk
Charanayedhalli	-	Shimoga taluk
Sakrebylu State forest	-	Shimoga taluk
Muthinakoppa	-	NR pura taluk
Siddamaga Houser	-	NR pura taluk

3.2 Need of UTP

Existing

The existing cropping pattern is rainfed wet crop in the initial reaches of the proposed command (semiarid zones) and in the remaining area rainfed crops consisting of Ragi, Jawar, Groundnut, Cotton, Pulses, Chillies. But this is not assured due to the uncertainty of rainfall during the crop period.

Requirement

The irrigation facilities proposed for providing irrigation to 2124 Ha.

Table 16 : Irrigation facilities

Sl. No.	Project	Master plan for	
		734 TMC	1050 TMC
1.	Tunga lift irrigation scheme	-	10
2.	Tungabhadra lift irrigation Bhadravathi	-	12
3.	Tungabhadra lift irrigation at Shingatalur	8	15
		8	37

Detailed investigations for the three lift irrigation schemes were taken as per the above allocations. In the meanwhile in view of the acute power shortage faced by the state and also due to the difficulties involved in the management of large scale lift irrigation schemes, an alternative to the above lift irrigation schemes called Upper Tunga Scheme was taken up for detailed investigation for an utilization of 70 TMC providing irrigation facilities to 3,20,000 acres.

Originally the Upper Tunga Project flow scheme comprised of the following:

1. Construction of an composite dam across Malathy river, a tributary to Tunga on the upstream of the existing Tunga Anicut at Gajanoor.

2. Modifications to the existing Tunga Anicut at Gajanoor and installation of 13.75 ft. crest shutters over the Tunga anicut and

3. Running a high level canal of a length 350 miles on the left bank taking off at R.L. 1920.00 ft. for providing irrigation to 3,20,000 acres utilizing 70 TMC.

The detailed survey and investigation of Upper Tunga Project was prepared and estimated to cost Rs. 37.42 crores and submitted to the Govt. during May 1969. Normal rainfall of Shimoga is considered for computing consumptive use requirement of Kharif Paddy since Kharif Paddy is proposed for command area in Shimoga taluk only. Normal rainfall of Honnali is considered to Kharif. Semidry crops are proposed for command area of Honnali Taluk. Weighted average normal rainfall of Hirekerur, Rannebennur, Haveri and Hangal taluks is considered for computing consumptive use requirement of Kharif semidry crops proposed in Dharwad District. Hence a guideline for preparation of detailed project report of irrigation and multipurpose project was published by Ministry of Irrigation 1980. An allocation was made for three life irrigation schemes or upper Tunga project in the Tunga and Tunga Bhadra sub basin under 734 TMC and 1050 TMC master plan respectively.

The crop pattern proposed is KSD followed by RSD with an utilization of 70 TMC. An extent of 25059 acres was likely to come under submission under the Malathy Dam. Out of which 17021 acres are forest lands. About 25 villages will be affected due to the construction of Malathy reservoir. Further, there is a lot of agitation from the public of the submersible area of the Malathy reservior for the construction of Malathy dam due to huge submersion of the forest land and the other crop lands.

In order to avoid heavy submersion a decision was taken to prepare the project for utilization, that is available from the run off the river without the support of Malathy reservoir. Several alternative proposals for different utilization of 50 TMC, 48 TMC., 37 TMC., 25 TMC., 12.5 TMC., 15 TMC., 10.8 TMC., 8 TMC., have been prepared and submitted to the Chief Engineer, Water Resource Development Organization (WRDO), Bangalore.

As per the recent instructions of the Chief Engineer, Water Resources Development Organization, Bangalore, Upper Tunga Project without Malathy Dam for an utilization of 12.24 TMC has been prepared and submitted to the Government during May 1989. It comprises of

1. Erection of 30 radial crest shutters of 9.79 M × 4.19 M (32.15' × 13.75') over the existing Tunga Anicut with modifications to the existing structure.

2. Running a high level canal on left bank from Tunga Anicut with off take level at RL 584.45 M (1917.50 ft.) for a length of 208.58 miles flow canal and 2.42 miles of tunnel near Ranebennur to irrigate 2,34,000 acres of land.

The present M.W.L. at R.L. 588.23 M (1930 ft.) of the existing Tunga Anicut will become the F.R.L. after modifications to the existing Tunga Anicut and erection of crest gates over the anicut.

Table 17 : Salient features of upper tunga project

Taluk : Shimoga		District : Shimoga

I. General

1.	Name of the Project	:	Upper Tunga Project
2.	Name of the River	:	Tunga river
3.	Location : Latitude	:	1385' 30"
	Longitude	:	750 31'-0"
4.	Purpose	:	Irrigation
5.	Means of approach	:	Approachable by an aspahlt road 11 kms (7 miles) from Shimoga on Shimoga Mangalore road.

II. Geophysical features

1.	Catchment area at the site	:	2240 Sq. Kms. (865 Sq. miles)
2.	Nature of C.A.	:	Good
3.	Climate	:	Moderately cold
4.	Annual rainfall	:	Min. 762 mm (30 inches)
5.	Yield available at		
	a. @ 50% dependability	:	6393.56 M. cum (325.938 TMC)
	b. @ 75% dependability	:	5251.46 M. cum (184.184 TMC)
6.	Allocation as per 734/757 Master plan for Krishna Basin	:	To be accommodated in the allocation of water as per KWDT award.

III. Technical details

A.	Reservoir	Existing	After Modification
1.	Average river bed level	RL.571.50 M (1875')	RL.571.50 M (1875')
2.	Lowest foundation level	RL. 566.00 M (1857')	RL. 566.00 M (1857')
3.	Dead storage level	LBCRL 582.1 M (1910.00') RBCRL 582.75 M (1912.00')	High level LBC proposed at RL 584.45 M (1917.50')
4.	Minimum draw down	RL 584.04 M. (1916.23')	RL 588.24 M (1930')
5.	F.R.L.	RL. 584.04 M (1916.25')	RL. 588.24 M (1930')
6.	M.W.L.	RL.588.24 M (1930')	RL. 592.80 M (1933.25')

Contd......

		Existing	After Modification
7.	T.B.L.	RL.590.06 M (1936')	RL.592.80 M (1945')
8.	Rigid crest level	RL.584.04 M (1916.25')	RL.584.04 M (1916.125')
9.	Gross storage capacity	37.89 M. Cum (1338 Mcft)	91.86 M. Cum (3244 Mcft)
10.	Dead Storage	23.47 M. Cum (829 Mcft)	41.62 M. Cum (1471 Mcft)
11.	Live Storage	14.24 M. Cum (509 Mcft)	50.17 M. Cum (1773 Mcft)

B. Dam (Earthen Dam)

1.	Type of Dam	:	Earthen dam with Central puddle core	
2.	Max. height of Dam	:	13.10 M (43')	15.84 M (52')
3.	Length of Dam	:	176.77 M (580')	176.77 M (580')
4.	Top level of Dam	:	RL 590.06 (1936')	RL 592.30 M (1945')
5.	Top width of the Dam	:	5.48 M (18')	5.48 M (18')
6.	Free Board	:	1.82 M (6')	3.58 M (11.75)
7.	Side slope of earthdam	:	Upstream 1:1 Downstream 2:1	2:1 2:1

C. Spillway

1.	Type of Spillway	:	High Co-efficient weir	High Co-efficient weir
2.	Length of Spillway	:	365.74 (1200')	364.67M (1196.5')
3.	Location of crest Shutters	:		Central 30 spans of 9.79 M X 4.19 M (32.15' x 13.75')
4.	No. of piers	:		29 piers of 2.44 M (8')
5.	Maximum flood discharge	:	7362.20 cusecs 260000 cusecs	7362.30 cusecs 260000 cusecs

D. Submergence

1.	Submergence/acres	:	NIL
2.	Villages affected	:	1
3.	Roads affected Shimoga-Thirthahalli Road	:	12.35 Kms

Contd......

		Existing	*After Modification*

E. Canals

1. Length of the canal : 331.45 kms (206.01 miles)
2. Length of Tunnel : 3.90 kms (2.42 mils)
3. Length of deep cut : 4.148 kms (2.57 miles)
        Total : 339.56 kms (211 miles)
4. Capacity at head reach : 68.25 cumecs
5. Bed width : 14.25 meters
6. Full supply depth : 3.50 M
7. Bed fall : 6 inches/miles
8. Value of No. : 0.018
9. Velocity : 1.00 M/sec (3.289 ft/sec)
10. No. of reaches : 5
11. Canal off take level : RL. 584.45 M (1917.50')

F. Tunnel

1. Type of tunnel : D Type
2. Length of tunnel : 3.90 Kms (2.42 miles)
3. Capacity : 45.70 cumecs
4. Bed width : 6.20 meters
5. Full supply depth : 3.50 m
6. Bed fall : 5.50 ft/mile
7. Velocity : 2.33 M/sec (7.64 ft/sec)
8. a. Depth of cut at Ch. : 27.15 m (89.08 ft)
        111 miles 1315 ft.
    b. Depth of cut at Ch. : 28.21 M (92.56 ft)
        113 miles 3556 ft.
    c. Length at approach : 2.01 kms (1,249 miles)
    d. Length at exit : 2.136 kms (1,327 mils)
    e. Total length of deepcut : 4.146 kms (2,576 miles)
    f. Bedfall in deep cut : 2.25 ft/mile

G. Irrigation

1. Gross command area : 1,19,872 Ha (2,96,203 acres)
2. Irrigation intensity : 9 per cent
3. Net Irrigable area : 94,698 Ha. (2,34,000 acres)
4. Cropping pattern
    kharif pady : 2124 Ha (5250 acres)
    Kharif semidry : 92,574 Ha (2,28,750 acres)

Contd......

			Existing	After Modification
5.	Crop water demand	:	454.639 M. Cum (16.065 TMC)	
6.	Evaporation losses	:	20.72 M.Cum (0.734 TMC)	
7.	Total	:	474.359 M.Cum (16.80 TMC)	
H.	**Cost**			
1.	Total cost of the project	:	Rs. 37987 lakhs	
2.	Modification to existing Tunga anicut	:	Rs. 3366 1akhs	
3.	Cost of the canal	:	Rs. 34,621 lakhs	
4.	Cost per ha/acre	:	Rs. 40,113/Rs. 16,233	
5.	Benefit cost ratio at 10%	:	1.48	
6.	Additional food crops grown	:	3.742 lakh tonnes worth 178.56 crores	

3.4 Submergence

Biological diversity, genetic resources, food security, sustainable agriculture, rural development and environment are complex issues which have to be considered seriously not merely to improve agricultural productivity for immediate sustenance, but also for long term growth for the future generations. Depletion of forests is directly related with the rate of depletion of the forest gathered food resources. This dam destroy the flora and fauna of the Western Ghats specifically, magnificent sal forests, virgin forests and cultivable land of high quality. This means that a small high quality part of forest is destroyed to serve arid and semiarid parts of the country. It is a normal practice, that people from arid and semiarid regions (i.e., drought prone regions) migrate to more favorable regions and not vice-versa. In the present case, people residing in favorable regions who have no complaint whatsoever against nature are first made destitute and then shown charity. This is a crime against science and humanity. High technology is there to serve science and humanity, it is not there to destroy well settled people. If the Government of Karnataka is so rich as to submerge a thick patch of virgin forest, relocate people who do not need in the first place relocation, why does it not relocate its own people in favourable regions within the region? This will be in the National interest. The National beauty of the country and lives of simple innocent folks living in such surroundings cannot be saved by all kinds of compromises in the name of national interest.

According to the data collected by Karnataka Forest Department, the area coming under the river Tunga catchment area is as under (Table 18).

Table 18 : Details of tunga catchment area

Sl No.	Name of the range	Area in ha
1	Agumbe range	12022.57
2	Hanagere range	1700.00
3	Mandagadde	9204.37
4	Rippenpet	2828.34
5	Sakrebyle	5392.87
6	Thirthahalli	17668.21
		48816.36 ha

4. Socio-economics

The sociocultural roots of our present environmental crisis lie in the paradigms of scientific materialism and economic determinism which fail to recognize the physical limits imposed by ecological systems on economic activity. The emergence of the concept of sustainable development in realization that societal perceptions must shift towards ecological determinism so as to achieve qualitative growth within the limits of ecosystem carrying capacity. As defined by the United Nations World Commission on Environment and Development 'Sustainable generation without compromising their own needs and aspirations"

The project should be fully integrated so as to avoid actions being perceived as extremely driven and becoming "subject to problems". Along with land and energy sources water has been the focus of disputes and in extreme cases, even wars!! The increasing value of water, concern about water quality and quantity and problems of access and denial have given rise to the concept of resource "hydropolitics". In the current socio-economic and political context creating new bird sanctuary (?) is an arduous and delicate task. It requires consultation with the eventual beneficiaries, who, conversely, can also be seen as the 'victims. It is then necessary to evaluate the local people's perception of the project and provide them with the means to define the strategies and actions needed.

An efficient structure also includes a programming committee, which is indispensable for continuous integration of project activities into regional development plans.

The western ghats are being subjected to a series of human impacts which endangered their very existence and sustainability. Since the majority of the problems stem from human activities, it is of prime importance to put in place a program of education, training and extension at all levels. The greatest threat to western ghats is caused by the so called Developmental Projects which practices agriculture, destroying the forest and forest resources. The insecurity of land tenure and the lack of incentives for the sustainable use of the forest resources have contributed to a feeling of apathy on the part of the communities towards natural resources conservation efforts, creating an environment which is not conducive to concrete action. An analysis of the present status of forest in Western ghats reveals that the key problem is the increasing rate of deforestation and deterioration of the remaining forests.

Entire mosaics of the ecosystems are being submerged by massive dam projects. While habitat loss is one of the most visible threats. Melodious songs and flashy colors are not all that is being lost as birds decline. The disappearance of bird species sounds an alarm that environmental life support system are flattering and that our future is also at stake. Many people have to understand at least intuitively, that continuing environmental degradation would eventually extract a heavy economic toll. Of the major economic sectors, the one most vulnerable to environmental degradation is agriculture, simply because it is so directly dependent on natural systems and resources.

The Western ghats are the very centers of natures economic cycle on the southern parts of India, thus deserves the utmost care and attention that they can be given. To destroy them is the cut out the heart and thus bring death to the whole structure. The number of families who will get ousted by this project ranged from 156 to 600 and the author doesn't know which figure she should take as reliable. Assuming the lowest possible figure, even then, human suffering involved is heart rending. To honestly document the aspirations and hopes of these families is not an easy task. This shows the lack of scientific temper among the people in power.

The connection between water and human life is most dramatic in arid regions, where crop irrigation is essential to food production. Sustainable development is one that meets the needs of the present without compromising the ability of future generation to meet their own needs. To quote an eminent meteorologist Prof. Pisharoty a reputed scientist in Gujarath has argued that the best solution for our water problems is what the Dravidians did several centuries ago, the construction of about 30,000 water tanks, each about 100m X 100m X 100m in each of the 300 and odd districts which have an annual rainfall of 50 cms or more. Such water ponds could improve our ecology, trees would grow around them and ground water will be recharged. By means of their alternative the vested interests now centered around major and medium irrigation works would loose the opportunity of pocketing and wasting hundreds of crores of rupees, but the agricultural produces and the nation would certainly benefit.

Several fundamental questions related to the project objectives had to be answered in order to shape the basic guidelines. The relationship between society and the environment is a metabolism that can be disturbed and is being disturbed. However, it is dialectical relationship. The environment is an open subsystem evolving together with the economic and societal subsystem. Neither subsystem is a static, determined machine. Both society and nature are necessary sources of sustained human existence. It is tragic that the threat of aggression resulting from environmental problem, posing highly dangerous additional chains on scarce environmental resources. The world food program (WFP) definition makes theoretical sense "To allow for future generations, requires that we preserve our remaining resources and that we heal or rehabilitate resources that have been treated carelessly in the past. To do these things systematical is to follow a path of environmentally sustainable development. Governments generally tend to organize and administer water sector activities separately. One department is incharge of irrigation another oversees water supply and sanitation, a third manages hydropower activities, a fourth supervises transportation, a fifth controls water quality, a sixth directs environmental policy and so forth. These fragmented bureaucracy make uncoordinated decisions, reflecting individual agency responsibilities that are independent of each other. This project by project, department by department, region by region approach is no longer adequate for addressing water issues. The integrated approach requires water managers to understand not only the water cycle (including rainfall, distribution, ecosystem interactions and natural environment and land use changes) but also the diverse intersectoral development needs for water resources. Like many public sector personnel irrigation managers must walk a fine line between a tighter control of finance, the need for more active leadership and better planning of resource allocation on the one side and the contradictory need for more idea from below (farmer customer) on the other.

5. Versions and Reality

Sedimentation in reservoirs has other consequences besides the loss of storage capacity. Sedimentation and delta formation in the upper part of the reservoir, known as Aggradation in the back water zone, raise the water level and may cause flooding if the backs are low. At the same time, the opposite process, known as degradation, may occur in the stream bed below the dam, as the stream picks up a new load of sediment to replace what was deposited in the reservoir. The biological effects are many and complex both above and below the dam. Extensive changes in biota are frequently observed below dams. Small changes in meteorological variables such as temperature, precipitation and wind may have a serious impact on human activities such as agriculture, transportation, constructions and urbanization.

The use of forest resources and their extent of state are inextricably linked. From the time human began to break branches of trees to create shelters and feed the fire that afforded warmth, energy for cooking and protection from animal predators, the adequacy of the forest resources has been crucial; indeed, a question of life or death. It is to recognize that mans interaction with interference with nature in the course of agriculture, industry and trade is by far, having the largest impact on the environment as compared to all living organisms. Therefore understanding the extent of magnitude and dimensions of each factor, will be helpful in studying the degeneration of the environment.

Every individual organism, communities/species have a right to exist in nature. How much of the forest may we destroy to provide timbers or fire wood or arable land for agriculture required for the fellow humans? How much of environmental damage is acceptable or permissible, at what cost? Who benefits from the activities that cause environmental impact and who suffers damage? Can there be an equation or adjustment? It is sometimes difficult to reconcile the interests of those who stand to benefit from a given project and the interests of others who are likely to suffer a loss from it. This conflict is particularly acute when the project affects communities of native people following a traditional way of life.

A combination of ecological understanding and sympathetic consideration of the feelings and aspirations of the people likely to be affected should go a long way towards the prevention of undesirable environmental and social consequences of the further development of South India's water resources.

Mr K.V. Shetty hails from an agriculturist family in Mandagaddae, had his house near the agriculture fishery camp. He had 12 acres of paddy, 2 acres of arecanut and 100 coconut trees. In that cost of holding is 10 lakhs (building) and 20 lakhs (land property). In addition he also had mango, jumbo, orange, lemon, guava, sapota and turmeric trees. He know about upper Tunga project since long back. He said because of the proposal, his village, i.e., Mandagadde is not developed. As per the survey, Mandagaddae would become an island and all his property would be nothing. It was a question of sentimental. He also claims that the survey by the government bodies was the only intimation. He was also in a confusion, that knowingly the UTP, permission was given to construct new buildings by the mandal panchayat. He was also worried, about the new land which would be given as compensation for the ousters. He was not hopeful of making it useful since it may demand decades of years together. With all the facilities the land given might not be an irrigated land. He also stated that each acre of his land cost Rs 4 lakhs, but the compensation which he would be getting would be not more than Rs. 10-15,000 per acre. He opined that it might bring about a radical change in the life style even if they have to take new profession.

According to Maji Mandal Pradhan, Mr Devadasan, the whole area which would be submerged might effect nearly 4-5 thousand people of which 70 per cent are either with small holdings or labourers. The whole area opposes the UTP with one voice. He was worried about the fate of the labourers. The inconvenience and pain is unfathomable and beyond any expression. He said that during Bangarappa's (The then Chief Minister of Karnataka) regime this UTP gained momentum. Mr Jayaprakash questions the every justification of the project. He was quoting the example of the neighbouring Varahi project, blamed the government that it was cheating. He has also pointed out that the correct figures of submergence area was not given by the government in order to get grants from the world bank. He claimed that Karnataka government has taken funds Rs.300 crores to preserve the 'fragile western ghats' and again Rs. 300 crores to 'submerge western ghats'. The main problem in the region according to him was that the cultivators are not able to organize and oppose the project because they leave in scattered remote region and hence need an effective common platform to rise their voice.

He explained how difficult it would be to take care of plants. Blaming the government of its stupid activity questioned about the outcome of the protest in connection to Kaiga and Varahi projects. Admitting that one is not bound to protest for such beneficial irrigational project, he demands with a raised voice in getting justification regarding the compensation. Pointing out fault of the land holders that is though they sell their land at a cost of 3 lakhs/acre, in order to get rid of tax and stamp paper, they just show it in thousands. And here the story goes wrong, government catch holds of such documents and assess the rate and thus takes advantage affecting the land holders very badly. Even the opponent leader in the parliament does not oppose the project which simply proves that there is no clear hand politician just a vote can do anything and everything, can submerge or dislodge and thus proving the POWER !!! Another aged person from the area receiving the blow for the 2nd time, earlier in 1956 he lost 4 acres at the cost of Rs. 250 per acre which was worth paying Rs.30 - 40,000 - acre. He is fed up with the Government bodies and calls them as cheats, bogus, finds no faith in them. He opine that instead of stopping the project it would be worth to justify the project giving justifiable compensation.

For others it is sheer agony. anxiety writ large on their faces they say, "They (the government) are going to rip the guts out of us. They are going to fluck out our souls. But lacking a leader to espouse their cause, the affected remain helpless and confused. In fact even the historians and archaeologists have seem to have abandoned the villagers, through Mandagadde has evidence of the presence of Shivappa Naik and Tipu Sultan. The Dynasty held its away in these areas. Traces of pottery and utensils have been found in the areas, which may unable to understand the history little better. Some stone sculptures have been found which finds links with the Vijayanagar empire. More of these will be swallowed if development is allowed to give its way.

6. Conclusions and Remarks

Scientists have differing views on the benefits and risks of irrigation, with some seeing it as a cure for the world's increasing food supply shortages and others viewing it as a threat to ecological systems and indigenous sps. As the 21st century approaches, the world is facing problems, many experts see as intractable. Irrigation water is in short supply in many parts of the global and erosion threatens the productivity of much farmland. While the amount of land under cultivation cannot be expanded greatly, almost 100 million people are expected to be added to the world's population each year for the next 30 years. If we fail to convert our self-determining economy into one that is environmentally sustainable, future generations will be overwhelmed by environmental degradation and social disintegration. Simply stated, if our generation does not turn things around, our children may not have the option of doing so.

Engineers propose 'solving water problems by building ever more mammoth river diversion schemes, with exorbitant price tags and damaging environmental effects. It would be prudent if the necessary studies were done and the data made available for informed decision making before future construction takes place. Meeting human needs while facing up to water's limits economic, ecological and political-entails developing a wholly new relationship to water. Historically, we have managed water with a frontier philosophy, manipulating natural systems to whatever degree engineering know-how would permit. Modern society has come to view water only as a resource that is there for the taking, rather than a life-support system that underpins the natural world we depend on. Instead of continuously reaching out for more, we must find ways to meet our needs while respecting water's life sustaining functions. More water devoted to human needs means less for sustenance of ecosystems and in many areas, nature is loosing out fast.

These dry lands present a formidable challenge to crop production. For them conservation for more efficient use of scare water is quite literally and matter of life and death. Most native peoples are bound to their land through relationships both practical and spiritual, routine and historical. The project takes toll on native habitats who have fragmentary protection. Those over, native peoples frequently aim to pressure not just a standard of living but a way of life rooted in the uniqueness of a local place. Indigenous people's unmediated dependence on natural abundance has its parallel in their peerless ecological knowledge. Most forest-dwellers display in utter mastery of botany and use innumerable techniques to husband their forest and wild life.

The romantic and uncritical espousal of tradition of ecological knowledge and management is an extreme and almost as unfortunate as that of dismissing it. The emergence of the concept of sustainable development in recent years has brought in the general realization that societal perceptions must shift towards ecological determination so as to achieve qualitative growth most ecological problems are the cumulative results of activity in the region, only environmental impact assessment would form a major instrument for the assessment of developmental activities in the context of the regional carrying capacity. It is still not too late to shelve the hare-brained scheme. Surely Malnad cannot afford to lose any more of its flora and fauna. We will have to take care of what nature has taken millions of years to fashion. Man has no right to tinker with nature for his own selfish ends.

Can one simply dismiss this as price one has to pay for development ? or can one really term 'development' something which enriches a few while impoverishing many more? Something often decided on without taking into account at all the local needs. Are we willing to shed our ignorance before it is too late?

Acknowledgments

The author thank Dr Jayachandra, former chairman, Department of Environmental Science, Kuvempu University, Shimoga for his guidance. The author is grateful to Dr. B.B. Hosetti, Department of Biosciences, Mangalagangotri, Dakshina Kannada for his suggestions in the preparation of the report.

V. IMPACT ON AIR QUALITY DUE TO CEMENT MAKING: A CASE STUDY OF THE A.C.C. LTD., MADUKKARAI, COIMBATORE DISTRICT

C.V.R. Indira, Rajammal Thirumalnesan and
M. Umamaheswari

Abstract

Cement industry is one among the energy intensive industries. The present study is concerned with the determination of suspended particulate matter (SPM) concentration at different locations of point sources, point emissions and existing pollution control means such as Electro Static Precipitator (ESP), Fabric Filter (FF), installed to control SPM concentration. The meteorological parameters like wind speed, wind direction, rain fall, humidity, temperature etc., were found to interact with the ambient air quality in the vicinity of the factory. The study was undertaken with M/s The Associated Cement Companies (ACC) Ltd., at Madukkarai, Coimbatore District, India. Results of the study revealed that,

(1) The emission concentration was maximum at New Kiln ESP stack whereas it was minimum at clinker cooler after ESP stack.

(2) Ambient air level cement dust was found to be highest at the location siding gate.

(3) The lowest was recorded during the month of October and highest during August.

Introduction

Human health is the paramount, lent to elusive justification for strong control programmes of air pollution. The present study is concerned with the determination of air quality monitoring with specific reference to stack emission monitoring coupled with meteorological parameters, have provided information about its fluctuations between different months, SPM (Suspended Particulate Matter) concentration levels and ambient SPM concentration. Comparative study of SPM concentration between ambient air and stack emission have brought to light the difference that occur at different sources of emission and their impact on air quality at Madukkarai, Coimbatore District, India.

The Madukkarai Cement Factory of M/s. The Associated Cement Companies Limited (ACC) is selected for impact assessment as one of the up-todate case study for representing air pollution due to cement industry. Madukkarai is located about

14 kms away from Coimbatore city in Tamil Nadu and the cement plant lies close to the National Highways (NH 47) Coimbatore - Palakkad (Figure 4.4). The plant has the production capacity of 1734 tonnes per/day (TPD). The emission standard for cement plant having the production capacity of >200 TPD, set upto 150 mg/NM^3 (in protected area) and (250 mg/NM^3 in other areas) (Environmental Statement for ACC, 1993-94).

The major control equipments installed for SPM are the Electro Static Precipitator (ESP), Fabric Filter (FF), Cyclone collectors and wet Scrubbers (Figure 4.5).The present study has been focussed on monitoring the SPM in and around the cement factory. The pollution control equipments installed in the cement factory and meteorological parameters like humidity, rainfall, wind speed, wind direction were recorded. Despite this, to assess environmental impacts and to identify specific sources of pollutants affecting the air quality in the factory area have been explained.

Methodology

The method followed for measurement of SPM are specified by Indian Standard (IS) 5182 (Part II) 1969 (Emission regulations, 1984). Observations were made for eight hours using High Volume Air Sampler for ambient air measurement and Thimble sampling train-assembly for stack emission monitoring isopinetically. Concentration of dust in the ambient air was calculated according to the following equation:

$$\text{Concentration of dust} = \frac{W \times 10^6}{V \times T}$$

where W = weight of dust

V = flow rate

T = period of sampling.

(Atkinssion Jr. and Faith, 1988).

Samples were collected from stacks of selected important inlets and outlets of the New Kiln ESP, Clinker Cooler ESP Outlet, Cement Mills I, II, III, IV, V, VI, and VII (Fig. 4.5). These tests were carried out for 6 months i.e., from June, 1993 to November, 1993 at important locations (Table 19).

To determine the capacity of pollution control equipments, the

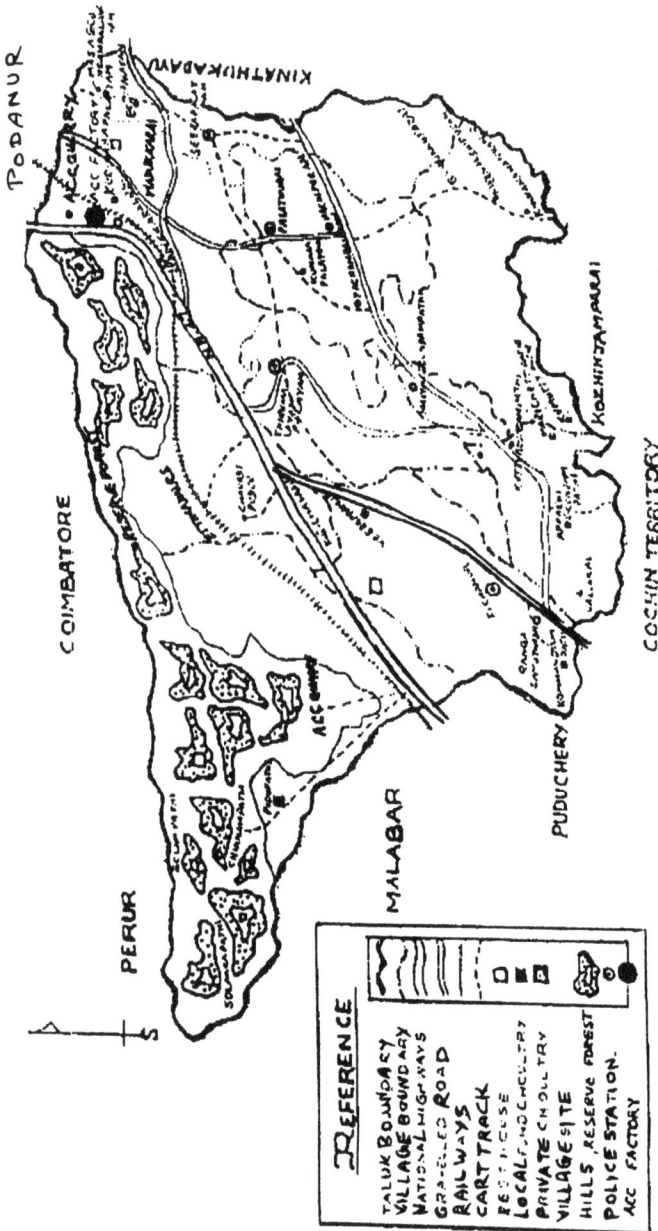

Fig. 4.4 : Site map, showing the A.C.C. Ltd.

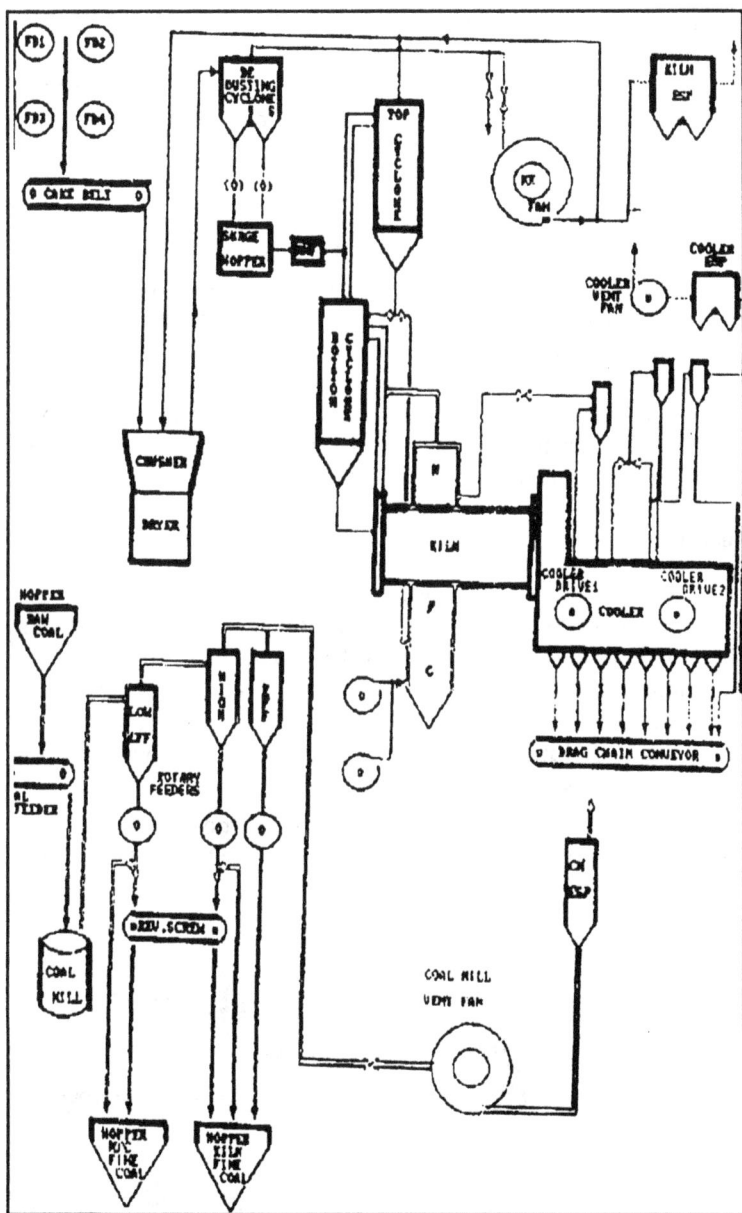

Fig. 4.5 : Flow chart of cement processing steps in ACC Ltd.

dust concentration before and after the collection by electrostatic precipitators has been calculated. The operation and maintenance of ESP, auxillary effect of ESP proper and its auxillaries, process parameters, collected dust evacuation systems on ESP performance and hot gas hazard were investigated by Kalyan Kumar Das Nag (1993).

In the present study, 't' test was used for comparing the SPM concentration of ambient air and stack emission. Depending upon the 't' value , the difference in SPM concentration between the ambient air and the stack emission is stated as significant or nonsignificant (NS). Analysis of Variance (ANOVA) method was also used in the present study to find out the variation between the samples and within the samples. Certain meteorological factors like humidity, rainfall, temperature, wind direction and wind speed were also recorded.

Results and Discussion

The results monitored are presented in Table 19, 20, 21 and 22. Table 19 gives a comparative account of mean SPM concentrations of ambient air and stack emissions. The stack emission concentration detected during different months, at various sources are compared with the standard limit of 150 mg/NM³, prescribed by Tamil Nadu Pollution Control Board (TNPCB). The concentration of SPM is expressed in mg/NM³, (Fig. 4.6).

In Newkiln ESP outlet, spread over a 6 months survey, the mean SPM concentration lies around 118.6±13.52 mg/NM³, which

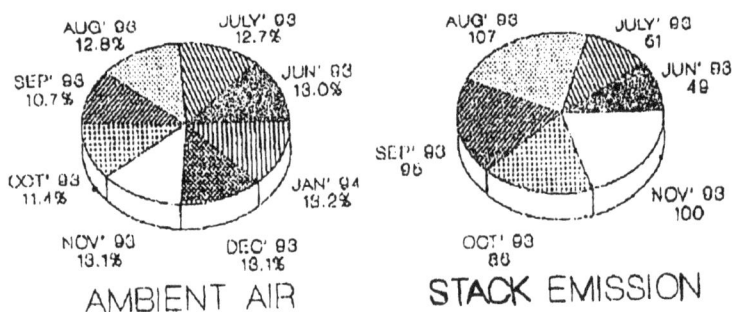

Fig. 4.6 : Comparison of mean SPM concentrations of ambient air and stack emission

was the peak mean SPM concentration out of all the sources. Earlier work of Westman (1989) have shown noticeable increase of pollutant concentration in air is because of the process method adopted and also the non-functioning of major operations and control equipments (Table 19).

Table 19 : Comparison and variation of mean SPM Concentration in ambient air and stack emission

Sl. No.	Comparison and Variation	SPM concentration mean + I.S.D.	't' value '	F' value
1.	Ambient air	91.05+9.4		
2.	Stack emission	82.30+21.3	0.552NS	
3.	Variation between the samples	24597.8		
4.	Variation within the samples	42326.3		0.581147NS

NS - Non significance ; n = 9 observations

From the data assembled in Tables 19 and 22 it can be noticed that,

(1) In order to compare the ambient air and stack emission SPM concentrations, 't' test was carried out and the 't' value was calculated to be 0.5522 N.S. (N.S. —Non Significant). The variation between and within the samples of ambient air and stack emission were analysed and calculated by using the Analysis of Variance (ANOVA) method. The table value of 4.3874 which was higher than the 'F' value 0. 58147 calculated justifies the accuracy of methodology followed for monitoring the air quality (Table 20).

Table 20 : Report of ambient air monitoring

Sl. No.	Location	Mean Concentration ug/m³ ± S.D.		
1.	Near laboratory	96.00	±	13.0.5
2.	Guest house	89.31	±	14.40
3.	Siding gate	115.90	±	9.20
4.	Quarters J–24	64.00	±	4.57

*S.D.–Standard Deviation

(2) The wind may be the cause for the increase of SPM concentration in ambient air at Quarters J. 24.

(3) The operation of major process and control equipments were found to dwindle the stack emission. In the new kiln ESP, the SPM concentration was recorded as about 102 ug/m³, 128 ug/m³, 56 ug/m³ and 49 ug/m³ in the clinker cooler after ESP. This reveals that the double fold increase in the concentration is because of the less running hours of ESP (Table 22, Fig. 4.7.)

Table 21 : Monthly mean of certain meteorological parameters recorded at Madukkarai

Elements								
Humidity (%)	86	81.5	86	84	83	85	85	85
Rain (mm)	—	—	2.4	35.7	—	—	—	—
Temperature	28	27	29	27	27	28	28	29
Wind direction	SW*	SW	SW	SW	SW	SW	SW	SW
Wind speed (Km/h)	14.7	14.8	15.3	14.6	14.7	18.5	14.5	14.6

* SW—South West Wind Direction

(4) It was also found that the meteorological parameters do not affect the stack emission.

(5) The meteorological parameters influence the SPM concentration in ambient air and also produce drastic changes in the working environment (Rajeev and Mohan 1990). During the period of survey, the rainfall occurred twice i.e., during Aug., 1993 (2.4 mm) and Sept., 1993 (35.7mm) have ultimately decreased the SPM in ambient air near laboratory, guest house, siding gate and quarters. No. J-24 to a half than the normal concentration. Because the rain droplets subsequently subsides the suspended particles in atmospheric air (Table 21).

Table 22 : Operation of major process and control equipments

Process and control equipment	Total running hours from Jun. to Jan., 1994 Hrs. Min.	Net hours of non-functioning of control equipment Hrs. Min.
Kiln	5500.30	—
Kiln ESP	5411.10	89.20
Clinker cooler ESP	5489.52	10.78

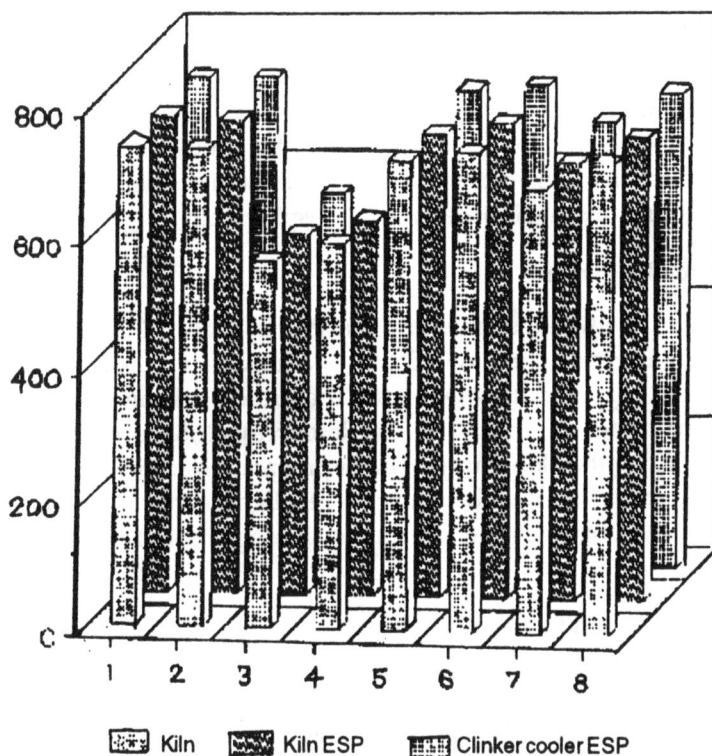

Fig. 4.7 : Operation of major process and control equipments

(6) The sampling station (quarters J. 24) is located nearer to the factory and exposed to heavy public and private vehicular transport.

(7) Sometimes due to constraints of storage or questionable kiln operating status, clinker is dumped from the dummy chutes in the clinker conveyor belt and released in the open space between Diesel generators building and cursher department. During this period high SPM concentration was observed due to fugative emissions from clinker falling from height in open area (Fig. 4.5).

118   Environmental Impact Assessment and Management

Recommendations

Studies of this kind may be extended to cover the modelling of pollutants and other parameters viz., carbon-dioxide (Stern, 1977), oxides of nitrogen (Strauss, 1977) and sulfur (Zutshi, et al., 1973), in the ambient air and aeromycological studies related to the monitoring of fungal spores from atmospheric air.

Acknowledgement

We acknowledge gratefully Mr. Pawan Malik, Ex. Superintending Engineer of the A.C.C. Ltd., Madukkarai for the constant help and support during the study. The authors acknowledge the British Council and Environmental Resource Unit, Salford University, U.K. for their kind support in the preparation and publication of this paper as a follow up of the Higher Education Link Programme between the two Universities sponsored by British Council.

References and Suggested Reading

Ambasht Lal, 1983. Environmental Engineering Division. *Journal of Instrumental Engineering,* India (5), 18.

Atkinssion Jr. A. and Faith, W.L. 1988. *Air Pollution,* Willey Inter Science Publication, New York, 380-394.

Bhave, V.R. 1983. *Instructions for Environmental Monitoring Protection Environment Aspects of Metullurgical Industry,* Nagpur, (4), 201-229.

Charles Komanoff, 1972. *The price of power: Electric utilities and the environment.* The council on economic priorities, 195.

Environmental Statement for A.C.C. , 1993-94. Madukkarai Cement Works, prepared by Associated Industrial Consultants (India) Private Ltd., Bombay,3-25.

Eriksson, E. 1970. The importance of investigating global background pollution. *World Meteorological Organisation.* 21. 724-732.

Howard, A.B. 1989. *Global air pollution problems for 1990's.* Belhaven Press, London, 7-127.

Joshi, 1989. Elemental Analysis of aerosol in work room atmosphere of factories. *Indian Journal of Environmental Health.* 26, 102-120.

Kalyan Kumar Das Nag, 1991. *Operation and maintenance of ESPS. National Seminar on ESPS,* Volume 12-14 (11).

Odum, E.P., 1971. *Fundamentals of ecology,* W.B. Saunders Co., Philadelphia, 574-582.

Perkins, 1978. *Air pollution,* McGraw Hill Co., 350-374.

Rajeev D. and Surendra Mohan, 1990. Monitoring of ambient air quality at Korba, *Indian Journal* of *Environmental Protection.* 10 (7), 29-31.

Rao, M.N. and Rao, H.V.N. 1990. *Air Pollution*. Ist Edition, Tata McGraw Hill Publishing Company, New Delhi, 16-170.

Stern, A.C. 1977. Air Pollution. Volume 3 - Measuring and monitoring air pollutants. Wiley - Interscience Pub., New York, 170-198.

Strauss, W. 1977. *Air Pollution* Volume 3 -Measuring and monitoring air pollutants, Wiley - Interscience, New York.

Westman, WE 1985. Eco-impact Assessment and Environmental Planning, John Wiley and Sons, New York, 532.

Zutshi, P.K., Mahadevan, T.N. and Banerjee, T. 1973. Trends of concentration of pollutants in Bombay, *Indian Journal of Meteorological Geography*. 24 : 61-64

VI. HEAVY METAL POLLUTION IN THE COOUM RIVER FLOWING THROUGH MADRAS CITY

Ayyadurai, K. Kadirvel, R. Swaminathan, C. and Krishnasamy, V.

Abstract

The Cooum river flowing through Madras city is polluted by industrial and domestic wastes. An attempt is made to monitor the heavy metal pollution in water and sediments in various seasons fixing five locations within the city limits. Mean concentration of Hg, Cu, Fe, Mn and Zn in water and sediments was 0.006, 6.42, 127.55, 54.81 and 15.58 mg/1 and 0.76, 390, 18589, 427 and 2282 mg/kg respectively. Significant differences were noticed between seasons as well as locations. The heavy metal concentration in water exceeded the limits prescribed by WHO except for Hg.

Introduction

The Cooum river is flowing through the heart of Madras city and was once a fresh water stream used for bathing and washing activities. In the course of time as a consequence of increased population, it become a sewage carrying canal. In addition to the sewage, this stream also received contaminants from Ambattur and Padi Industrial complex. The water in this stream is not free flowing and only during rainy seasons the floods will wash off the pollutants into the sea. It is noticed that the river is highly polluted and the fish population was practically nil (Jhingaran, 1988). Due to high load of heavy metals, the possibility of their accumulation in ground water by seepage is definite, which will lead to heavy metal pollution in the neighbouring wells. An attempt was made to study heavy metal concentration in water and sediments collected at 5 different locations (Fig. 4.8) (Vide Fig. 4.9 Lat 13^{-0}N Long 80^{0}15'E.). The source of contamination are shown in Table 23.

Methodology

Programme of Work Done

Collection of water, sediment and fish samples was done over a period of one year (January to December 1993) comprising of four seasons. The classification of seasons in South India was reported else where (Ayyadurai *et al.*, 1994).

Table 23 : Seasonal periods

Season	Month	Mean Temp. (C) in Tamil Nadu	
		Max.	Min.
Cold weather	Jan. to Feb.	29	22
Hot weather	Mar. to May	37	28
South West Monsoon	Jun. to Sep.	34	26
North East Monsoon	Oct. to Dec.	30	23

Water

The polythene sample bottle of 2 lit. capacity was thoroughly washed with detergent and tap water. It was then rinsed successively with 1:1 conc. HNO_3, tap water, 1:1 conc. HCl, tap water and finally with deionized glass distilled water. The water samples were collected following the method of Taylor (1958). The bottle was plunged neck downwards to a distance of about one foot below the surface of water, the mouth being directed towards the current and was gently rotated. When the bottle was completely full, it was raised rapidly above the surface and the stopper was placed immediately. The water samples were preserved by adding 3 ml of redistilled HNO_3 (1:1) per litre and were brought to a pH 2.0 The holding time for metal analysis of these samples was six months (EPA,1974).

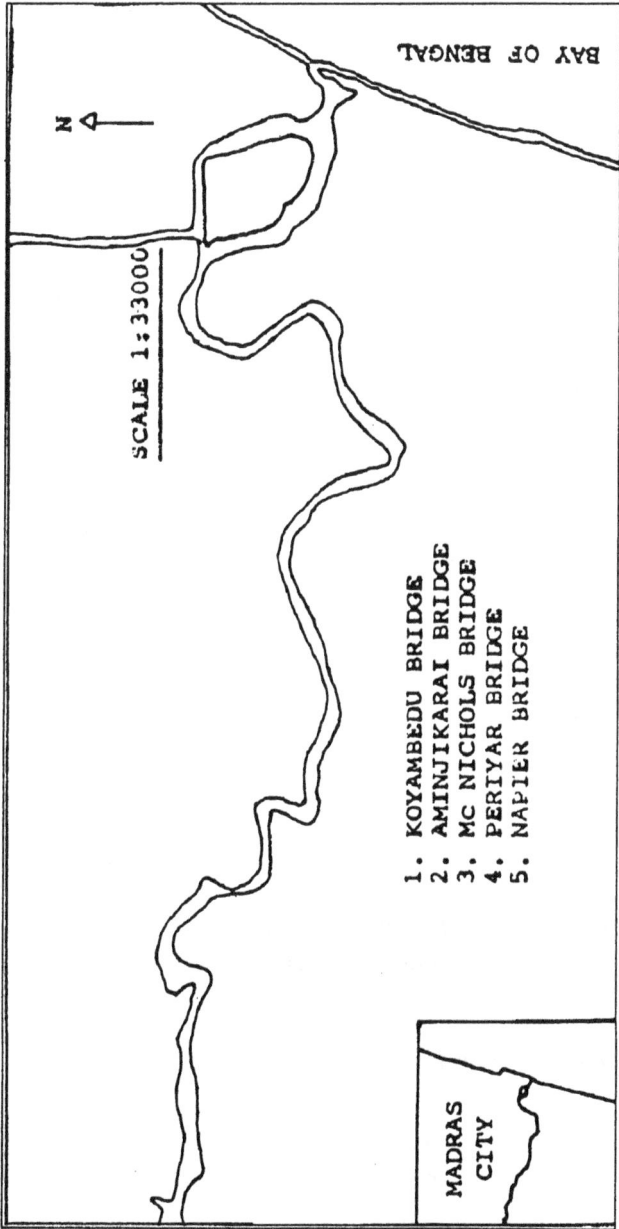

Fig. 4.8 : The course of the cooum river in Madras city

Sediment

Sediment samples were collected using a hand grab. These samples were placed in a plastic bag and transported to the laboratory in an ice box. At the laboratory the samples were preserved in a deep freeze.

Pretreatment of Samples

Pretreatment of water sample for analysis of heavy metals (Cu. Fe. Mn and Zn)

100 ml of the sample was taken in a beaker and 3 ml analar conc. HNO_3 was added. The beaker was kept on a hot plate and evaporated continuously to dryness without allowing the liquid to boil. The beaker was cooled and 3 ml of analar conc. HNO_3 was added to the residue. It was covered with a watch glass and kept on the hot plate. The temperature was increased slowly untill a clear solution was resulted. The walls of the beaker and watch glass were washed with glass distilled water and the insoluble material was removed by filtration. The volume was made upto 10 ml, labelled and used for analysis.

Pretreatment of water sample for analysis of Hg.

100 ml of water was taken in a conical flask to which was added 5.0 ml of dil. H_2SO_4 2.5 ml of conc. HNO_3, 15 ml of KMnO4 and 8 ml of 5% KH_2SO_4 and heated at 95⁰C for 2 hours. The contents were cooled and treated with 1:1 solution of sodium chloride-hydroxylamine hydrochloride in order to reduce the excess of $KMnO_4$. This solution was used for analysis.

Pretreatment of sediment for Cu, Fe, Mn and Zn

The sediment sample was wet seived through nylon screen to remove the coarse particles. Then the sample was homogenized in a blender and stored at 4⁰C on sealed polythene bag. Before analysis, the sample was dried at 105⁰C, pulverised with mortar and pestle and thoroughly mixed.

The pulverised sediment (0.59) was weighed, transferred into a Kjeldahl's flask, treated with 8 ml of conc. HCl and 2 ml conc. HNO_3 and kept overnight at room temperature. The digestion was continued next day at 105-⁰C for 1 hour and then at 140-⁰C until the sample was dry. After cooling, 12.5 ml of 20% (by volume) HCl was added and the mixture rewarmed at 80-⁰C for a few minutes. After cooling, the solution was filtered and made upto 50 ml glass distilled water and used for analysis.

Pretreatment of sediment for analysis of Hg

The sediment sample was wet seived through nylon screen to remove the coarse particles. Then the sample was homogenized in a blender and stored at 4-⁰C in a sealed polythene bag. Before analysis, the sample was dried at 60-⁰C and pulversed using pestle and mortar and thoroughly mixed. 0.1 g of this dry powder was accurately weighed, to which 5.0 ml of glass distilled water, 2.0 ml aqua regia and 15 ml of 6% $KMnO_4$ were added and heated in a water bath at 95-⁰C for 30 minutes. After cooling the contents to room temperature, it was treated with 1:1 solution of sodium chloride-hydroxylamine hydrochloride to destroy the excess of $KMnO_4$ and filtered. The filtered solution was made upto 25 ml in a volumetric flask and the solution thus obtained was used for the analysis of mercury.

Estimation of Heavy Metals

Estimation of Cu. Fe, Mn and Zn by using Atomic Absorption Spectrophotometer (Varian Techron AA 120)

Preparation of reagents and calibration was done as per Varian Techron Cook Book.

Preparation of standard solutions of strength 1000 mg/1

1 g of respective metal was dissolved in 1:1 HNO_3 and made upto 1 litre.

Calibration and estimation

The required halo-cathode lamp was inserted in the AAS and the required wavelength for various metals adjusted. Air-acetylene flame was used for atomising the metal. Initially the blank solution (the respective solutions used for digesting water, sediment) was aspirated and the instrument was adjusted for zero absorbance. Subsequently the metal solution of different concentrations, prepared from the standard solution were atomized one after the other and the readings were noted. A standard graph was plotted between concentration and absorbance. Similar procedure was followed for the samples to be analysed. From the absorbance, the concentration was worked out using the standard plot. The working conditions of the AAS instrument for different metals are furnished in Table 24.

Estimation of Hg by cold vapour technique using Mercury Analyser (ECIL.1800 MA)

Mercury analyser works on the principle of cold vapour technique. The mercury present in the pretreated sample was reduced to elemental state by using stannous salt in an acid solution. The working wavelength was 253.7 nm and the sensitivity 0.001 ug. Hg solution of concentration 100 mg/1 was prepared by dissolving 0.1354 g of $HgCl_2$ in 2% HNO_3 made upto 100 ml using 2% HNO_3 and calibration was done as indicated in ECIL book.

Calibration and Estimation

The apparatus was set up and connection was checked for leakage. The required aliquot (2-4 ml) of the blank solution (respective solutions used for digesting water and sediment) was taken in the reaction vessel. 18 ml of 10% HNO_3 and 2 ml of $Sncl_2$ were added and the stopper was replaced immediately. Magnetic stirrer was switched on and the contents stirred vigorously for about 5 minutes. The pump was started to allow the air to purge through the reaction vessel. Absorbance was recorded as quickly as possible. The above procedure was followed for various contentrations of standard Hg solutions and the respective absorbance recorded. A standard graph was plotted between concentration and absorbance. The above procedure was followed for the various sample solutions and the corresponding absorbance noted (after deducting the absorbance value for the respective blank solution). The Hg concentration was calculated using the standard plot.The correlation matrices were carried out using Magnum Multiprocessor System (Bassett *et al.*, 1986) multiple correlation was worked out as described by Snedecor and Cochran (1967).

Table 24 : Sampling station and sources of pollution in the Cooum River

Station	Location	Distance from the sea (KM)	Major source of Contamination
1.	Koyambedu Bridge	18	Cattle and Lorry washes
2.	Aminjikarai Bridge	13	Domestic wastes
3.	Mc Nichols Bridge	9	Cattle and sewage waste
4.	Periyar Bridge	5	Metal workshops and bus depot.
5.	Napier Bridge	1	Chloride from sea

Foot Note : Apart from the major sources of contamination mentioned, untreated domestic wates are let out at various points throughout the entire route within city limits of the river.

Results and Discussion

Heavy metal concentrations in water and sediment samples during the four seasons are depicted in Table 25.

Mercury

Mercury content in water ranged from Nil to 0.022 with an average of 0.006 mg/1. The concentration of Hg in water was below U.S. Tolerance limit of 0.05 mg/1 (Zajic, 1971). Hg concentration in the Cooum river ranged from 0.29 to 1.75 with an average of 0.0768 mg/kg dry weight. The results revealed a highly significant (P<0.01) difference for Hg in sediment between different seasons. Highest concentration of 1.33 mg/kg weight was observed during hot weather which may be due to evaporation effect (Brosset, 1981).

Copper

Highly significant difference were observed between seasons with respect to Cu content in water. A high mean value of 9.29 mg/1 during hot weather and low mean value of 3.15 mg/1 during north east monsoon was registered. The low concentration during the north east monsoon may be attributed to the free flow of water in the Bay of Bengal. This is supported by the observation of Baghylakshmi (1989). Cu level (4.75 mg/1) in the Cooum water crossed the WHO permissible limit (1 mg/1) (Wong et al., 1980). Copper content in the sediment was 390 mg/kg dry weight which is well below the 1830 mg/kg dry weight as reported by Boyden et al. (1979) for Restrongent Creek (UK). Highly significant difference (p < 0.01) was noticed between the seasons for Cu content in the sedimenet. As in the case of Hg, maximum value of 435 mg/kg was registered in hot weather season.

Iron

Iron content in water (127.55 mg/1) crossed the permissible limit stipulated by WHO limit of 0.3 mg/1 (Wang et al., 1980). The very high value of Fe in this rivulet may be due to the result of Fe ore tailings from metal workshops and mixing of untreated domestic and industrial wastes. Iron concentration in sediments ranged from 5000 to 36333 mg/kg dry weight. Highly significant difference (P< 0.01) was noticed between seasons.

Manganese

Manganese concentration in water ranged from 11.6 to 122.7 mg/1. The high values of Mn in the Cooum Rivulet may be from sewage input and effluents from different metal workshops. The high concentration could also be due to chelation of this metal with soluble organic compounds presented in less oxygenated waters. Or their presence in a reduced soluble form (Mn^{2+} or due to both reasons (Saad et al., 1981). Mn concentration in water exceeded the WHO level of 0.1 mg/1.

The average concentration of Mn was 427 mg/kg dry weight. Which is close to the average value of (495 mg/kg dry weight) reported by Jones and Jorden (1979) in surface sediments of the estuary, river Liffey in Dublin, Ireland. The very high value of Mn in the Cooum must be due to the result of sewage addition and formation of Mn complexes with organic matter.

Zinc

Zinc content of water in the present investigation ranged from 460 to 29.77 with an average of 15.58 mg/1 which is in agreement with the reported values of Cu (31.5 mg/1) in the Cooum (Bernice *et al.*,1987). Concentration of Zn was above the permissible limit (5.0 mg/1) fixed by WHO indicating a high sewage and industrial water pollution. As in the case of Cu, highly significant (p<0.01) difference was noticed between seasons in both sediment and water.

Heavy metal concentration in water and sediment at different locations on the Cooum River are shown in Table 25. The results revealed that the concentration of heavy metals in the Cooum waters were significantly high. A steady increase from station 1 to 4 was observed and declined thereafter. The reason may be sea water interaction since station 5 is the point of entry into the sea. The highest concentration at station 4 must be due to washings and wastes from a huge bus depot and metal workshops located nearby.

Statistical Modelling

The data collected on heavy metals in water and sediment were analysed for statistical interpretation. Correlation matrices were worked out by taking the combination of any two metals at a time keeping the sample size as 20 for water and sediment. Zn and Cu had significant positive correlation with other metals at 95 per cent level of probability and the rest of the metals do not show significant correlations.

R1 (2.....5) = 0.8582 (Mn Vs other metals)

R2 (1....5) = 0.8777 (Fe Vs other metals)

R3 (1....5) = 0.9316* (Cu Vs other metals)

R4 (1....5) = 0.9513 * (Zn Vs other metals)

R5 (1....4) = 0.7312 (Hg Vs other metals)

Multiple regression analysis was done with the view to develop system of equations keeping, one of the metal as a dependent variable multiple regression co-efficients were determined to

Table 25 : Mean concentration of heavy metals in the water (mg/1) and sediments (mg/kg dry weight)

Metal	Cold weather	Hot weather	Southwest monsoon	Northwest monsoon	Mean
Hg W	0.009[a] ± 0.003	0.007[a] ± 0.002	0.004[a] ± 0.002	0.004[a] ± 0.002	0.006 ± 0.001
S	0.372[a] ± 0.031	1.366[c] ± 0.180	0.926[b] ± 0.100	0.408[a] ± 0.040	0.768 ± 0.110
Cu W	4.75[a] ± 0.003	9.29[b] ± 1.49	8.49[b] ± 1.29	3.15[a] ± 0.55	6.42 ± 1.27
S	296[a] ± 58	435[a] ± 94	455[b] ± 70	373[ab] ± 73	390 ± 31
Fe W	130.78[ab] ± 21.97	152.37[a] ± 94	121.77[b] ± 21.84	105.29[a] ± 17.67	127.55 ± 8.50
S	18700[ab] ± 3443	25400[b] ± 3768	16650[ab] ± 3258	13067[a] ± 2884	18589 ± 3165
Mn W	55.44[a] ± 17.84	59.62[a] ± 13.79	59.83[b] ± 23.10	44.33[a] ± 11.47	54.81 ± 3.15
S	506[b] ± 100	487[b] ± 98	353[a] ± 58	363[a] ± 55	427 ± 35
Zn W	16.94[b] ± 1.75	20.36[b] ± 2.27	17.00[b] ± 2.31	8.02[a] ± 1.24	15.58 ± 2.29
S	2300[b] ± 336	2663[b] ± 473	2329[a] ± 388	1834[a] ± 347	2282 ± 148

W = Water   S = Sediment

Means within each category bearing atleast one common superscript do not differ insignificantly.

Table 26 : Heavy metal concentration in the water (mg/1) and sediments (mg/kg) at different locations

Metal		Station I	Station II	Station III	Station IV	Station V	Mean
Hg	W	ND	$0.0052^a \pm 0.002$	$0.004^{ab} \pm 0.001$	$0.013^b \pm 0.003$	$0.038^c \pm 0.0004$	0.006 ± 0.092
	S	$0.48^a \pm 0.092$	$1.628^a \pm 0.164$	$0.90^{ab} \pm 0.276$	$1.003^b \pm 0.261$	$0.833^{ab} \pm 0.232$	0.768 ± 0.20
Cu	W	$3.08^a \pm 0.53$	$6.13^{ab} \pm 1.06$	$6.57^{ab} \pm 1.77$	$0.94^b \pm 1.86$	$6.39^{ab} \pm 1.26$	6.42 ± 1.27
	S	$171^a \pm 23$	$275^{ab} \pm 31$	$389^b \pm 27$	$649^c \pm 54$	$456^{bc} \pm 36$	390 ± 31
Fe	W	$31.63^b \pm 2.24$	$143.8^a \pm 9.97$	$152.19^a \pm 16.30$	$165.34^a \pm 16.30$	$148.81^a \pm 6.06$	127.55 ± 8.50
	S	$8604^b \pm 1590$	$19708^{ab} \pm 2314$	$25296^a \pm 3251$	$26917^a \pm 2825$	$12421^b \pm 1988$	18589 ± 3.15
Mn	W	$20.20^a \pm 3.83$	$33.80^a \pm 5.43$	$37.26^a \pm 9.72$	$95.20^b \pm 13.91$	$87.57^b \pm 11.67$	54.81 ± 3.15
	S	$201^a \pm 27$	$416^b \pm 38$	$465^b \pm 64$	$668^c \pm 74$	$338^{ab} \pm 44$	427 ± 35
Zn	W	$18.48^a \pm 1.39$	$14.22^a \pm 2.50$	$14.68^a \pm 2.98$	$22.26^b \pm 3.43$	$16.26^{ab} \pm 1.56$	15.58 ± 2.29
	S	$959^a \pm 78$	$1942^b \pm 130$	$2557^b \pm 252$	$3556^c \pm 188$	$2394^b \pm 207$	2282 ± 148

develop linear equations using which, the level of any one of the metal can be forecast if the concentrations of others are known (Bassett *et al.*, 1986).

Multiple Regression Analysis

Water

(Mn) $x_1 = 17.3002 + 0.2438\ X_2 - 2.1368\ X_3$
$$+ 3.7286\ X_4 + 240.8908\ X_5$$

(Fe) $X_2 = 70.5301 + 0.56965\ X_1 + 9.1581\ X_3$
$$- 3.3961\ X_4 + 2936.1499\ X_5$$

(Cu) $X_3 = -1.8466 - 0.0086\ X_1 + 0.0158\ X_2 +$
$$0.4620\ X_4 + 39.4340\ X_5$$

(Zn) $X_4 = 4.8170 + 0.0455\ X_1 - 0.0178\ X_2$
$$1.4005\ X_3 + 167.7917\ X_5$$

(Hg) $X_5 = 0.00354 + 0.00001\ X_1 + 0.00004\ X_2$
$$-0.00033\ X_3 + 0.000456\ X_4$$

Sediment

(Mn) $X_1 = 191.6760 + 0.0113\ X_2 + 0.7868\ X_3$
$$+ 0.1649\ X_4 - 51.7346\ X_5$$

(Fe) $X_2 = 1787.2305 + 24.0522\ X_1 - 6.1061\ X_3$
$$+ 3.5441\ X_4 + 5757.6836\ X_5$$

(Cu) $X_3 + 110.3438 - 0.3889\ X_1 - 0.0014\ X_2$
$$+ 0.1956\ X_4 + 39.7957\ X_5$$

(Zn) $X_4 += 279.9443 + 1.653_3\ X_1 + 0.01660\ X_2$
$$+ 3.9490\ X_3 - 11.3839\ X_5$$

(Hg) $X_5 = 0.03111 - 0.0067\ X_1 + 0.00003\ X_2$
$$+ 0.00104\ X_3 + 0.00001\ X_4$$

References and Further Reading

Ayyadurai, K., C.S. Swaminathan and V. Krishnasamy, 1994. Studies on Heavy metal pollution in the finfish. *Oreochromis mossambicus* from river Cauvery, *Ind. J. Environ Health, 36* (2), 99-103.

Bassett, E.E., J.M. Bremner, I.T. Jolliffe, B. Jones, B.J.T. Morgan and P.M. North, 1986. *Statistics.*, Edward Arnold Publishers Ltd., Baltimore, U.S.A., pp-177-201.

Baghyalakshmi, V., 1989. *Heavy metal pollution and its effects on the micro and macro fauna of Adyar River Madras.* Ph.D. Thesis, University of Madras, Madras.

Bernice, A,V. Baghyalakshmi and R. Lakshmi, 1987. Limnology of River Cooum with special reference to sewage and heavy metal pollution, *Proc. Indian Acad. Sci. (Anim. Sci)* 96, 141-149.

Boyden, C.R., S.R. Aston I. Thomton, 1979. Tidal and seasonal variations of trace elements in two cornish Estuaries. *Estuarine Coastal Mar. Sc.* 9, 303-317.

Brosset, C, 1981. The Mercury cycle, *Water Air Soil Poll.*, 16, 253-255.

Environmental Protection Agency, 1974. *Methods for Chemical Analysis of water and wastes* No. 625/6-74-003.

Jhingran, V.G. 1988. *Fishes and Fisheries of India*, Hindustan Publishing Corporation (India), New Delhi. 127.

Jones, G.B. and M.B. Jordan, 1979. The distribution of organic material and trace metals in sediments from the Liffey Estuary, Dublin, *Estuarine Coastal Mar. Sci*, 37-47.

Saad, M.A.H. , A.A. Ezzot, O.A. EI. Rayis and H. Hafez, 1981. Occurrence and distribution of chemical pollutants in Lake Mariat, Egypt, 11 Heavy metals, *Water Air Soil Poll.*, 16, 401-407.

Snedecor, G.W. and W.G. Cocharn, 1967. *Statistical methods*, 6th Ed., and IBW Publishing Co., Calcutta.

Taylor, E.W., 1958, *Examination of waters and water supplies*, 7th Ed., F & A Churchill Ltd.

Wong, M.H., Kuin-Chung Ho and T.t. Kwok, 1980. Degree of pollution of several major streams entering Tolo Harbour, Hongkong, *Mar. Poll. Bull.*, 11, 36-40.

Zajic, J.E, 1971. *Water Pollution, disposal and Reuse Vol 1*, Marcel Dekkar, Inc, New Yorks.

CHAPTER 5

SOLID WASTE MANAGEMENT IN INDIA : GENERAL ASPECTS

B. B. Hosetti

Q.1. What is refuse or Solid waste?

Q. 2. How to handle these wastes?

Q. 3. How to recover and reuse the valuable components from solid wastes?

Q. 4. What can we do with hazardous waste?

Q. 5. How to dispose these wastes without side effects?

Q. 6. Which is the filthiest city in India?

Definition

Any useless, unwanted discarded material that is not a liquid or gas is classified as solid waste. For example it may be yesterdays news paper, junk mail, todays dinner table scrap, pieces of chapati, roti, waste rice, racked leaves, dust, grass clippings, broken furniture, abandoned materials, animal manure, sewage sludge, and industrial refuse or street sweepings etc. For the first time in India the Municipal Corporations worked overtime to remove the filth produced due to the threat caused by the outbreak of plague -like disease in Gujarat in the year (1994).

(a) Refuse: This is all putricible and nonputricible waste except body waste. It includes, all types of rubbish and garbage.

(b) Rubbish: This refers to that portion of the refuse which is non putricible solid waste.

(132)

(c)   Garbage: This refers to that portion of the refuse which is putricible component of solid waste. These are produced during cooking and storage of meat, fruits and vegetable.

Introduction

Any refuse or unwanted material which is not liquid or gas is referred to as solid waste. That is yesterday's newspaper cotton pieces, food stuff, skin, saree, leather, your dress, fish etc. anything of solids produced by man will be termed as solid waste some time, some where and some how. It means waste material is produced as a result of human activity. The quantity of this material is increasing due to increase in human population and increase in standards of living. Eg. in Bombay, 100 tonnes of animal dung about 2250 tonnes of city refuse and organic waste are being produced every day. All this is contributed by kitchen refuse, markets, slaughter houses etc. These wastes have to be disposed off so that environment remains clean and healthy to live in. Solid waste management includes the process of generation, storage, collection, transport and disposal or reuse of refuse. These are useless unwanted or discarded materials that arise out of human's activity and are free flowing materials (WHO, 1971).

Until 1950 solid waste disposal had not posed any problem. However during the period between 1953-1955 the spread of viral disease to hogs has attracted the attention of several sanitary engineers and farmers. Since then feeding of garbage to hogs was banned in U.S.A. However in other parts of third world countries, garbage feeding to hogs and cattle is continued unabatedly. It is probably for the first time that scientific studies on refuse were started and published in Chicago by the public administration Department (APHA, 1980). The EPA of US also published its fourth report on the resource recovery from solid waste in 1977. After 1970, several people started to work on this topic. Winkler and Wilson (1973), Rimer (1981), Alter (1981), Bruce (1984) and. Howard et al., (1985).

Classification

Typical classification of solid waste was suggested by Hodges (1974) as follows:

(1) Garbage: Putrecible (decomposable) wastes from food, slaughter houses, canning and freezing industries.

(2) Rubbish: Non putreciable wastes either combustible and non combustible. This includes wood, paper, rubber, leather and garden wastes as combustible components whereas non-combustible are glass metal, ceramics, stones and soil.

(3) Ashes: Residues of combustion solid products after heating and cooking or the incineration of solid wastes by municipal, industrial and apartmental areas.

(4) Large wastes: Demolition and construction wastes, automobiles, furnitures, refrigerators and other home appliances, trees, fires and other items.

(5) Dead animals: Household pets, birds, rodents, zoo animals etc. Also anatomical and pathological parts from hospitals.

(6) Sewage treatment process solids : Screenings, settled solids and sludge.

(7) Industrial solid wastes: Chemicals, paints, sand and explosives.

(8) Mining wastes : Tailings, slug ropes, culmpiles at coal mines.

(9) Agricultural wastes: Farm animal manure, crop residues and others.

Techobanoglous (1977) has categorised sources of solid wastes into three types viz.,

(1) Municipal (2) industrial and (3) Hazardous wastes.

The locations where wastes are generated from residential areas include single family and multifamily localities. In commercial areas wastes are generated from stores, restaurants, markets, office buildings, hotels, shops, medical facilities and institutions etc. Industrial wastes include rubbish, ashes, demolition and construction wastes.

Bhide and Sundaresh (1983) have classified solid wastes into six categories.

(1) Physical : paper, glass, metal pieces, polythene bags etc.

(2) Chemical : Synthetic organics, Inorganic metals, Salts acids, alkalies, inflammables, explosives etc.

(3) Biological: (1) Hospitals, malignant tissues contaminated material like hypodermal needles, bandages (2)Wastes from biological research facilities.

(4) Flammables : Mostly liquid forms alongwith solid chemicals. Eg. Organic solvent oils, plastics and organic sludges.

(5) Explosives : From factories.

(6) Radio active : (1) Regulated by atomic energy commission and separately disposed off.

Currently waste is commercially classified into 4 groups.

(1) Residential

(2) Commercial

(3) Industrial, and

(4) Municipal

Mismanagement and Side Effects of Solid Wastes

Solid waste management is an important facet of environmental hygiene and needs to be integrated with total environmental planning (WHO, Expert Committee, 1971). Its storage, collection, treatment and disposal can lead to short term risks. In the long-term there may be dangers arising particularly from the chemical pollution of water supplies. Wilson (1974) said that the problems connected with refuse storage in buildings were, Insects, rats, fire and odour. These problems are also associated with other problem of human health and aquatic systems.

Insects

A common transmission route of bacillary dysentery, amoebic dysentery and diarrhoeal diseases are from man's faces by flies to food or water and thence to man. Flies thrive on food wastes and in USA 90 per cent of house flies in cities breed on open garbage. If night soil and unprotected latrines are close to refuse dumps, the disease routes are widespread, as the flies can fly upto 10 kilometers. Refuse dumped on ground results in infestation with fly eggs, and larvae have been found upto 50 mm below the surface. The breeding of mosquitoes in discarded tyres, tins, jam, juice are reported.

Rats

The main source of food for rats and other smaller rodents is refuse and rubbish dump where they quickly proliferate and spread to neighboring houses. The rats become vectors to histoplasmosis, rat bite fever, salmonellosis, tularemia, trichinosis etc.

Fire

Ashes added to combustible refuse pose a great danger at the source and fire in uncontrolled tips have been known to burn for months or even years. Usually the fire starts with unsustainable practice of open dumping of refuse and it can spread accidentally. Occasionally fires begun spontaneously from the heat given off by decomposition or by glass on an open tip acting as a lens for sunlight. Flammable industrial waste increase the danger of fire and can convert old tyres into toxic ones.

Odour

While passing through a crowded city in tropical areas a traveller will experience a bad smell. It is due to the combination of rottening vegetation and fecal matter and other solid wastes indiscrimately discarded. When this stink persists all day and night it causes major environmental nuisance. This bad smell is also due to the release of H_2S during decomposition.

Other Effects on Health and Environment

Apart from diseases of which insects and rats are carriers, the handling of it causes illness to workers. A survey in India conducted by CPHERI, 1973 (Central Public Health Engineering Research Institute) showed that at Bhopal upto half of the sample of refuse in the slum areas contained roundworm ova. The accident rate amongst refuse workers is also high as a result of lifting heavy load of waste and dealing with mechanical equipments.

Atmospheric Pollution

When refuse is burnt in an open area, a dense smoke often covers the site and neighbouring land. Old fashioned incinerators without air   cleaning equipments are little better than open burning. Apart from particulate matter that constitutes smoke, the gaseous discharge from the incomplete combustion may include SO_2, NOx and various gases. If PVC is a constituent of the refuse, the gases may include hydrogen chloride.

Tourism/Ethics/Recreation

Uncollected refuse and insanitary tips in full public view are eyecores. If tourism is important, in such cases esthetic nuisance may reduce the number of visitors with resultant economic loss. There can be depreciation of the value of property nearer to a garbage area or incinerator spillage from vehicle, bad smell, increase of flies and rats and wind blown dust, paper and plastics all of which are, harmful to the locality. On the other hand, refuse can be used in a well planned controlled tip to improve low lying and derelict lands, and property values may then increase in the vicinity.

Water Pollution

When rain run-off from an open dump joins surface waters there is an inevitably pollution of floating solids. Organic matter exerts high oxygen demand and pathogen can create a health danger to down streams users. Unless the water table is not high or underlying, rock is fissured, ground water will hardly be affected. Dumps should not be close to shallow wells. A distance of 12 m is suggested. On the other hand, avoidance of ground water pollution is of paramount importance, in the siting of refuse.

Waste Resources

During the last few years there has been an increasing awareness in industrialised countries that vast growing quantities of refuse are an indication of wasteful use of resources. Separation of valuable material from refuse is supported by conservationists. In developing countries everything of value are already separated for recycling. The components of municipal refuse are as follows.

Components   (eg. Municipal waste)

```
            ┌──────────────┴──────────────┐
```

Putricible   Non Putrecible

Food, Wood, Paper, Leather   Metal, Glass, Plastics
Flower, Vegetable, Ash   Hair, bone

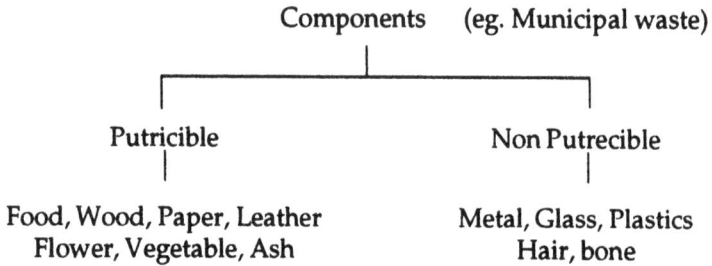

The management strategy includes the following steps:

(a) Material flow in a society

(b) Reduction in raw material usage

(c) Reduction in solid waste quantity

(d) Reuse of materials.

(e) Energy recovery

(d) Day to day solid waste management.

Characterization of solid waste is important in evaluating alternative equipment needs, systems and management programmes and plans, especially with respect to the implementation of disposal and energy and resource recovery options. It depends on a number of factors such as food habits, cultural traditions, socio economic and climatic conditions. Refuse characteristics vary not only from city to city but within the city itself and also seasonally. Quality of refuse should be assessed taking into account seasonal variations and zonal characteristics etc. NEERI (1983) reports that per capita production in the metropolitan cities like Bombay, Calcutta, Madras and Delhi range from 0.45 to 0.6kg/day. In other townships it ranged from 0.15 to 0.33 kg/day. It means the density value of Indian cities varies from 300-660 kg/km^3. High per capita refuse production indicates high standard of living.

Management of Solid Wastes

Management of solid wastes from the point of generation to its final disposal have been grouped into six functional steps. Those are,

(a) Waste generation

(b) Onsite handling storage and process.

(c) Collection

(d) Transfer and Transport
(e) Processing and Recovery
(f) Disposal.

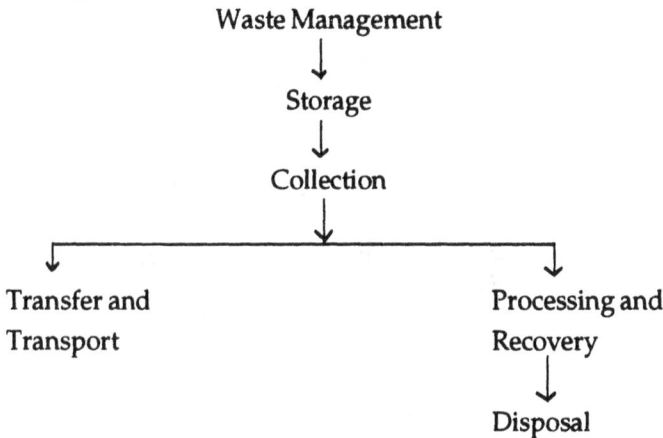

Waste Management
↓
Storage
↓
Collection
↓

Transfer and Transport	Processing and Recovery
	↓
	Disposal

Onsite Collection, Storage and Transport

The ideal arrangement is for each household to have its own cabinet with a tight fitting lid. Refuse should be removed regularly and frequently. In the humid tropics waste food decays quickly and hence it requires daily collection. It is also proposed that two collections in a week is a better option. The cost of house to house collection is costly and must be inevitably be beyond the means of many municipalities.

Since the early 1950s there has been considerable discussion about the merits of using paper or plastic for household storage of refuse. The disadvantage is increased cost, increased volume, unless the refuse is compressed in collection vehicles and damage to sacks by hot ashes and by sharp edged objects.

Refuse Dumps

For high density housing, it is useful to provide refuse dumps. Refuse is scattered by winds, goats and children. Even after collection of the main mass of refuse some amount still remains as a breeding site for flies whose eggs and larvae penetrate deep in the ground. Various types of community refuse containers are designed in India.

Collection of Solid Wastes

Waste produced from individual households is removed initially by the owner or an employee and later by municipal staff. In the case of community bin system adopted in most of the developing countries, waste is collected and taken to the community bin by the house owner or an employee, from where it is removed by conservatory staff. Wastes from the streets were collected and removed by the conservatory staff. Wastes from industries are collected and taken to the disposal site by the industries themselves.

In house to house collection system, as adopted in most of the developed countries, the workers collect the waste from individual premises where it is stored by the owner in standardised containers. If wastes are to be stored in individual house, then every premises should be provided with specific containers having specific size and air tight lids. When these are provided and located at specific points in individual premises the house to house collection system would be successful. The house to house collection system is successful only in such areas where daily collection of refuse per unit is more and where better civic sense prevails as in high income group areas and having predominantly commercial and industrial activity. This system is thus feasible to a few areas of metropolitan cities like Bombay and Calcutta. Thus house to house collected wastes are carried in smaller vehicle, preferably manually operated which can then be transferred to main collection vehicles (Techobanoglossus, 1977).

If the dwelling is small or in clusters and outside the storage space, under such conditions daily collection is essential. In the community bin system, workers sweep the roads and collect the material at specific points. The capacity of community bin should be at least 50 per cent in excess when collection is daily and 100 per cent in excess for 6-7 days. The spacing of the containers has to be accordingly fixed, that is they should be placed more than 100 mt apart. In addition to the waste generated in individual premises, wastes are generated on streets also, the collection of which is the responsibility of civic authorities. In most of the developing countries mechanical equipments are also used, which sweep the road, collect the waste and take it away for disposal.

In manual methods, the collection from the streets is deposited in common storage places from where a separate vehicle collects it for transport to processing or disposal site. In manual operation, it is desirable to have a group of not more than 3 persons, one person should clean the foot path, the second one is to sweep the road and the third to collect the material in order to transfer it to a wheel barrow. Large number of persons or groups may be needed when quantities to be cleared are large and the work has to be done in a short time. Refuse collection service involves a wide variety of tools and equipment facilities. For optimum performance of the system proper selection of equipment and utilization are essential.

In house to house collection waste is collected in containers, bins or plastic sacks. Those bins and sacks collected are directly emptied in the collection vehicles. Individual households having a service of daily collection may store wastes in used metal tins or plastic buckets provided with tight lids. For commercial establishment bins of 50-100 lt capacity with handle and wheels would serve better.

The refuse from roadside dust bins is collected by the road sweeper and transported to disposal/processing site by using a variety of vehicles and carts driven by bullocks and by municipal workers. A study was conducted in 33 cities by a team of scientists of NEERI in 1971-1973. Out of these in 10 cities, waste from narrow streets were collected in bullock carts. Now -a-days as a substitute for bullock carts small capacity transport vehicles like tractor, three wheeler auto rickshaws and trucks are used.

Disposal

1. Open Dumping

In this method, the solid waste collected from the town is deposited in low lying land, usually on the outskirts of the town. Since the open dumps are uncovered these attract flies, insects, rodents and emit odours. This method is unscientific and causes nuisance to the public and is subject to fire and fire hazards. At the same time it causes health and pollution hazards and is not suitable aesthetically. Yet this method being the easiest, it is adopted in many urban places.

2. Sanitary Landfilling

This is a modified form of open dumping. Waste is deposited in 0.9-4.5 m thick layers in depressions and then compacted and covered at least once a day with bulldozers. The covering prevents breeding of flies and rats etc. While selecting sites for land filling it is necessary to examine if any underground potable water source in the vicinity will be polluted and if the neighbourhood habitation would be affected by odour or fire. Sanitary land filling is a cheaper method of refuse disposal.

Advantages

(1) The process is completely sanitary.

(2) Highly skilled personnel are not required.

(3) Land can be safely overloaded without any problems.

(4) Converts low lying wastelands into useful area.

(5) Natural resources are returned to soil and recycled.

Disadvantages

(1) A large area is required.

(2) Since land available for this is usually away from townships, transportation becomes costly and causes fire hazard due to release of methane in wet weather.

3. Incineration

In this method, the refuse is burnt in a controlled manner. Incinerators are built with lined furnace, large area for burning air, blowers for helping combustion and oil burners to provide additional heat to burn wet garbage are needed. The primary products of combustion are carbon dioxide, water vapour and nitrogen and solid residue of glass, ceramics and mineral ash. 90 per cent reduction in the volume is attained and left with only 10 per cent ash. Air is necessary to keep pollution under control.

Advantages

(1) The residue is only 20-25 per cent of original waste, the clinker can be used after treatment.

(2) It requires very little space.

(3) Cost of transportation is not high as these incinerators are located within the city limits.

(4) Safe from hygienic points of view .

Disadvantages

(1) It is capital oriented and operating costs are high.

(2) Needs skilled persons

(3) Air pollution may occur.

Composting

Composting is a biological decomposition of organic substances available in the waste under controlled conditions. Rottening, putrefaction etc. are natural processes that take place in a controlled manner. The compost thus formed under controlled conditions is a brown peaty material. For centuries refuse has been used for producing manures. However systematic ways of preparing compost was developed by Lord Howard and is known as indoor process. This involves laying of alternate layers of the decomposable materials like night soil, animal dung, garbage etc. The mass is usually turned double during the compost process. The indoor process is anaerobic i.e., the compost is formed by anaerobic bacteria in the absence of oxygen. There are two systems by which compost can be formed.

(1) aerobic (2) anaerobic.

In the anaerobic system, anaerobic bacteria perform the work in the absence of oxygen, hence oxygen has to be supplied. The disadvantage in anaerobic system are,

(1) The process is slow (4-12 months)

(2) It is a low temperature process/also warm.

(3) It produces offensive odour.

In the aerobic process the compost is produced by aerobic bacteria. High temperature is produced, but bad odours are absent and the compost is formed rapidly. There are two methods by which compost is prepared in the aerobic process.

(1) Non-mechanical method.

(2) Mechanical method.

In both the methods of composting the following steps are involved :

(a) Sorting (b) Grinding or Shredding (c) Windrow making (d) Turning (e) Aerating.

(a) Sorting

The material like empty tins, metals, glass rags, plastics etc. have to be removed before grinding, as these are dangerous when pulverized. These are removed manually except tin and ferrous materials which are removed by ballistic separator.

(b) Shredding of Refuse

The refuse is chopped into pieces of small size ranging from 25 to 50 mm. The shredding/cutting into thin long pieces increases the rate of decomposition and aeration.

(c) Windrow System

The refuse is stacked into piles or mounds in rows of 1.5 to 1.8 meter height and a width of 2.4 – 3.0 meter. The windrows are kept in the open or in shreds. The bacterial activity results in an increase in the temperature of the biomass. The colour of the refuse changes from green to brown or bluish gray. It takes 11-31 days to prepare the compost.

Turning

In the windrow system, composting is turned frequently so that aerobic process is continued. The mixing is effected from both sides of the pile and is normally done at intervals of 2-3 days.

Bottom Aeration

Several methods are adopted for aeration but the one most commonly adopted is stacking the material on a perforated floor and forcing air into the pile by way of bottom aeration. Non-mechanical methods are adopted when the quantity of refuse to be handled is small. In India, compost plants have been set up in Ahmedabad, Bangalore, Calcutta and Bombay. The Delhi composting plant has eliminated a number of processes like prefermentation. The refuse is transported and unloaded by tipping trucks. Windrow are formed in regular rows and shape by the turning and separating machine and moisture is added with a hose pipe. A beater is fixed to the machine and the refuse is shredded. The machine removes rags, stone, plastics etc. The windrow are turned on 5th, 10th and 15th day with beaters. Aeration is also given through this machine. The garbage is then removed to mill process on the 31st day. It is fed into hopper and rags are removed. The garbage next falls on a vibrating screen and from there again through another high speed screen the dense materials like glass pieces, stones are removed from compost.

(1) This compost when added to the soil, increases water and ion exchange capacity of the soil.

(2) A number of industrial solid wastes can also be treated by this method.

(3) It can be sold thereby reducing the ash of waste disposal.

(4) Recycling of nutrients occurs.

Disadvantages

(1) When mechanized, it involves high cost.

(2) The non-consumables have to be disposed of separately.

(3) Use of compost has not yet caught up with farmers and hence it has no assured market.

Status of Refuse Production in India

For decades in India sanitation has been treated like a dirty world. Municipal corporations are ridden with corruption and inefficiency. The spread of unauthorized slums in already congested areas and a visible lack of civic sense among most of the Indians have added to growing mounds of filth and diseases which have turn the country into a big cesspool. According to the UNICEF nearly 3 lakh urban children die annually due to diarrhoea mainly in the slum areas. The National Institute of Urban Affairs (NIUA) estimates the infant mortality rate at 123 per 1000 births. The major causes are diahorrea, diptheria, tetanus and measles and poor sanitary conditions. One third of the Indian population live in urban areas. The number of cities with more than 10 lakh population or class I cities increased from 216 in 1981 to 300 in 1991. In fact the incidence outbreak of plague-like disease in Surat, in the year 1996 has initiated in many places a process of cleaning up of their mess of refuse.

Ministry of Urban Development is the key agency for urban affairs in the country. At the planning level, there is a lack of understanding of what sanitation means and as well as a poor coordination.The ministry of non-conventional energy sources is looking for power generation using garbage in incinerators. The Ministry of Health looks into the health related problems of sanitation only. Centre will not fund for garbage cleaning. It is the duty of the respective municipalities. But they have not been able to raise taxes and are poorly funded and poorly managed.

Table 27 : Refuse production rate in metro cities

Name of the City	Garbage Generated (tonnes)	Garbage cleared (tonnes)	Sewage generated (million litres)	Sewage collected	Annual Budget (crores)
Bombay	5800	5000	1800	1450	2436
Delhi	3888	2420	—	1260	1016
Calcutta	3500	3150	800	675	250
Madras	2675	2140	250	240	145
Bangalore	1500	1000	400	250	48
Ahmedabad	1500	1200	400	240	270
Patna	1000	300	141	83	15
Surat	1250	1000	130	70	170

Bombay : The world's sixth largest metro city contains 16460 persons/sq km and a total of 10 million people. 5 million live in slums.

Delhi : 4300 slums do not have sewage facility. Only VIP areas are cleaned.

Calcutta : 150 years old city. 4 million people live in 140 km^2 area. This city is overflowing with filth and sewage.

Bangalore : One of the so called cleaner cities of India. It is fast expanding. There are 419 slums.

Lucknow : This city after plague incidence has evidenced a refuse clearing drive.

Ahmedabad : Sanitation equipments are seen. Out of the required 600 containers for garbage storage, only 250 are present.

Patna : Out of 3,400 sanitation staff, half do not turn up, the rest do not work. The municipality does not have money to buy brooms, buckets etc.

Surat : This plague city is the diamond capital of India. The municipality does not have deficit. But 10 lakh people living in slums are neglected.

Mangalore : In the study conducted by MCC (Mangalore City Corporation and EMPS by DANIDA during December, 1993), it is estimated that 73000 tonnes/year or about 200 tonne/day of solid waste is produced in Mangalore city.

The origin of municipal solid waste is estimated as follows:

Household	-	29%
Commercial	-	19%
Industrial	-	4%
Street sweepings	-	24%
Market	-	17%
Building	–	2%
Hospital	-	4%

The composition of waste is estimated as

Putreciable vegetables	-	60-70%
Recyclables	-	22-34%
Others	-	Upto 18%

In Mangalore city dust bins are located at a distance of 600 m. These storage bins are made of cement concrete and are 430 in number. Waste from street are collected by paur Karmikas. About 400 workers collect the refuse through 10 open trucks and at the rate of 60 tonnes/day. The collected refuse is dumped in a place called Moodushedde, 10 km away at outskirts of the city. Garbage is also dumped in Yemmekere, Gujarakere and in Nethravathi river banks. There are absolutely no efforts to reuse and recycle the refuse, except that the rag pickers are active in collecting recyclable materials like metal pieces and plastics. The organic components and paper are wasted or burnt.

CHAPTER 6

CASE STUDIES ON USWM

I. URBAN SOLID WASTE MANAGEMENT WITH REFERENCE TO HYDERABAD CITY

K. Jeevan Rao

1. Introduction

Upgradation of environmental quality has focussed on problems caused by urban agglomerations. The wastes generated by urban centres and the associated environmental problems have revealed that greater emphasis must be given to the management of urban solid waste and their by-products. Consumerism evident world wide, has magnified the problems of waste generation, handling, transportation and ultimate disposal. 'Waste' is any material disposed off as no longer useful to the normal human activities. A waste product which is not free flowing is recognised as a solid waste. Urban Solid Wastes (USW) also called by the name municipal solid wastes, comprise commercial, industrial, agricultural and mining wastes.

Solid wastes comprise countless number of different components : dust, food wastes, packing in the form of paper, metals, plastics or glass, cloth and furnitures, garden wastes, and hazardous substances.

In the most cities of India, the decomposable organic matter is the major component of the municipal refuse and makes up nearly 70 to 80 per cent of it. It is this component of solid waste which when left uncleared, starts decomposing, emitting foul odour and also causes breeding and proliferation of undesirable and disease causing organisms.

(148)

In the industrialised countries the environmental health problems have been solved at the stages of storage and collection of solid wastes, although major problems remain in the matters of resource recovery and disposal. The technology of wastes handling is now highly developed. On contrast, India and many other developing countries are suffering due to the problems caused by urbanisation. The solid wastes generated from different sources are highly heterogenous in their physical, chemical and biological properties and are called for different methods for their safe disposal. The waste has often been able to provide developing countries with technical guidance in environmental matters but solid wastes may be exception ; there are too many climatic, economic and social differences for systems to be successfully transplanted.

The most obvious environmental damage caused by solid wastes is aesthetics and environmental hygiene. A critical factor in the evaluation of alternative systems of solid wastes handling is the quality of management.

2. Population Growth Rate

India's decennial population growth rate dropped by a little over point two per cent from 24.7 in 1971-1981 to 23.5 in 1981-1991. The average annual exponential growth decreased from 2.2 to 2.1 per cent in the same period. As on March 1, 1991 the Indian population stood at 844.0 millions. In real terms India had added a Japan to its population over 1981-91 or an Australia every year. Indians constituted 16 per cent of the world's population. Popultion density increased from 261 per sq km in 1981 to 267 in 1991. The figure stood at 117 in 1951.

An eight-fold increase of Indian urban population recorded in the past 90 years from 25.85 millions in 1901 to 217.18 millions in 1991. In India 217.18 million people that is 25.72 per cent out of 844.00 million people live in 3768 urban agglomerations or towns.

The National Commission on Urbanisation (1988) predicted that the urban population would be of the order of 35 per cent of total population by 2001 AD as against 33.06 per cent projected by the Experts Committee on population projections. Urban populations continued to multiply in Bombay topping the list with a population of 12.57, Calcutta 10.86, Delhi 8.38, Madras 5.36, Hyderabad 4.27 and Bangalore 4.11 millions. As on March 1, 1991 the total population of Andhra Pradesh was 66,304,854. The urban population was 17,764,266 and the rural population was 48,540,588, the percentage of urban population to the total population is worked out to be 26.79 per cent.

The Hyderabad agglomeration accounts for roughly one-fourth of the urban population of the Andhra Pradesh State, which is 17,764,266 as against 12,487,576 at 1981 census. The hike in urban population as compared to the total population standing at 26.79 per cent as against 23.32 per cent at 1981 census. In Andhra Pradesh cities with more than one lakh population, numbered 32 as against 20 in the 1981 census. The growth rate of class I cities in Andhra Pradesh has been the highest recorded among the major states.

3. The Urban Challenge

Urbanisation is one of the many crucial challenges being faced by modern society. More than half the world's population will be living in cities and towns by the year 2000 AD. Significantly, 17 of the world's 20 large agglomerations will be in developing countries. More people in the developing world will leave their villages to start a new life in the city. Nearly 40 per cent of Indian urban population lives in slums.

Today, 217.18 millions out of India's total population of 900 millions live in urban settlements and the level of urbanisation is expected to reach 35.00 per cent in another 10 years. The concentration of populations poses several social problems, but it also means better opportunities for civic development, gender awareness, information and participation.

The relationship between national income and degree of urbanisation is now well established from the economic point of view. World Bank data shows that 60.00 per cent of the gross domestic product of developing countries are produced in towns and cities. The mushrooming of smaller cities are another significant factor in the urbanising process. In 1981, there were 218 cities of over the lakh population, by 1991 the number had risen to 316.

4. Urban Solid Waste Generation

In India, the total urban population of 240 million produce three lakh tonnes of USW every day. The percapita waste generation is estimated to be 0.33 kg. The country is already spending Rs. 23.00 crores annually on disposal of this USW. Most of it is produced by urban houses and industries. The problem is typically "Dump and Forget".

4.1 USW Problem in Hyderabad City

Hyderabad is fifth largest Indian city with a population of >5 million. The population of Hyderabad generates 1500 to 2000 tonnes of wastes every day. The enormity of the disposal can very well be gauged by the huge quantities of wastes generated in Hyderabad. As the Municipal Corporation of Hyderabad (MCH) has resorted to dumping USW even near residential areas and the citizens are now exposed to health hazards the situation demands immediate attention.

The situation has now came to such a state that even the USW which is lifted has no place to be disposed off. The residence colony near Autonagar, Vanasthalipuram, where a landfill/dumping site is located, are offering still resistance to the MCH lorries bringing in the USW from the city with houses having come up in the areas, the residents oppose the dumping off USW here as it is a health hazards for them. There are problems in other places also. As a result of new problems, the MCH lorries are disposing off the USW in low lying areas near residential areas in the city (Fig 6.1). While the transport problem continues, there is no place to dispose off the lifted USW, the residents of the twin cities, Hyderabad-Sikandarabad are forced to live in the filth.

Several political groups protested against the dumping of USW by the Corporation at Autonagar, on the city outskirts in L.B. Nagar Municipal limits. The activists wanted the Municipal authorities to stop dumping USW at Autonagar, as the environment in the surrounding colonies was badly affected. The students and staff of the Sanjay Gandhi Technical Academy were unable to concentrate on their studies because of the bad smell. The USW dumping had affected poultry farms, forcing their closure. Physical composition of USW generated in Hyderabad (%) is given below (Table 28)

Fig. 6.1 : Garbage site near residential area

Fig. 6.2 : Garbage collector and turner

Table 28 : Physical composition of USW generated in the Hyderabad City (%)

Gravel	12.5
Glass	1.0
Rags & Textile	7.0
Paper	2.0
Metals	1.0
Leather	0.2
Plastic and polythene	1.0
Earth and Ash	35.0
Compostable matter	20.0
Miscellaneous	18.4
Moisture	25.0
Density kg m^{-3}	305.0

Source : Jeevan Rao, 1991.

4.1.1 Threat to Deer Park

Even as the population of the black buck and the spotted deer is going up at the Mahavira Harina Vanasthali Deer Park on the Hyderabad-Vijayawada Highway, a threat to its inmates is looming large. The threat is manifested in the shape of a chain of USW dumps skirting a side and extending all the way to the second phase of the park. People in the neighbourhood and the forest officials alike complain of the stink eminating from the USW heaps along the road margins. A large number of vultures hover over the park and prey on carcasses dumped. Stray dogs making a further mess of it. Hyderabad is turned 400 years old, but the occasion serves only as a grim reminder of how much civic conditions in the State capital have been deteriorated.

5. Management of Urban Solid Wastes

Various methods are available to treat and dispose of the refuse generated in cities. The feasibility of those methods, is dependent upon the local conditions, including the availability of land, the market conditions for refuse generated products, transport, equipments and availability of skilled labour. Following mentioned environment friendly methods can be used.

5.1 Recycling

Recycling is only the first step in a development process. Cities that recover their materials to avoid paying disposal costs. But the real benefit to the local economy comes from converting scrap into useful products, processing aluminium into ingots and paper into pulp. How far a city can move in this direction is a function of its size and density, industrial mix and political will.

Waste recycling principally of paper, plastics and metal item in USW came into vogue in developed countries recently, the 6 lakh rag pickers in India have unobtrusively been performing this function all along. In today's highly environment conscious society, effective waste management is a key factor. Recycling is basically a means by which we can regain and reuse materials from select wastes which are usually discarded. Paper, aluminium and plastic are commonly recycled.

The following constituents are commonly extracted for industrial re-uses :

— paper, for re-pulping,
— textiles for paper making,
— metals for remelting,
— glass for re-melting,
— plastics for production of inferior grade plastic.

Recycling has become a necessity, as many countries are unable to locate sites suitable for USW disposal, since land is scarce. Further, dumping the wastes is delayed, the emissions thus produced are harmful to living beings. The dwindling supply of raw materials also implies that we must extract the maximum from the existing resources. The financial element also attracts countries like ours to take to recycling as we can save on valuable foreign exchange required to import commodities such as oil and other raw materials.

The recyclable item has greater value if not mixed with the rest of the USW. A tremendous amount of emphasis is therefore laid on "source separation". That is, the trash is sorted out into glass, plastic, paper and the like by the consumers themselves. The recyclable articles can then be sold at "buy back" centres or at times it is collected from the individuals and then sold to recyclers. Source separation is advantageous as it involves the direct participation of society in quicker 'resource recovery' which aids both the country's economy and environment.

In India, source separation is disliked by many. So persons are employed to separate USW. People should be educated on the salient features of source separation. Buy back and collection of waste programmes which are already prevalent can be developed into a more organised industry.

Paper is one of the most universally recycled products. Paper recycling was in vogue from the time paper production was mechanised. The process includes removing the ink and pins from the waste paper, before transforming it to the regular pulp used in paper manufacture. Paper recycling is favoured, as the need for virgin wood pulp is reduced, this resulting in the conservation of trees. Plastic recycling is widepsread, as plastic waste is significantly responsible for environmental pollution. Some waste plastics can be melted into new plastic articles. Certain other types are ground into a fine powder and used as fillers or shredded to provide insulation in avilted jackets or sleeping bags.

Aluminium recycling is receiving great importance, especially in places like the U.S. where canned drink and food consumption is high. Aluminium recycling saves a phenomenal amount of energy, upto 95 per cent of th energy spent in the making of aluminium. The U.S. and Japan are among the leaders in the recycling trade. In Japan especially, recycling is very methodical due to a high level of civic awareness. Roughly about 35 per cent of Japan's paper comes from the waste paper basket. Their combustible waste is employed to create electricity. Even whatever little is not recycled is invariably used in the reclamation of land.

Raw materials which are bio-degradable or those that can be easily decomposed are advocated in the production of goods, as they facilitate recycling. As in foreign countries, more Indian commodities should start displaying a label which indicates that the item is recyclable or is a recycled one, and we as conscious citizens should go in for such items.

5.2 As a source of plant nutrients and organic matter (Composting)

The city refuse is often referred to as a resource out of place. It contains plant nutrients such as nitrogen, phosphorous and potash.

The data on the potential supply plant nutrients from solid wastes is presented below :

Table 29 : Nutrient supply from USW in India

Total urban population of India (1991)	240	million tonnes
Total solid waste generation	29	million tonnes
Nutrient quantity (Annual total tonnes)		
Nitrogen	1,45,000	
Phosphorus	70,000	
Potash	70,000	
Total NPK	2,85,000	

Source : Jeevan Rao, 1992

Compost from city refuse can be used with chemical plant nutrients as it improves the soil structure and water holding capacity. Under a central scheme for solid waste disposal, liberal grants-in aid are provided for installation of mechanical compositing plants. Such plants have already been installed in Bangalore, Baroda, Bombay, Calcutta, Delhi, Jaipur, Kanpur and Hyderabad etc., each producing 75 to 175 tonnes of compost daily.

Importance of agricultural authorities

The agricultural authorities who should determine the extent of which compost may be required for organically deficient areas, or as a primary source of nutrients or for agricultural use in general.

Precautions to be taken while using USW as a manure

The USW is highly heterogenous, alkaline in pH and varied widely in their chemical composition. The studies conducted by the author revealed the following important findings.

1. The composition of USW should be considered before application on agricultural land.

2. Compostable and non-compostable components of USW should be separated at source itself.

3. Indian Standard Institution 9533-1980 Guidelines should be followed to dispose off USW.

4. There is a need to control run-off from landfill sites.

5. From landfill sites upto 200 meters groundwater sources should not be used for drinking purposes.

6. To avoid the serious problem of weed infestation due to

use of USW as manure, it is recommended that stabilized (composted) wastes should be used.

7. Root crops should not be grown when USW are used on agricultural land, to prevent uptake of heavy metals.

8. Staggered application of USW minimizes the build up of soluble salts, sodium and heavy metals in soils besides improving soil fertility.

9. Disposal by dumping of USW to reclaim abandoned wells, tanks, low lying areas should be stopped.

5.1.1 As organo-chemical fertilizers

Organised efforts to improve process and technology of composting were initiated in India in 1930's. A significant advance towards the qualitative improvement of compost was the production of the material supplemented with phosphatic fertiliser. Studies revealed that (Mutatkar, 1985) compost quality would be greatly increased by incorporation of rock phosphte in the composting system. Two such products (organo-mineral) by blending organic and mineral fertilizers viz., 9 : 9 : 5 and 7.5 : 7.5 : 7.5 grade were produced by Agricultural Research Division, Rashtriya Chemicals and Fertilizers Ltd., Bombay. This is fortified compost granulated with off grade fertilizers and is therefore, very similar to commercial fertiliser in chemical content and appearance and contains about 25-30 per cent organic matter.

Recommendations : Based on the research conducted at Agricultural Division, Rashtriya Chemicals and Fertilizers Ltd., Bombay, the following recommendations were given by V.M. Mutatkar of RCFL.

1. The semi-mechanised compost plants should be set up with low capital cost in periphery of towns having less than 1,00,000 population so that they would be economically viable.

2. Locally available materials like sludge, biogas slurry, press mud, sugarcane waste along with fungi, azotobacter etc. should be used for enrichment.

3. The granulation plant of 50 tonnes day^{-1} capacity should be set up at the same site of compost plant and off grade fertiliser material should be used for enrichment.

Enrichment can be done by bulk blending according to the need of the soil and crop conditions of the area is also possible.

4. Organo-mineral fertiliser thus prepared should have :

(i) Sufficient nutrients with good rate of mineralization

(ii) product should be biologically active

(iii) product should be free from undesirable components

(iv) Preparation or product should be chemically feasible locally compatible, technically simple and environmentally non-polluting type.

5.2.2 Composting

Composting is more popular in Europe and United States of America. Many European Compsting plants, such as the one at Frederikssund in Denmark, compost a mixture of refuse and sewage sludge. In the U.S. pressure to find alternative ways of disposing of sewage sludge has also encouraged the development of such mixed plants. Portland in Oregaon may some have the biggest solid waste composting plant in the U.S. It is being build to handle 800 tonnes of waste a day in conjunction with a recycling scheme for metals, plastics, paper and glass. There are atleast three composting plants functianing in Egypt.

Table 30 : Composting of municipal solid wastes in different countries

Country	Amount K tonnes/yr	Composting % by wt.
Austria	2,800	18
France	20,000	10
Portugal	2,650	15
Spain	13,300	17
USA	177,500	2
Germany	25,000	2
Canada	16,000	2
The Netherlands	7,700	5
Italy	17,500	7
Switzerland	3,700	7

Source : Warmer Fact Sheet, 1992

In Duisburg, West Germany, a 31 year old plant composts 100 tonnes of domestic waste per day, a plant at Bad kreuznach handles twice this amount. In some parts of West Germany, however, concern over toxic metal contamination has limited the use of compost from mixed refuse to use as a filler and in sound reduction barriers on motor ways.

In Europe, early systems for composting refuse were based on mechanical separation and the compost often contained impurities. More recently, there has been a strong drive towards improving its quality. The best way is by house holders separating compostable material, such as food scraps and vegetable peelings, from non-compostable material such as vacuum cleaners, bagass, disposable nappies, sanitary towels and shoes.

In Munich, a one-year scheme is being set up in which 40,000 households will be given a "Biobin" that is emptied every fortnight. Munich council expects to divert about a one third (50,000 tonnes per year) of the waste going to two-incinerators and landfill sites in this way, with sales of the compost offsetting costs.

Similarly in Hyderabad there is a mechanized composting plant at Autonagar with the capacity of 200 t/day. There is a need to start 'Biocycle' as an extension to the existing recycling scheme. Almost every household should separate their compostables into a strong paper bag for weekly collection and composting by the Municipal Corporation of Hyderabad. (Fig. 6.3)

5.3 As a source of energy

Urban solid wastes contains a good deal of cellulosic matter in a form which can be burnt to produce energy. However, recycling plants producing energy from USW have been working in Europe since a long time. In the United States alone there are around 130 working waste-to-energy power plants. The largest of them processes 12,000 tonnes of refuse (fuel) every week generating 68 MW of power. The Union Governments Department of Non-conventional Energy Sources decided that this approach could kill two birds with one stone, that is get rid of USW by incinerting it and produce much needed electricity in the process. It followed this decision with an import of a Danish Plant at a cost of Rs. 18.0 crores which was installed in the mid-eighties at Timarpur near Delhi. The plant never worked. Technically, there was nothing wrong with the

Fig. 6.3 : Composting plant of the Hyderabad city

plant. It turned out that Indian USW was different in composition from those found in developed countries.

Indian USW is generally low in recyclable components (which probably would have been collected by rag pickers) but has much higher moisture content and large quantities of inert materials like sand, grit, stones and small metal scraps. As a result, raw Indian USW has low calorific value and does not burn easily. Unlike USW of developed countries, raw Indian USW cannot be directly incinerated to generate power (Timarpur plant was designed to do just that). To increase the USW's calorific value before using it as fuel it had to be processed, since its composition is different, the Indian USW has to be processed differently from those in the west where the intention is mainly to separate recyclables. So a continuous process should be used to remove sand, grit and metal pieces from the raw USW, drying, shreding and finally compacted into fuel pellets so as to use it as a raw material for power generation.

Technology

The technology needed is simple. The USW is passed at various stages over metal meshes to sieve out soil particles. Shredders chop the USW clumps into small pieces for easier processing. The USW is passed through a kiln to dry up the moisture with hot air circulation (It also serves to kill pathogens in the USW). Iron can be separated using magnets while non-magnetic particles like glass, ceramics, iron-magnetic metal pieces and stones are removed by sucking the USW vertically upwards in a strong stream of air.

Suggestions

Mix agricultural wastes like paddy husk and saw dust into the processed USW to further increase its calorific value. The mixture is then made into pellets using modified cattle feed pelletiser. This can increase the calorific value from 937 per kg for the raw USW to over 4000 K calories per kg for the blended fuel pellets (with agricultural wastes mixed in).

Working Examples

A 60 tonnes capacity/day plant in Bangalore can process about 200 tonnes of raw USW every day. It is in the proximity of dumping yard. The country's first large scale plant to produce fuel pellets from municipal USW began trial runs in Bombay in November, 1991. The plant was set up by the Department of Science and Technology with the assistance of CMC Limited. The two crore plant is designed to convert 150 tonnes of USW (Bombay generates 4000 tonnes every day) into 80 tonnes pellets which can be used for local industries. Nearby slum dwellers too can use it as a cooking fuel.

Procedure

The USW was first spread out and sun-dried in an open yard for a day This alone removes a 30 to 35 per cent of the moisture in the USW. Some items like chappals and tyres are picked out by hand. One hectare is enough to dry 250 tonnes of USW. The light combustible part of the USW is separated from the denser fractions by ballistic separation where a conveyor belt drops USW from a height while a blower provides a stream of air so that the lighter particles fall clear of the heavier ones. A part of the unprocessed USW, after a day's drying, is burnt to run the kiln in which the USW is dried. Some critical equipment like grinder and pelletiser are run in two streams. With proper separation of the non-compostible part of the USW and adequate drying, it is possible to get pellets with a calorific value of 3500 K Cal per kg.

Precautions

Such plants should be designed only after careful examining the kind of USW available locally. There could be seasonal differences even in the same place. Moreover, USW from road sweepings is entirely different in character from vegetable market wastes. In addition, climatic changes should also be considered. For instance, the plant could not be operated during the three monsoon months.

The calorific value of municipal refuse can be utilised by burning it in excess of air as in incineration. In the absence of air as in pyrolysis or by bio-chemical decomposition to produce biogas. For Indian waste characteristics, auxillary fuel will be required for incineration, making it costly. The process will, therefore, have limited application as for some hospital and industrial wastes are concerned. Pyrolysis yields number of products which can be utilised as sources of fuel. However, the process is economical only for certain commercial and industrial wastes.

Biogas can be produced by anaerobic decomposition of the organic portion of refuse. The gas produced has a calorific value of $1950/KCal/m^3$. If all the available organic matter in the city refuse is converted into biogas, the heat energy made available can satisfy 27 per cent of the domestic energy requirement of the Indian population. This will be an immense contribution especially as it also reduces pollution by solid waste. The digested solids can be used as a source of organic manure. It is, therefore, necessary to standardise the method and popularise it so that at least the domestic energy requirement can be taken care of.

5.4 Sanitary Landfilling

At present, 90 per cent of the municipal waste is dumped in low lying areas without proper precautions. This results in breeding of rats, flies and other vermin and pollution of surface as well as ground water. It is necessary to adopt sanitary land filling techniques so that the sites can be put to beneficial use much earlier. In some developed countries, the biogas produced during decomposition of the organic matter in the landfill is used as a source of energy. (Table 31).

Table 31 : Landfill gas utilization projects in Europe

Country	Total plants	Country	Total plants
West Germany	74	France	4
UK	33	Norway	1
Sweeden	20	Switzerland	7
Italy	7	Belgium	1
Netherlands	7	Spain	1
Denmark	5		

Source : Warmer Fact Sheet, 1993, Landfill gas

This is possible only when the landfill has been constructed by adopting certain design criteria and the site is sufficiently large.

Landfill Site—Precautions

1. Special public information programmes detailing all the environmental hazards arising from the landfill/dump site are needed.
2. Methane meters and various testing devices should be used to test environmental safety in the dumping site.
3. A de-gasification plant is to be put up and special drains to collect seepage water are to be laid.
4. In the future at least the philosophy of waste recycling should be integrated in USW disposal.

5.5 Refuse derived fuel (RDF)

The Department of Science and Technology, Govt. of India has set up a pilot plant at the Denar dump yard in Bombay to process nearly 120 tonnes per day (TPD) of municipal waste to produce 80

TPD of Refuse Derived Fuel (RDF) pellets. Presently the plant produces one tonne pellets per hour. Whereas, modifications are being worked out to raise this to 1.5 tonnes per hour, according to CMC Ltd., who is associated with the designing of the plant.

A similar project has been established by M/s. Shivshankar Engineering Company Pvt. Ltd. in Bangalore and got regular production of fuel pellets since October 1989. Compacting 50 tonnes of garbage per day converting into 5 tonnes of fuel pellets. The fuel pellets can be designed both for industrial and domestic uses.

The combustible pellets which have a 10 per cent ash content (against 35 per cent in coal) and do not contain sulfur but only carbondioxide, will be a suitable alternative to coal for textile mills and chemical industries. They can be used as a domestic fuel also. Already some industries have shown interest in the pellets which have a calorific value up to 3500 KCal/kg as compared to 4000 KCal kg^{-1} in coal, at half the rate.

Problem of monsoon : The only disadvantage is that the plant cannot be operational during the monsoon months because the USW is to be dried under the sun. Any other process would not be energy efficient to overcome this, it is better to plan to dry and bale the USW for use during the wet months.

5.6 Light Weight Aggregate

An Australian Research and Development Company had evolved a process of converting municipal waste into light weight aggregate. Here municipal solid and liquid waste are mixed with clay to produce a ceramic rock called neutralite graded for use in construction and building.

5.7 Vermitech

Vermicomposing of the wet waste by using earthworms is an ideal method of compost making. One can use two different types of earth worms. One is surface feeder and the other burrower. The efficiency of these two types in our conditions should be determined before using them. The worm castings (vermicompost) is an ideal pot culture medium, and it converts the wet wastes into a useful value added product vermicompost, costs Rs. 2000/tonne.

Vermiculture : Vermiculture means culturing of earthworms as versatile natural bioreactors for cleaning up the environment with cost effective waste management technology, developing of sustainable agriculture, waste land development. Vermiculture biotechnology can be used to handle biodegradable organic wastes, in three ways :

1. Bio-processing under a tree
2. Bio-processing in a container
3. Collection of waste and central vermi-processing

Bhawalker Earth Worm Research Institute, Pune, evolved a vermiculture technology. The package consist of :

1. Applying first a layer of vermi castings at the rate of 2.5 t ha^{-1} (cost about Rs. 4500 to 5000 per tonne at present)
2. Application of water to keep sufficient moisture
3. Watering every 15-20 days. This process helps to improve soil pH, in the soil structure etc.

5.8 Other uses of USW

The USW can be gasified at temperatures of 500 to 1000°C in the absence of oxygen. This process (called pyrolysis) reduces the waste to gases comprising hydrogen, methane, carbonomonoxide and carbondioxide, tarry liquids containing wastes and organic compounds (such as acetic acid and methanol) and carbonaceous char. Some of the gas can be burnt to run the reactor itself. The surplus gas and liquids are available as fuel products. The solid char can be processed to yield various gardes of useful products such as activated carbon or an inert slag which can be disposed off easily. Cellulose in USW can be converted into glucose enzymatically or through chemical hydrolysis. The glucose can be fermented to produce ethanol. Other useful chemicals like acetone, organic acids, glycerol can also be extracted through the hydrolysis of USW.

6. Health Hazards

In most of the Indian cities, the scavengers handle USW at a number of stages where they either come in direct contact with the infected material or the dust gets air-borne and is inhaled by them. Clinical examination supported by stool and blood testing of refuse workers as well as a control group carried out by the NEERI in 1971 in Trivandrum to study the health impact on the workers, showed that nearly 98 per cent of the test group was infected with parasites as compared to only 33 per cent in the control group. The clinical examination also showed a higher incidence of respiratory and skin diseases as well as jaundice, trachoma and eosinophilia. Thus, it is imperative that the scavengers handling the waste manually should be provided with protective garments, gloves and gum boots.

For the general public, the main risk to health are indirect and arise from the breeding of disease vectors, primary flies and rats. The present by-laws do not take into account all the aspects. Hence more comprehensive and stringent laws are necessary to ensure safe disposal of USW and to keep the environment clean.

7. Environmental Standards

There are international organisations for standardisation (ISO) and International Electrotechnical Commission Standards in the field of environment ranging from the quality of air, water and soil. While the international standards prescribe the methods of analysis, the national standards set the actual limits. The ISI has so far prescribed 262 such standards. The ISO/TC No. 146 pertains to air quality, No. 147 is on water quality, No. 190 on soil and No. 43 on acoustics.

The first legislation in India was the 1974 Act relating to water (Prevention and Control of Pollution Act). The second one was on air and Act of 1981. An umbrella Environmental Protection Act covering all the respects was also brought in 1986. Prescribing standards is easy, the real problem lies in enforcing regulations. The Pollution Control Board have their own limitations and are subject to too many pressures. Legal steps alone will not suffice. The real control has to come out from public awareness and the citizens.

8. Regulatory Measures

Public health and sanitation are state subjects. Collection and disposal of wastes in the urban areas is entrusted to the local urban body. The municipal laws lay down detailed lists of obligatory and discretional duties. Due to the direct bearing of sanitation on the health of the people, it has been listed as obligatory duty and therefore, the civic authorities are required to take adequate measures.

The old regulations suffer from the defect that the various categories wastes for which they would be applicable are not covered in sufficient details and are made applicable for domestic and to some extent trade wastes. They do not provide sufficient powers to the civic authorities for prosecution of offenders with the result that the enforcement has become ineffective.

Lack of suitable system for collection, transportation and disposal and complete networking is the major constraint for proper disposal of municipal solid wastes. The local bodies which are responsible for proper additional responsibility of much higher priorities like immunistaion, food sample testing, street cleaning, etc. Waste disposal is relegated and placed at the end of the list. The reasons for poor operation and maintenance of refuse are :

   (i)   Inadequate financing
  (ii)   Multiplicity of agencies for operation and maintenance
 (iii)   Inadequate training of personnel
  (iv)   Lack of performance monitoring
   (v)   Inadequate emphasis on preventive maintenance
  (vi)   Lack of management
 (vii)   Lack of appreciation of the importance of the facilities by the community etc.

The burden of environmental degradation fails largely on the poor who do not have the means to choose their living conditions as they become the victims of development. As the country proceeds on the path of industrial and urban growth, the generation of solid waste in cities would continue. Efficient ways and means for the collection, treatment and disposal of the re-cycled products and are of by products have to be found out composting is one of the cost effective solution in the presentation especially applicable to the type of solid waste generated in Indian cities.

At present the involvement of the Government of India, is limited to sharing the subsidy component for low-cost sanitation for the economically weaker sections only. Apart from this, there is no other centrally assisted programme for urban waste management.

The role of the Ministry of Urban Development, Government of India is largely limited to advising the urban local bodies, supporting research and development and providing technical assistance to the states and urban local bodies.

9. Conclusions

1. The immediate need is to establish a network of research efforts to closely monitor the generation, availability and characteristics of USW of large and small urban communities.

2. Out of operational expediency, it is very common to observe in cities that the USW is dumped to retain abandoned wells, tanks, low lying areas for reclaming land. Such reclaimed land is being utilized for establishing housing colonies and industries. The soil and ground water pollution is imminent due to such practice of disposal of large quantities of USW. This practice therefore, needs to be discouraged because once the soil profile gets polluted with pollutants like salts and heavy metals, it serves as a store house for polluting the ground water.

3. The urban centres are turning into a places of rubbish with the steady rise in living standards. There is a need to amend the USW disposal law for considering the refuse as a resource to be saved and reused.

4. Action programmes have to be formulated and the policy makers have to be sensitised on these issues to obtain their support at appropriate fora. Once these recommendations are implemented, there would be a substantial improvement in environmental friendly integrated utilization of USW in future.

5. Public awareness and cooperation and a sense of sanitation are the prerequisites for proper functioning of any solid waste disposal system. These can be ensured only through proper sanitation and health education. Continuous efforts by government and non-governmental agencies along with the peoples participation can make any city clean.

6. The technical know-how and other capabilities in this field, the organisations like the National Environmental Engineering Research Institute (NEERI), Nagpur, can be utilised by municipal administrations to good effect.

7. With option as varied and effective as the ones discussed, it would be a crime to just dump USW instead of looking at it as a valuable resource. We do have a long tradition of recycling and reusing something that the West is now relearning.

Literature Cited

Golueke, C.G. 1977. *Biological reclamation of solid wastes.* Rodale Press, pp. 1-3.

The Hindu, 1991. BJP Corporator suspended, 21st June.

Sundary Chronicle, 1993. Dispose of USW scientifically, May 2.

Rodney, L. 1968. Effects of land disposal of solid wastes on water quality. USDHEW. Bureau of solid wastes, U.S.A.

The Hindu, 1993. Energy from Garbage, 21st June.

Dr. Rashmi Mayur, 1983. Environmental pollution in India. Global Futures Digest, Vol. I, No. 2.

Radia Khatib, 1990. Evaluation of recyclable materials in municipal waste from Karachi, *Biological Wastes*, 31 : 113-122.

The Hindu, 1993. From Waste to Wealth, 16th January.

Jeevan Rao, K. 1994. Garbage disposal problem in Hyderabad. Paper presented at UGC sponsored *National Symposium & Workshop on Environmental Education in University Carricula* held at JNTU, Hyderabad from 21 to 23rd March.

The Hindu, 1991. Hyderabad 5th Largest City, 26th March.

The Hindu, 1991. Light weight aggregate from municipal waste, August.

The Hindu, 1990. Little to be happy about, 29th May.

The Hindu, 1991. Loss of soil fertility, threat to food production.

Kala, J C and Khan R R, 1994. Management of Municipal Solid Wastes–*Yojana*, Vol 37, No. 24, Jan. 15, pp. 18-20.

World Health Organisation, 1976. Management of Solid Wastes in Developing countries. SEAS No. 1, New Delhi.

The Hindu, 1991. Meeting the urban challenge, 6th October.

Gotaas V B H, 1962. Municipal solid waste disposal, part V American City 77, 6 : 190.

The Hindu, 1991. Noval approach to environmental hygiene, 24th May.

The Hindu, 1991. Novel garbage clearance in Bombay, 3rd June.

The Hindu, 1992. Oldest Dumping Ground, 12th April.

National Academy of Sciences, USA, 1991. One Earth, One Future.

The Hindu, 1991. Population growth rate dips, 25th March.

Mutatkar V K, 1985. Prospects of organo-inorganic fetili-zers. Fertilizer News, December.

The Hindu, 1988. Recycling USW to save money in America. 20th January.

Bhardwaj, K K R and Gour A C, 1985. Recycling of organic wastes. AICRP on microbiological decomposition and recycling of farm and city wastes. Fifteen years of research 1968-1982, ICAR, New Delhi.

Vinal, O P and Talashilkar, S C 1984. Recycling of urban solid wastes. *Journal of Science and Industrial Research, 43* : 65-67.

The Hindu, 1991. Role of standards for a better environment, October.

Jeevan Rao, K 1992. *Soil and Water Resource Characteristics in relation to land disposal of urban solid wastes.* Ph.D. Thesis, A.P. Agricultural University, Hyderabad-30.

World Health Organisation, 1971. Solid waste disposal and control. Geneva.

Sundaresan, B.B., A.D. Bhide, S.K. Titus and A.V. Shekdar, 1983. *Solid Waste Management–A Course Manual.* NEERI, Nagpur.

The Hindu , 1986. Solid Waste Management and recycling in urban areas, 17th September.

Solid Wastes Engineering Principles and Management Issues–George Techobanoglous 1977–McGran Hill, K.O. Gukusha Ltd. Tokyo.

Bhide, A.D. and B.B. Sundaresan 1983. Solid Waste Management in developing countries. INSDOC State of the first Report. Series 2, New Delhi.

Nandkishore V, 1980. *Studies on urban solid waste composting*. MSc. (Ag) Thesis, University of Agricultural Sciences, Bangalore.

The Hindu, 1991. The biggest compost heaps in the world, 31st October.

Deccan Chronicle, 1990. The USW city 26th November.

The Hindu, 1992. The Pollution Bomb. 7th June.

The Times, Guide to the Environment, 1990. U.S.A.

The Hindu, 1989. Threat to Deer Park.

The Hindu, 1990. Threat to Deer Park. 2nd March.

The Hindu, 1994. Turning Garbage to glass. 12th January.

The Hindu, 1991. Twenty three per cent point eight two (23.82) per cent rice in population. 26th March.

The Hindu, 1993. Urbanisation slows down. 31st March.

Shantaram M.V. and Jeevan Rao K., 1993. Utilization of urban solid wastes in agriculture problems and prospects. Paper presented at *National Conference on Biofertilizers and Organic farming* held at Madras, Tajcoro mandal on 25 to 26th December 1993. Organised by Ministry of Agril., Govt. of India and Dept. of Agril., Govt. of Tamil Nadu.

Bhide, A.D., 1984. *Urban Solid Waste Management—An Assessment*. NEERI, Nagpur.

The Hindu, 1990. You can help.

India Today—Our filthy cities can be clean The Mess—1994, October 31, pp. 62-79.

Warmer Fact Sheet 1993, Land fill gas Warmer Campaign, Bridge House, High Street, Tonbridge, Kent TN 9 IDP, UK.

Warmer Fact Sheet 1992, Composting Warmer Campaign. Bridge House, High Street, Tonbridge, Kent TN9, IDP, UK.

II. URBAN SOLID WASTE ASSESSMENT AND MANAGEMENT FOR SHIMOGA CITY

B.B. Hosetti, P.Y. Latha, R.S. Sabhahita

Abstract

A study on the status of refuse management in Shimoga city was carried out during July, 1993 to November, 1993. The refuse produced in lower and upper class residential, commercial, industrial and municipal area was collected and the samples were analysed for physical (loose density, compact density, organic matter, moisture content & pH) and chemical (BOD, Chlorides, Phosphates and Organic matter) parameters. Percapita values 0.15 kg and 3.6 kg and 16.11 kg was calculated for residential, commercial and industrial sectors of the city respectively. On the basis of this information total production of solids in the city was calculated to be 61 tons/day. A part of the city refuse (about 10 tons/month) was used for manuring orchard gardens and vermiculture practices. Major portion of the city refuse was disposed on the low lying areas. About 100 members of rag pickers were active in collecting recyclable materials like glass, metal, paper and plastic in the city all the year around. The study on solid waste indices revealed that moisture content was high in the residential areas. The present system of refuse handling and managing is not satisfactory and hence a systematic regular collection and hygienic management through sorting of components and their recycling is suggested.

Introduction

Until 1950's solid waste disposal has not posed any problem, however during the period from 1953 to 1955 the spread of a viral disease to hogs has attracted the attention of sanitary engineers and farmers in USA (Trivedi and Gurudeepraj, 1992). Since then feeding of garbage to hogs was banned in USA. However, in most part of the third world countries garbage feeding to hogs and cattles is still going on.

It is probably for the first time the scientific research on refuse management was started and published in Chicago by Public Administrative Department (APHA, 1980). The Environmental Protection Agency of U.S. has submitted the fourth report to the scientific congress on resource recovery and reduction in the volume of solid wastes of Washington city in 1977. However the studies on analysis of chemical components of solid wastes was reported earlier by Kaiser in 1966. Further, research on solid wastes aimed to study their quality and economic management (Winkler and Wilson, 1973; Howard *et al.*, 1985; Alter, 1983; Burce, 1984). Among the various components of solid wastes the inorganic substances like salts of heavy metals are most objectionable. Studies on removal and recovery of heavy metals from refuse and their accumulation in soils and plant materials was given more importance (Wagner *et al.*, 1990).

Perhaps the most difficult task facing anyone concerned with design and operation of solid waste management is to predict the composition of solid wastes. The problem is complicated because of the heterogeneous nature of waste materials. The studies carried out in USA revealed that repetitive sampling procedures should include at least ten grab samples collected from refuse dumps.

Characterization of solid wastes is important in evaluating alternative equipment needs, systems, management programmes and plans, especially with respect to the implementation of disposal, resource and energy-recovery options. It depends on a number of factors such as food habits, cultural traditions, socioeconomic and climatic conditions. Refuse characteristics vary not only from city but even within the city itself and also seasonally. Quality of refuse should be assessed taking into account the seasonal variation and zonal characteristics.

Scope

The present investigation is undertaken with an intention of evaluation of the status of solid wastes produced in the city both qualitatively and quantitatively in terms of their physico-chemical characteristics and to suggest possible abatement measures. The solid wastes arising from residential, commercial and industrial sources was collected and analysed for their physico-chemical factors. The solid wastes production, its management and recycling process are also discussed in this paper.

Shimoga City

Shimoga is a district place located on Bangalore-Honnavar road, lies at a distance of 272.00 km from Bangalore. It is situated at the latitude of 13°15' North and at longitude 75°35' East. The city is situated at 578.98 meters above the mean sea level on the bank of river Tunga. It is one of the fast growing centre, known for its scenic beauty. The total population of Shimoga city as per 1991 census was about, 1,92,647. Shimoga city is conventionally differentiated into three major areas.

(1) Residential area which includes
    Upper class and lower class ....417.35 hectares
(2) Commercial area   ....59.60 hectares
(3) Industrial area   .... 80.47 hectares

The main upper class residential area comprises Gandhinagar, Ravindranagar, Vinobhanagar, Old post office area, B.B. street, Bapujinagar etc. The lower class residential area comprises Harakere, Gopalnagar, Segahatti, Gurpura etc. The major commercial areas comprises Gandhinagar, B.H. Road, Nehru road, Durgigudi, Savarkarnagar (Fish market area) etc. The major industrial area are Garden area and industrial estate.

From the above mentioned areas, old post office area, B.H. road, Harakere and industrial estate in R.M.C. yard were selected for the study. Representative cross sectional samples were collected after an extensive survey of these areas. Zonal characteristics with seasonal variations and per capita income were given prominent considerations to analyse, evaluate and to obtain the requisite data.

Sampling

Polythene bags were distributed to individual houses (upper and lower class) in residential area and individual shops in commercial area. The refuse accumulated for 24 hours was collected on the consecutive days. All the refuse samples were stored at 4°C till analysis was over. Samples from five houses in upper class, seven houses in lower class residential area and sixteen shops of different business in commercial area were analysed on two samples in triplicates. Since industries produce huge amount of wastes, only a composite sample from six types of industries was collected for analysis.

Analysis

About 1 kg of refuse sample was weighed on a physical balance with handgloves, all the components were sorted manually, weighed and the percentage was calculated. The leachate of waste samples were analysed for physical (loose density, compact density, moisture content and pH) factors (Howard *et al.*, 1985) and chemical (BOD, Chloride, Phosphate and Ammonical nitrogen) factors (Saxena, 1990).

Results and Discussion

The average values of various components of residential and commercial wastes and their percentage are given in Table 32-34. The quality of wastes in terms of physico-chemical factors are presented in Fig 6.4 to 6.7. Based on the percapita waste produced by individual persons, shops and industries the total daily and annual production was calculated (Table 36).

Sorting the Components

The solid waste procured from five houses in upper class area are sorted out into individual components. It was characterized by high percentage of vegetables (39.9 per cent) and ash (19.8 per cent). The other components include dust 52 to 54 g/d, rice 18.9 g/d, paper 25.5 g/d and fruits 93.5 g/d respectively. The non-decomposable factors like cloth, metal and glass were almost nil or negligible. The glass 0.2 g/d, plastics 14.4 g/d and hair 1.6 g/d were recorded (Table 32).

The study on classification of refuse collected from lower class area reveals that ashes (236.4 g/d) and vegetables (302.9 g/d) were more. The metals, glass and wood were recorded in negligible amount. An intermediate amounts of dust, rice, paper and hair were noticed. Fruit components were completely absent. From Table 32 and 33 it is evident that the population in upper class use fairly large amount of vegetables, food, flowers and other groceries and discharge relatively large quantity of solid wastes. This indicates, the higher living standard of the residents.

The solid waste collected from sixteen different shops from commercial area are mixed and a representative sample was prepared. Sorting of the samples revealed that the decomposable matter constituted thirteen components and non-decomposables

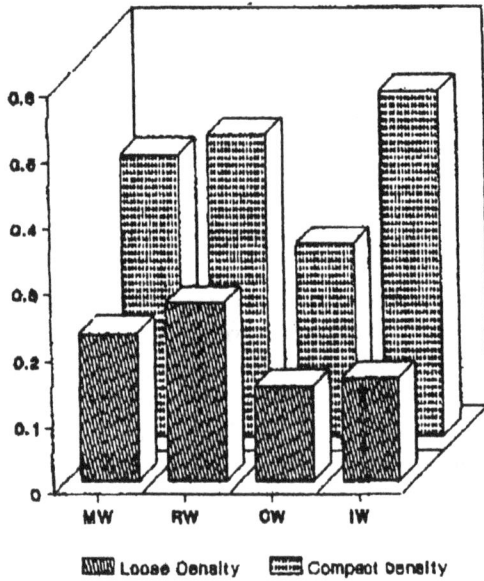

Fig. 6.4 : Variations in loose and compact density

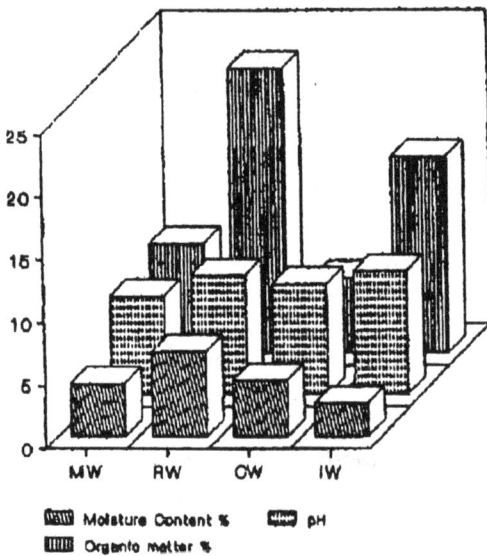

Fig. 6.5 : Variations in moisture content, pH and organic matter

Table 32 : Quality of lower class residential solid waste (Average values in grams)

	Dust	Rice	Paper	Veg.	Fruit	Ash	Flour	Metal	Glass	Plaster	Wood	Hair	Total
Average (gms)	61.5	26.42	17.79	302.85	—	236.42	27.7	—	3.78	8	Nil	2.57	673.5
±SE	8.99	3.29	3.33	35.40	—	19.12	6.6	—	0.55	1.40	Nil	0.3	65.03
%	9.13	3.92	2.64	44.97	—	35.1	4.1	—	0.56	1.19	Nil	0.38	—

— Indicate absent.

Table 33 : Quality of upper class Residential solid waste (Average values grams)

	Dust	Rice	Paper	Ashes	Flour	Fruit	Veg.	Cloth	Metal	Glass	Plastic	Wood	Hair	Total
Average values	53	65	25.5	225	27.5	93.5	45.2	—	—	0.2	14.37	37.5	2	1133.7
⓪ ± SE	1.14	5.7	3.73	38.4	2.6	23.1	16.18	—	—	0.08	0.71	25.16	0.1	87.7
%	4.67	5.7	2.34	19.8	2.42	8.2	39.86	—	—	0.002	12.7	33.07	—	—

— indicate absent.

Table 34 : Components of commercial refuse (Average values in grams)

Total No. of shops	Dust	Paper	Cloth	Grains	Flowers	Rice	Vegetables	Fruits	Ashes	Skin	Plaster	Tablets	Cotton
A. Decomposable Components													
Average values	80.93	2113.5	77.95	275	111	5700	8850	525	10000	55	1500	7	900
%	14.81	3.86	1.42	0.50	0.20	10.43	16.20	0.96	18.30	0.10	2.74	0.01	1.64

Total No. of shops	Plastics	Metals	Hair	Glass	Wood	Rubber	Animal wastes	Leather	Total
B. Non-Decomposable components									
Average Values	1748	235	790	487.5	1875	250	10205	125	54624.5
%	3.20	0.43	1.44	0.89	3.43	0.45	18.68	0.22	

* Animal wastes : Scales, bones, claws and feathers.

All parameters are expressed in grams.

in seven components (Table 34). In all the three sectors i.e. upper and lower class residential and commercial areas decomposable components like vegetables, ash were most conspicuous. The metals, glass and cloth were absent or rarely recorded in resident areas. The components varied every day and month depending upon persons life style, type of business, crop pattern, cultural and ethical practices (Miller, 1985). Since Shimoga is malnad (tropical forest belt) area and vegetable are available at low price in all parts of the year. Hence vegetable wastes dominated the samples.

Among the industries rice mills and lathe machines were most conspicuous. Survey on refuse production in various industries is carried out in the main industrial area and also isolated industrial units. The industries surveyed include metal, plastic, wood, tyre retreading, oil, rice mills and paper industries (Table 35). The rice mills produce wastes in the form of rice husk and boiler ash. Both wastes are economically reused. Husk is used as fuel to boilers and boiler ash is used as manure for the fields. There is a single paper industry within the city limites which utilizes packing and other paper wastes as raw materials and manufactures kraft paper. There are seven tyre retreading industries which produce small pieces of rubber wastes. These are collected for months together and sold for reuse. The wood industry produce large amount of saw dust and wood pieces which are reused as fuel. The plastic industry and metal industry produce cut piece of plastics and metal scrap, both are reused elsewhere. It is interesting to note that invariably all the industries have developed a potential to reuse their refuse. Except few metal industries and lathe machines which produce large amounts of wastes, rich in lead and other heavy metals, harmful to many organisms. At present there are about 1199 industries in Shimoga city. Still many more industries are yet to come up in the areas like RMC yard and Mechanahalli village.

Physical Factors

The density of solid waste varies markedly with the geographic location, seasonal variations, photoperiod and storage or retention periods. In the present study also both loose and compact densities

Table 35 : Survey of solid waste production in industrial area of Shimoga City

Sl. No.	Name of Industry	Raw material Used	Type of produce	Type of Solid Waste Produced	Total Amount	Whether recycled or not
1.	S.G.K. Metal Industry	Iron alloy	Suspension Cylinders	Metal Scrap	100 kg/month	Recycled. The scrap is solid, remelted & reused elsewhere.
2.	Indian Plastic Industry	Plastic Rolls	Plastic Bags	Plastic pieces	3.33 kg/day 3 Kg/day	Recycled. The waste are sold & reused in plastic making.
3.	Sanatana Wood Industry	Wood logs	Furniture	(a) Saw dust (b) Wood pieces	(a) 10 kg/day (b) 8 kg/day	Recycled. Sold, used as fuel
4.	Karnataka Tyre resoling	Rubber rolls	Tyre Retreading	Rubber powder	70 kg/month 2.33 Kg/day	Recycled. Sold and reused in rubber making.
5.	Sharada oil Industry	Castor seeds	Castor oil and cakes	Boiler ash	30 kg/day	Ash is used as manure by farmers
6.	Amruth Rice Mill	Paddy	Rice	(a) Rice husk (b) Boiler ash	(a) 300 kg/day (b) 50 kg/day	Reused as fuel used as manure
7.	Paper Packaging Ltd.	Waste Paper	Kraft Paper	(a) Pins (b) Plastics (c) Clay (d) Tar (e) Boiler ash kg/day	(a) 15 kg/day (b) 15 kg/day (c) 10 kg/day (d) 30 kg/day (e) 10 × 10	Pins and Plastics are recycled Boiler ash is used as manure

showed marked variations. These two densities were comparable in municipal and residential wastes. Loose density was 0.22, 0.27, 0.14 and 0.15 g/cm³ in municipal, residential, commercial and industrial wastes respectively (Fig. 6.5). Determination of density of a refuse indicates its organic content, paper and other components. The high density values of municipal and residential wastes indicate that the refuse was rich in decomposable matter and can be economically used for composting or energy recovery (Alter, 1983) from decomposition of refuse by microbes. Usually residential and municipal refuse are rich in coliform flora derived from the fecal matter of human beings and excreta of animals. Under high humidity condition garbage absorbs more moisture and in tropical areas moisture content will fasten the decomposition of wastes. However in temperate areas though moisture content is high due to low ambient temperature composition takes place very slowly (Miller, 1985). The moisture content of solid wastes is usually expressed as the weight of moisture of unit weight of wet or dry material to the total volume. Moisture content also varies in accordance with ambient temperature and precipitation. Among the four types of refuse samples analysed residential wastes contain maximum water content (6.7 per cent) and minimum in industrial (2.6 per cent) waste. The moisture content of commercial and municipal wastes were 4.4 per cent and 4.2 per cent respectively (Fig. 6.5). The lowest amount of water content reported in industrial wastes is attributed to low levels of organically rich components. The rich decomposable matter like food and vegetables have contributed higher moisture to residential wastes (Park and Bhargava, 1992).

Organic matter indicates the components of the refuse originated from plant and animal sources. The organic matter anaylsed from municipal and commercial wastes were minimum and maximum in residential wastes followed by a comparatively higher value in industrial wastes respectively (Fig. 6.5). The organic content was 8.6 per cent, 22.7 per cent, 5.9 per cent, 15.6 per cent in municipal, residential, commercial and industrial wastes respectively. Comparatively higher values of organic content in residential wastes are owing to the utilization of maximum amount of vegetables and food materials by the residents. It is well established that the residential and industrial wastes concerned to food, breweries, milk and paper production discharge, are the organically rich solid wastes. Such organic wastes are also rich in bacterial and fungal cells, and hence these wastes should be disposed off immediately, otherwise they impart offensive odors and create unhygienic conditions. The pH values of four refuse samples were determined by testing the leachate samples obtained from the refuse collected. The leachate of industrial and residential wastes was highly alkaline whereas it was slightly alkaline in case of commercial and municipal refuse samples. The pH level were 7.8, 8.7, 9.4 and 9.7 in the leachate samples of municipal, commercial, residential and industrial wastes respectively. Since most of the enzyme activities in the bacterial cell are pH dependent and the alkaline pH levels are not favourable for the decomposition activities of bacteria in the refuse (Trivedi and Gurudeepraj, 1992). However most of the oxido-reductive enzymes are active over a wide range of pH i.e. from 6 to 9.

Chemical Factors

The BOD values are directly influenced by the organic content of refuse samples. The BOD values obtained for municipal and residential waste were lower than industrial wastes. The industrial refuse always exert higher demand for O_2. The commercial wastes possessed lowest BOD values (Fig 6.6). During our analysis an average BOD of 376.5 mg/litre in industrial, 173.8 mg/litre in municipal, 98.15 mg/litre in residential and 30.9 mg/litre in commercial wastes were recorded respectively. The organic content in the commercial wastes was minimum. Hence the BOD values analysed from its leachate were also minimum. The BOD is considered as an important parameter in the design criteria for solid waste management.

Chlorides in the refuse samples are derived from sewage, chlorinated tap water and waste discharges from animals, human beings in the domestic areas. During monsoon season most of the nutrient levels namely chlorides, nitrates and amonia nitrogen are leached out from the refuse dumps and flow into natural streams thereby increasing the nutrient levels in the natural waters leading to eutrophication.

The average values of chlorides in reside:.tial wastes was maximum. It was 2.29 mg/g, 0.5 mg/g, 0.42 mg/g and 0.12 mg/g in residential, municipal, commercial and industrial samples respectively (Fig. 6.7). Residential wastes were rich in chlorides, it is because they constitute the chloride rich substances like detergents, sludge and tap waters.

The inorganic phosphorous determined was 1.4×10^{-3} mg/g, 2.7×10^{-3}, 2.6×10^{-3} mg/g in commercial, industrial, municipal and residential wastes respectively. The phosphate level was also responsible for eutrophication in water. The phosphorous level in the sludge leachate is derived from the soaps and detergents used in individual houses. Hence the refuse collected from municipal dumps were characterized by higher levels of nutrients especially phosphorous and chlorides.

The quantitative study on total solid wastes produced in Shimoga city limits, amounts to be 61 tonnes/day. It constitutes 29.28 tonnes, 12.39 tonnes and 19.33 tonnes of residential, commercial and industrial wastes respectively (Table 36). The pollution survey conducted by scientists of NEERI reveals that always the waste production in the form of liquid and solid is more dominant in residential areas as compared to others. Similarly our study also presents that higher amount of refuse was produced in residential area followed by industries and commercial sectors.

At present Shimoga city holds a population of 1,92,647. The per capita production of waste by each individual is worked out to be 0.15 kg/day. A total of 10608.05 tonnes of wastes was produced in a year with the above population in residential area. Similarly the waste produced per shop in commercial area was 3.6 kg/day. The total wastes produced in commercial area was 4522.8 tonnes/ annum for a total of 3442 shops. According to the information obtained from the Office of Industry and Commerce there are 1199 industries in Shimoga. An average refuse production per industry per day was 16.11 kg and the total waste production estimated to be 7052.8 tonnes/annum (Table 36).

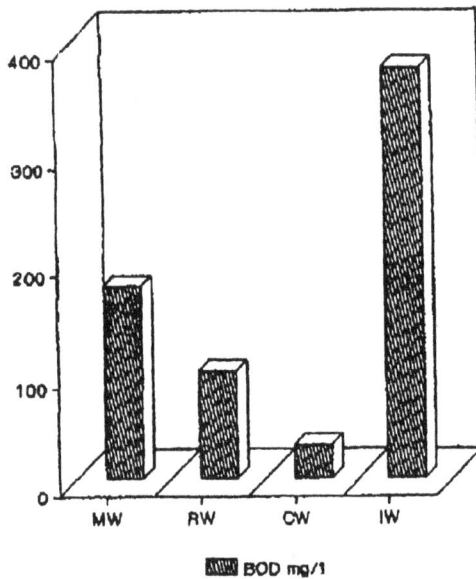

Fig. 6.6 : Variations in biological oxygen demand

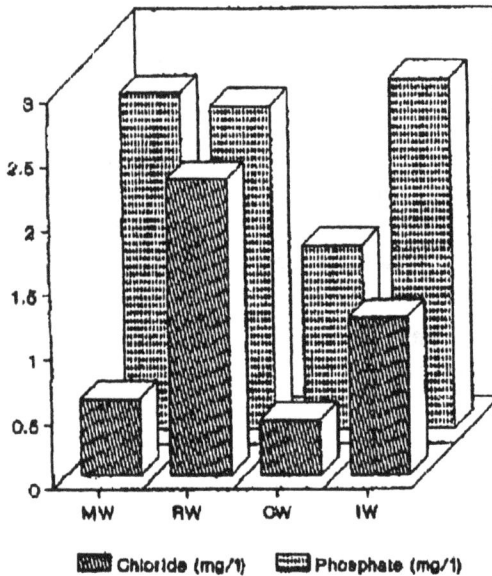

Fig. 6.7 : Variations in chloride and phosphate (10^{-3}) values

MW = Municipal Waste   CW = Commercial Waste
RW = Residential Waste   IW = Industrial Waste

Table 36 : Details about solid waste production in Shimoga city (1993-94)

Sector	Waste Produced Per Individual or Industry or Shop	Total
Total population—1,92,647	Per capita waste produce 0.152 kg/day	Total residential waste 10688.05 tonnes/annum 29.28 t/day
Total No. of industries—1199	Waste produced/industry 16.11 kg/day/industry	Total industrial waste 7051.8 tonne/annum 19.33 t/day
Total No. of Shops—3442	Waste produced per shop 3.6 kg/day	Total commercial waste 4522.8 tonnes/annum 1239 t/day

Management of Wastes

NEERI studies (1983) indicated that the per capita refuse production in metropolitan cities like Bombay, Calcutta, Madras and Delhi range from 0.45 to 0.6 kg/day. In other townships it ranged from 0.15 to 0.33 kg/day. It means the density value of Indian cities varies from 300-560 kg/m^{-3}. The above factors clearly indicate that with the high standard of living (Predominant in metropolitan cities) the per capita value of refuse produce also increases.

The investigation carried out on the solid waste production and its prospects of management reveals that an average of 61 tonnes of refuse is produced every day in the Shimoga city. The management practices are inadequate, improper and irregular. Many people do not use spittoons and waste bins and in some areas most of the dust bins are broken or over-filled. Usually the waste bins and waste storage dumps are disturbed or scattered by garbage pickers, dogs, cattles, pigs and birds. This may lead to the spreading of refuse all around the bins and around the storage places. The bins are supposed to be situated in every street depending on the quantity of generation of waste. But in actual practice the bins are provided without proper assessment of the population density and their behaviour. Also the collective bins are not collected by municipal authorities at regular intervals as planned. The frequency of wastes collection by the corporation workers is not followed according to the plan.Therefore the transporting vehicles are often over loaded, due to this over loading and haphazard road conditions, vibrations are much more and the refuse was often spilled out to the roads. This can be prevented by utilizing completely covered dumpers as in advanced countries.

The disposal of solid waste at the same time is out of plan. Major portion of the refuse was dumped in low lying areas and also on barren land surfaces. A part of it was also dumped into river Tunga. This particular activity and sewage has enriched the nutrient level in the river. It was evidenced by the complete darkening of the water and growth of aquatic weeds in the polluted reach near the old bridge. The present dumping grounds like the proposed auto complex site and Shankar Mutt might have contaminated the ground and surface waters in that area. This has also lead to the bad smell and mosquito growth, pigs, rodents activity in the area. Fire hazards are also possible. The proposed auto complex site is being utilized as a dumping ground has a high water table and if the dumping of refuse in this place is continued, it may become totally unsuitable for any construction activities.

A little portion of garbage (approximately 8-10 tonnes/month) was used as manure for orchard plantations by the localities. There are obsolutely no efforts to recycle the refuse. However the rag pickers were always active to collect the non-combustible components like glass, plastics, metals and sell them to local agents for recycling. There is a great potential for the reuse and recycling of refuse in Shimoga city. It was also evidenced that in the rural or suburb parts of Shimoga solid waste namely animal dung and other decomposable garbage was stored in earthern pits and allowed to decompose for 6 to 8 months. Such earthern pits are popularly known as oxidation ditches and the decomposed matter was used as manure for agricultural crops. Application of sludge to soil increases the soil fertility. To avoid damage to crop land and the contamination of ground water, sludge should be applied in small amounts at short intervals to balance the sludge infiltration and evaporation (Tietjen, 1973). The city refuse can also be used for energy production by means of biogas plants. Energy can be obtained by anaerobic digestion of refuse and through sludge pyrolysis. Both in the case of pyrolysis and anaerobic digestion the manure obtained will be aesthetically acceptable, essentially pathogen free, easy to handle and valuable soil conditioner to farmers (Wagner et al., 1990) or sanitary land filling should be undertaken regularly.

References

Trivedi, P.R. and Gurudeepraj. 1992. Management of Pollution Control. *Encyclopedia of Environmental Sciences. Vol. 8-9.* Ashadeep Publishing House, New Delhi-12.

Kaiser, E.R. 1996. Chemical Analysis of Refuse Compounds, *Proc. Nat. Conf. ANSE.* New York.

Winkler, P.F. and Wilson, D.G. 1973. Size characteristics of municipal solid wastes. *Compost Science.* p 14.

Alter, H. 1983. *Material recovery from Municipal wastes.* Marsel and Dekker Inc. New York.

Bruce A.M. 1984. Assessment of sludge stability. Cited in *Methods of Characterization of Sewage Sludge*. Ed. Casley. T.T., Hermit, P.L. and New Man. P.T. Reidal Publishing Co., U.K. pp. 131-143.

Wagner, D.J., Bacon, G.D. Knocke, W.R. and Switzinbaum, M.S. 1990. "Changes and variability in concentrations of heavy metals in sewage sludge." *Environmental Technology*, 11, pp. 949-960.

Howard, S.P., Donald R.R., George T. 1985. *Environmental Engineering*. McGraw Hill Book Co., New York.

Saxena, M.M. 1987. *Environmental Analysis of Water, Air and Soil*. Agro Botanical Publ., Bikaner, India.

Miller, G.T. 1986. "*Environmental Science : An Introduction*. Wadsworth Publ. Co., Belmont, USA.

Park, H.J. and Bhargava D.S. 1992. "Environmental Impacts of Solid wastes incineration". *Indian Journal Environ. Protection*. p. 12.

Tietjen, C. 1973. European experience in applying sludge on small farmland" *Compost Science*, 4-6, 1973.

III. SOLID WASTE GENERATION AND ITS DISPOSAL IN THE STEEL PLANT, ROURKELA, ORISSA

Swayam Prakash Rout

Introduction

This is a report of solid waste generation and its disposal from Rourkela Steel Plant, Rourkela Township and hospital. The paper outlines the composition and the amount of solid waste generated with the mode of disposal from various industries. Management of solid waste plays a vital role in environmental protection. The pollution control measures adopted in effluent treatment, dust collection from emissions and during raw material handling and processing result in solid waste generation. The improper handling of solid waste creates water, air and soil pollution. The solid waste may get air borne causing air pollution or may be washed out by rain water thereby soluble chemicals from solid waste may also leach through the soil thereby contaminating the soil and ground water.

The best way to manage solid waste is to evolve methods to recycle solids in the process to maximum extent possible or utilize the waste in producing saleable products. The solid waste generated from different industries in Rourkela area vary in composition and quantity. The indiscriminate disposal of solid waste in the form of garbage from various urban and municipal bodies also pose serious concern. The garbage includes hospital rejects, sludge from sewerage drains and sewage treatment plant.

1. Rourkela Steel Plant

In the process of steel manufacture a number of operations generate solid waste. Some of these solid waste are recycled while others are dumped on land. The major solid waste generated from various units of Roukela Steel Plant (RSP) is shown in Table-37. The reuse and chemical composition of solid waste generated from various units of RSP are discussed below.

Blast Furnace Slag

Total slag generated	4,95,274.00 tonnes
Total B.F. slag granulated	1,95,840.00 tonnes
Total B.F. slag dumped	2,99,434.00 tonnes

The chemical composition of B.F. slag is given as,

Silicates	31 to 37 per cent
Iron Oxides	2 to 2.5 per cent
Calcium oxide	26 to 33 per cent
Magnesium oxide	4 to 10 per cent
Aluminium oxide	0.1 to 0.6 per cent

The granulated slag is sold to cement manufactures and utilized in the production of slag cement. The rest of the slag is the molten state is transported by ladle and poured in the dumping area. The huge dump looks like a hillock around the boundary wall of RSP.

Blast Furnace flue dust and sludge

The total quantity of B.F. flue dust arrested by dust catchers generated per annum is 70,044.00 tonnes. The chemical composition of B.F. flue dust is given as :

Total iron	31 to 38 per cent
Aluminium oxide	12 to 13 per cent
Silicates	10 to 11 per cent
Calcium oxide	05 to 07 per cent
Magnesium oxide	02 to 2.5 per cent
Carbon	14 to 18 per cent

Presently all the B.F. flue dust is dumped in the dumping yards.

Steel Melting Shop Slag

Steel Melting Shop slag is generated both from the L.D. Converter and Open Hearth Furnace. The total SMS slag produced annually is 2,34,792.00 tonnes. The chemical composition of SMS slag is given below :

Iron oxides	15 to 30 per cent
Silicates	12 to 18 per cent
Calcium oxide	30 to 50 per cent
Magnesium oxide	02 to 08 per cent
Aluminium oxide	02 to 2.5 per cent
Manganese oxides	08 to 14 per cent
Phosphorus	Upto 02 per cent

The molten SMS slag from Open Hearth Furnace and L.D. Converters is carried in slag ladles by rail and poured in the slag dumping yard.

Lime Dolomite Kiln Dust from SMS

The dust is collected from ESP of Gas Cleaning Plant of L.D. Converter. The total L.D. dust generated is 7338.00 tonnes per annum. The chemical composition of L.D. dust is,

Silicates	02 to 35 per cent
Calcium oxide	08 to 9.8 per cent

Aluminium oxide	2 to 2.21 per cent
Magnesium oxide	01 to 1.8 per cent
Total Iron	54 to 56 per cent
Manganese oxides	2.2 per cent

Presently all the L.D. dust are disposed off into the dumping yards.

Fly Ash

Fly ash is generated from coal fired boilers of the Captive Power Plants and medium pressure boilers. Captive Power Plant (CPP–I) does not have an ESP and consequently no ash pond. However the fly ash is collected partially by the mechanical dust collector. Presently the fly ash from Captive Power Plant (CPP–II) is located in two number of ash ponds of capacity of 4,53,537 m^3 and 7,55,485 m^3 respectively.

The fly ash generation from Captive Power Plants is as follows:

CPP–I	80,000 tonnes/annum
CPP–II	2,40,000 tonnes/annum

The chemical composition of fly ash is given below :

Silicates	45 to 51 per cent
Calcium oxide	02 to 03 per cent
Aluminium oxide	20 to 25 per cent
Magnesium oxide	1.1 to 15 per cent
Iron oxides	7.3 to 9.5 per cent
Loss due to ignition	18 to 26 per cent

Acetylene Sludge

The sludge is generated during the reaction of calcium carbide with water. 1500 tonnes of sludge is generated annually.

The chemical composition of sludge is given below :

Silicates	04 to 06 per cent
Calcium oxide	60 to 70 per cent
Aluminium oxide	01 to 1.5 per cent

Table 37 : Generation and disposal of solid waste from various units of RSP

Units	Waste Generated	Quantity in tonnes/annum	Mode of disposal
Coke Oven	Coke breeze		Sold as coke briquette
Blast Furnace	(a) Slag	4,95,274.00	Slag granulation for cement partly and dumping on heaps.
	(b) Gas Cleaning Plant (GCP) flue dust from mechanical separator.	70,044.00	Dumped outside the industrial premises.
	(c) GCP Clarifier sludge	8,000.00	Dumped alongwith B.F. slag
Steel Melting shop	(a) SMS Slag	23,4045.00	Dumped on heaps.
Fishing Mills	(a) Mill Scale	36,116.00	Utilized in sinter making
	(b) Ferrous sulfate from pickle liquor	6,420.00	Used as coagulant
Captive Power	Ash from		
	(a) CPP I	80,000.00	Dumped in ash pond.
	(b) CPP II	3,20,000.00	
Refractories	Salvaged refractories	40,000.00	Partially reused but dumped on land.
Lime/Dolomite kins	Fine rejects	3,500.00	Reused
Foundry	Sand (Used)	1,530.00	Dumped on land
Acetylene Plant	Process sludge	1,448.00	Dumped on land

All these figures pertain to calendar year 1988-89.

Iron oxides   0.1 to 0.3 per cent

Loss due to ignition   25 per cent

The sludge is dumped inside the industrial premises.

Calcium Carbonate by-product of Fertilizer Plant

Carbon dioxide is removed by sodium hydroxide in the gas fractionation plant. Sodium hydroxide is regenerated by adding hydrated lime, which also produces calcium carbonate as a by-product. 2,400 tonnes of calcium carbonate is generated annually and is dumped outside.

The chemical composition of this by-product is,

Silicates   38 per cent

Calcium carbonate   60 per cent

Iron oxideds   02 per cent

Rejected Refractory Materials

40,000 tonnes of rejected refractory materials are generated from various units of Rourkela Steel Plant. Out of them only 3,000 tonnes are sold and the rest 37,000 tonnes are dumped inside the premises.

Beside the above wastes, solid waste is generated from Hot Rolling Mills as mill scales and pickling waste was ferrous sulfate.

Other Sources of Solid Waste Production in the Rourkela Area

The garbage collected from steel and civil township is of the order of 30 tonnes per day which is very substantial to cause environmental pollution. The solid waste generated from Ispat General Hospital is 8 tonnes per day which along with solid waste from steel township is dumped in low-lying areas near village Jhirpani. The disinfection is done by spraying disinfectant which is inadequate.

The solid waste from the township and other local municipal bodies are indiscriminately dumped in available low-lying areas in the township. However it has been observed that no specific area has been earmarked for solid waste disposal.

Orissa Industries (ORIND)

The major solid waste from ORIND is from rejected refractory materials and clinker from gas producer plant. 1400 tonnes of rejected refractory materials are generated annually which is used for construction purpose whereas 80 tonnes of clinker generated annually from coke is used in Gas Producer Plant.

Kalunga Industrial Estate (KIE)

The solid waste generated from the various industries in KIE varies widely. There are chromium bearing solid waste from sodium dichromate plant as well as inert materials like bottom ash from coal fired boilers and furnaces. The solid wastes are dumped inside their premises or are dumped outside the factories in low-lying areas.

Discussion

Solid waste handling and disposal cause severe environmental problems. A pragmatic approach for the disposal of solid waste has become absolutely necessary. Environmental pollution can be minimized when the solid waste are recycled again and again. The proper disposal of these solid wastes is equally important to ensure minimum leaching of ground and surface water.

The solid waste disposal is a complex operation in Rourkela Steel Plant. Slag dumping area has been earmarked and railway track is extended upto that point as and when necessary. The molten slag are brought by ladles and are poured on the heaps. The flue dust from LD. Converter and wastes from Gas Cleaning Plant of Blast Furnace are also dumped in the area. A lot of fugative emissions have been observed during the dumping of these fine dust. The fine dust also makes its way to the atmosphere during prevailing winds.

The recycling of solid waste will bring about better resource utilization thereby minimizing handling costs. (Behra et. al., 1990; Rout and Rout, 1990). The reuse and recycling of wastes, will ensure environmental protection and simultaneously, minimises environmental pollution. When 100 per cent solid waste are re-cycled or sold so that no facilities will be required to handle these wastes in dumps, the situation with regard to waste disposal should be considered as ideal. Thus in this situation, the environmental pollution will be contained to a large extent. However solid waste disposal in the major industries in Rourkela area are not well organized and therefore a fresh approach has to be made. (Behra, 1992).

The hospital wastes from Rourkela is simply dumped in an abandoned area. It should be incinerated as they contain toxic and pathogenic waste components.

References

Behra, D.K. Patnaik, D. and Rout, S.P. 1990. *Indian J. Environmental Protection*, 10 (4) 299p.

Behra, D.K. 1992. Ph.D. Thesis, Utkal University, Bhubaneswar.

Rout, S.P. and Rout, M.K. 1990. In *Environmental Impacts of Industrial and Mining Activities*, edited by L.N. Patnaik, Ashis Publishing House, New Delhi.

CHAPTER 7

COASTAL ZONE MANAGEMENT WITH SPECIAL REFERENCE TO KARNATAKA

K.S. Jayappa

Introduction

India has a long coastline of about 9,000 km including its major indentations and shores of islands. Coastal zones are receiving an increasing importance in view of their high productivity and growing use. There is an intense pressure to develop coastal areas in view of dense population, exploitation of living and non-living natural resources such as fisheries, aquaculture, seaweed, minerals and tidal energy. The increasing load on harbours, location of waste-effluent disposal sites, development of various chemical, petrochemical, fertilizer and allied industries, spurt in recreational activities and above all petroleum exploration activities have necessitated to recognise the importance of these fragile zones.

Coastal zone is the interface of the land, the sea and the atmosphere. It starts from seaward margin of the continental shelf and extends upto inland limit of the coastal plain. Over exploitation of its resources is led to an unbalanced developmental scenario. An improper planning and the conflicting demands, uses and problems add to the choas and destruction of the coastal environment. Any sustainable development of the coast should involve the maintenance of a beach (Baba, 1979), which is the most important component of a natural coastal dynamic aquatic system.

What is coastal zone management ?

The coastal zone management is a multidisciplinary task, defined as "the activities, maintenance and other matters con-

(196)

cerned to be necessary to protect the amenity of the coast and thereby to minimize damages to property from erosion or encroachment by tidal water". Or integrated coastal zone management involves prevention of marine pollution from all sources, protection from erosion, resources sharing, maintaining the quality of marine/coastal resources and meet the needs of present and future generations.

How to proceed ?

A major part of coastal zone management is public education by the way of holding training courses, preparing audiovisual packages and publications on 'beach conservation' and display of information leaflets. Various informations required for coastal zone management are as under.

1. Reliable knowledge on coastal processes

2. Effects of beach sand mining, and decrease in sediment supply to the beaches.

3. Conditions of existing wetlands and changes that may take place through time and space.

4. A reliable sand budget for coastal "littoral cells". This requires quantitative information on sand inputs and outputs.

These ideas have to be conveyed to the public so as to motivate them to work for coastal zone management programmes, in line with those already existing in the developed countries like USA, USSR, UK, Japan, etc. for the purpose of proper exploration, planning and utilisation of coastal resources.

Problems of Coastal Zones

The problems associated with the management of coastal zones of Karnataka are : coastal erosion due to anthropogenic activity and its prevention measures, siltation in navigable estuaries, fishing activities, flooding of low lying areas, pollution problems and recreation potentials. Of these, the man made erosion and steps taken for its prevention are important ones to be attended immediately.

Emery and Neev (1960) have estimated that beach sand was

mined in Israel (until 1960), at the rate of 10 to 20 times greater than the natural annual replenishment, and that this mining caused sand deficit along many beaches causing an accelerated erosion of the coast. Similar is the case in different beaches of India, sand is removed for the purpose of construction and extraction of economic minerals as well as silica sand. Illegal sand mining is being carried out at various places, like Someshwar (Fig. 7.1), Mulur, Padukere on the coastline of Karnataka. Legal mining from Tekkatte, Uliargoli, Kapu, Mulur, Hejamadi and Tonse for extraction of silica sand which is used in foundries as moulding sand and the manufacture of sodium silicate was reported (Narayana *et al.*, 1991). This is being a direct loss of material affecting the dynamic equilibrium of the beaches.

Oradiwe (1986) reported that about 382, 000 m³ of sand is mined every year from Montery bay of California. In many regions sand carried by rivers to the coast is a major source of beach sand. This is deposited in river mouths during floods as an ephemeral delta, with waves and currents gradually move much of it onto and along adjacent beaches. Mining of sand from the estuaries and rivers is being carried in Karnataka state for construction and other purposes and is a common practice. To our knowledge the volume of sand mined is not quantified anywhere in the State. Whatever may be the quantity of sediment mined, it has to be checked, otherwise further depletion of sand on the beaches may upset the existing sediment balance and it may result into further coastal erosion. Thus, the river inputs and mining output should be balanced in such a way that the coastal erosion is controlled or minimised.

It has been found in recent years that coastal communities are taking advantage of newly accreted beaches, recently evolved spits and those areas protected by seawalls for their settlement. For e.g. Bengre and Ullal near Mangalore which have been accreted and provided with seawall respectively and a long narrow spit near Udipi grown in recent years are occupied with thick population. If such illegal migrations are not checked on time, coastal zone management would continue to allure the country for ever. This is both an administrative and engineering problem which calls for proper management.

Another problem associated with coastal erosion and subsequent construction of hard structure and its negative effects on

Fig. 7.1 : Illegal sand mining from Someshwar beach, Mangalore

fishing activity. The coastal residents who mainly depend on fish catch for their livelihood, face a major problem when a beach which they were using for fishing is disappeared. If a structure like seawall is constructed to stop the beach erosion, they have to face further problems because structure dampen their shore-based fishing activity. In addition, siltation in the estuarine mouths due to longshore drift and river discharge causes severe problems for fishing and navigation.

Sediment accumulation in estuaries and mangrove areas has been steadily increasing due to natural and anthropogenic activities. Continuous reclamation of margins of these areas for industrial and urban agglomeration, has resulted in reducing the buffer zone between the land and the sea, increasing the pollution, flooding the lowlying areas and reducing the spawning ground for marine organisms. This may often lead to ecological imbalance and a conspicuous decline in biodiversity. There are only two alternatives for the pollutant disposal. It is either into a nearby aquatic system or onto land. A general feeling is that the rivers, estuaries, beaches and the adjoining continental shelf are the cheapest, convenient and safe areas for dumping of industrial wastes. This feeling has come into practice in large scale and lead to the degradation of coastal waters.

Objectives

The broad techno-economic objectives of the coastal zone management are :

1. Prevention of coastal erosion through prevention of human interference with coastal processes.
2. Estimation, conservation, proper exploration and utilisation of the coastal resources like ground water, fishery, agricultural products for the well being of mankind.
3. Improvement of navigation, dredging and proper utilisation of dredged sediments.
4. Maintenance of recreational potentials of beaches and development of tourism.

Why people like to live in coastal zones ?

Many people like to live in coastal zones because of availability of water and other resources and facilities like trade and commerce. At present about 2/3 of world's population (or nearly two billion people) live within an average width of 10 km from shoreline, which constitute only about 3 per cent of the total land surface (Chavadi, 1996). In addition, millions of people visit the coasts to enjoy the beauty of the regions, to run along a beach, to play in the surf, to ride a surf board, to sail on the sea, or simply to sunbathe. More fundamental reasons for attraction of people in large numbers to these areas are job potential in seaports, shipping, thermal power plants and other industrial activities. These places have attracted industrialists, mainly because of the facilities like sea port and water as a coolant.

Why are we interested in this subject ?

We are interested in this subject because about 60 per cent of humankind live in the coastal zone and is expected to increase further within the next 2-3 decades. The number of tourists are in the tens of millions annually. For e.g. in 1985, Spain had about 43 million tourists. In Japan, recreational use of beaches has been increased from 110 million in 1975 to 220 million by 1990. 24 million people use rockway beaches on the Atlantic coast of New York city annually. In India, the state of Goa alone, registered 2.0 lakh tourists in 1975 which was increased to 9.7 lakhs by 1993. In some areas of the world, tourists visiting beaches are responsible for a major segment of local jobs and income. This is true with the state of Goa in India, where the income from tourism is about 7 per cent of the net domestic product of Rs. 550 crores and about 17 per cent of its population earn their livelihood either directly or indirectly. As the coastal zones are so densely populated and a large number of tourists visit them, they need greater attention of scientists, planners, administrators and the public to protect and upgrade these fragile areas.

Coastal erosion management

Coastal erosion control is a subset of coastal zone management. The degree to which such management is needed depends upon the density of population, extent of development of the region and the intensity of the problem. The main functions of coastal erosion management are to alleviate these problems associated with waves and beach erosion on the downstreams of backwaters. One of the solutions for downstream erosion is nourishment by bypassing the sediment dredged from harbor and approach channel areas and letting natural processes to transport it further down the coast. Sand bypassing is a well recognized coastal erosion management tool. The management includes management of structures, beach nourishment, retreat and setbacklines. But so far the coastal erosion prevention methods practiced in India, are mostly defensive ones, namely seawalls and groins.

Structures

Structures include groins, seawalls, revetments, bulkheads etc. Though these are used as erosion management tools, they often complicate the erosion problem or shift the erosional sites to adjacent areas. In most of the cases, the cost of seawall construction and their maintenance is greater than the value of the property saved. In order to save the beach and the property behind it, we would destroy them. In some cases they have been destroyed by huge waves thereby recreational potentials of beaches are also lost and posed a severe threat for shore-based fishing activity. Collapse of seawalls is experienced in Kerala and at some places like Ullal and Mukka in Mangalore, Polipu and Kapu in Udipi, Manki in Honavar and Harawada in Ankola taluks of Karnataka state (Fig. 7.2). In order to overcome this problem of erosion seawalls are constructed at some places away from the reach of storm waves. In such cases, the walls remain intact but do not protect the beach. The guiding principle is that interference with the coastal system should be minimized so that coasts can achieve a natural equilibrium.

Beach Nourishment

It refers to management of coastal erosion by artificial addition of suitable quality sediment to a beach area in order to rebuild and

Fig. 7.2 : Collapse of sea wall in Manki village due to erosion

maintain the width of that beach which provides storm protection and recreation in the area. The first recorded beach nourishment project took place in 1922 at Coney Island, New York. From that time to late 1980's, over 640 km of US coastlines have been nourished at a total cost of about $8 billion (Dixon and Pilkey, 1991). In overseas, beach nourishment has become very popular, particularly in developed countries such as Denmark, Netherlands, Germany, Australia, and Great Britain and also in developing countries like Brazil (Vera Cruz, 1972) and Nigeria (Ibe *et al.*, 1991). Many environmental groups feel that beach nourishment is a costly affair (Dean, 1989). The conclusions drawn on the basis of an assessment of more than 150 beach nourishment projects revealed that :

1. Beach nourishment projects excepting a few have short lifetimes–less than 5 years,

2. Nourished beaches typically erode much faster (2 to 10 times) than their natural counterpart,

3. Nourished beaches do not recover from storms like natural beaches.

4. The storm waves are main factors for the failure of these projects.

In a well developed, high populated area it can be economical to dredge sand from offshore, harbor entrances and inshore regions and place it on the shore to build a wide beach. Examples for this are Miami beach, Florida, where 10.5 miles of beach was built with 10.7 million m³ of sand. 6.2 miles length of beach of New York has been developed with about 5.4 million m³ of sand. Likewise in Belgium, England, Hawaii, California, Morocco, Holland, South Africa, Singapore, Japan the beach nourishment was practiced. It is also done at low populous and isolated areas where large scale industries, highways, harbors and beach resorts exist.

Dutch Government has recently decided to maintain the position of its existing coastline with beach nourishment. Similarly in Britain, beach nourishment is expected to form a major element of coastal management in the future (Lacey, 1991). European countries such as Netherlands (Anon, 1990) and Denmark (Moller and Swart, 1987) are presently leading in the field of integration of beach nourishment with coastal zone management.

Fig. 7.3 : Position of a building during calm (A) and retreated during storm (B) wave conditions (after intergovernmental panel on climate change, 1990)

India, being a developing country is unable to afford to practice beach nourishment as a coastal erosion management tool. But in certain places like Visakhapatnam harbor, in Andhra Pradesh and Old Mangalore Port in Karnataka and at few places in Kerala where navigational channels and harbor areas are dredged, the dredged material is pumped on to the downdrift sides which are subjected to erosion due to longshore currents.

Retreat

Retreat refers to abatement of coastal erosion by relocating buildings and other infrastructure in a landward direction (Fig. 7.3). During 1980's heavy criticism was received in US because of failure of numerous beach nourishment projects (Gilbert, 1986). During this time, Leonard *et al.* (1990b) began to contradict traditional coastal engineering methods used to design and evaluate such projects. Such criticisms were supported by coastal environmental groups which advocated planned retreat as the only true and sustainable solution to coastal erosion control (Dean, 1989). Dean (1988) opined that, when maintenance of shoreline becomes costly, under such cases relocation is more attractive.

Setbackline

It is an imaginary line drawn roughly parallel to the shoreline. The zone existing between this and low water line is called as buffer zone which is required for waves to play on it. If man doesn't interfere with natural coastal processes, nature itself will protect this zone. Buffer zones would include sloping beaches, berms, sand-dunes and longshore bars which can absorb most of the incoming wave energy. Hence it is essential to have setbackline and beyond this line constructions should be stopped. This can be done by strict execution of the Coastal Zone Regulation Act (CZRA) 1991 of India. Then maintenance and management of natural buffer zones may become easy.

Planning and Management

An important aspect of a coastal management is planning for beach front landuse. Planned use of beachfront and island area sets the stage for development and subsequent demand for services impacting on the coast. Coastal engineering research centres, Research institutes and individuals who have undertaken studies related to coastal process, coastal geomorphology and coastal engineering are the primary sources of data. They can prepare comprehensive plans, landuse maps, and provide information on the areas subjected to beach erosion, kind of hazard, magnitude and frequency of hazard that can occur along the coastal zones. Major differences can exist from location to location, and we must take care to recognize the reasons for these differences while preparing plans. The beach management provides an overview of existing coastal planning, strategies, including perceptions of local beach problems. The planning and management include the following elements :

(a) **Beachfront landuse** : The objectives of this, is to demarcate the lands according to their uses. Condensed standard landuse classification includes five categories. They are :

(1) low/medium density residential, (2) high density residential, (3) commercial, (4) recreation/conservation, and (5) vacant. This classification may be carried out by referring to topomaps, satellite images, and landuse landcover maps available in Corporations and Municipalities.

(b) **Physical public beach trends** : The primary objective of this trend is to provide a general overview regarding the number of public beaches, their position, cumulative linear distance and also their association with various geomorphological landforms. Identification of primary and secondary sources of erosion and accretion of public beaches is secondary objective of this trend. This information is useful for coastal managers to find out the causes of beach erosion and to estimate the funds required for erosion control. One has to remember that beaches are public property and not belonging to those beachfront land owners only.

(c) **Recreation** : It is an inventory of :

1. The number and area of all recreation activities

2. The number of access points to these public beaches

3. The parking capacity.

Capacity of the beach area, parking and the number of access points are the limiting factors for the use of public beaches for recreation. This information is important for allocating funds to improve existing public beaches, develop parking or beachfront area. Innovative landuse measures to expand the area available for public use should be explored at both state and central government levels.

Organisation of physical, cultural and cartographic information, state wise in a format is useful for the researchers, planners, managers, government and funding agencies. This information is useful to act over the decisions like :

1. Coastal planning and recreation needs of each maritime state.

2. Determining new areas for public beach needs.

3. Setting priorities for beach protection.

4. Locating beach areas suitable for development.

5. Identifying erosion control needs.

6. Identifying storm hazard areas and expected losses.

7. Assessing coastal growth management needs.

Hawaii, for example, was one of the first to develop coastal zone management tools in order to alleviate problems associated with tsunamis, as this was perceived to be a major public problem including public safety. US congress passed the Coastal Zone

Management Act in the early 1970's, which presented a national policy to preserve, protect, develop and where possible, restore or enhance, the resources of the nation's coastal zone for the present and future generations.

Australia has established its Beach Protection Authority in the year 1968. Japan, in 1956 enacted its coastal law and recently formulated three, 5-year shore protection projects (1970-1985). Egypt has got a coastal zone management plan under its Shore Protection Authority. Malaysia has completed a national level coastal erosion study in 1986 which is a major step towards the development of its coastal zone management plan. Denmark too has a Shore Protection Plan. The main functions of these authorities are to provide advice to local authorities, harbor boards and river authorities. So many studies or programmes are planned, but they are funded for first 1 or 2 years, later no funds are released. This leads in abrupt ending of these programmes without any ultimate result.

Most of the beaches which were nice places to live, have become uninhabitable and some of them even unsuitable to visit. A beach is simply not a suitable place to build a home, a hotel, or anything else. Beaches unlike rocky coastlines and to a lesser extent mangrove-stabilised coasts, are shifting, eroding, accreting and ever changing; with passages opening and closing, sandbars building up, washing away; and with every hurricane producing almost instantaneous and massive change.

There are places without beaches, but possess resistant rock cliffs. These serve as excellent sites for resort development. Many people do not like to sit on a beach or to swim in the surf; rather, they relax on deck chairs and swim in pools. They like to be in a scenic area. Antalya, on the south coast of Turkey and the Turquoise coast are good examples of this activity. In India too, there are some locations in Uttara Kannada district of Karnataka state and on the coastlines of Goa, Maharashtra and Andhra Pradesh which can be utilised. Thus coastal erosion management may also includes the encouragement of developmental activities in erosion resistant regions where there are no beaches.

Coastal zones of India are also developing exponentially like many developed countries. Some of them have been already developed to such an extent that even the most radical environmentalist would not suggest to eliminate what has already been built. But planning for the future should include some very carefully drawn up restrictions on further development along the coastal zones. In fact we must make every effort to ensure that any further construction in the coastal zones of India takes place at least as far back as the permanent vegetation line behind the dunes. In addition to this we have to practice to follow guidelines as a part of coastal erosion management activity.

1. Buildings and seawalls that were built too close to the sea should not be replaced if they are threatened or damaged by sea erosion, instead they should be demolished and removed.

2. Even where seawalls have been placed too close to the sea, beach plants should be encouraged to grow on any dunes that still exist.

3. Shorelines stabilisation by planting casuarina, mangrove and other salt tolerant plants.

4. Only in extreme cases, beach nourishment may be used as a purely temporary solution to erosion.

The question is whether erosion can be reduced by natural ecological processes or man-made devices. The superior recuperative abilities of unaltered coastal ecosystems affected by storms and erosion are widely recognized. Some ecologists, landscape architects and members of environmental interest groups have suggested that public money would be better spent by discouraging investment in areas of critical erosion and by leaving them to achieve a state of dynamic, stable, equilibrium condition under natural processes. (Chapman, 1980; Carmel et al., 1984).

In comparison to other public works projects, coastal protection schemes have provoked relatively little opposition from environmentalists. When disputes are arised, they tend to focus on one of the two issues :

1. Dredging for sand to supply beach nourishment

2. Aesthetic and amenity impacts of large scale engineering structures.

Because the practice of pumping sand from wetland lagoons draws widespread reaction from fishermen, environmentalists, conservationists and others who contend that such activities damage extremely rich wildlife habitats. Although its precise environmental effects are unclear, this is a costly form of dredging appears less damaging and is certainly less controversial than removing sand from lagoon beds (US Corps of Engineers, 1973). US Coastal Zone Management Act of 1972 served as a vehicle for controlling accelerated growth of population and investment in maritime areas. The legislation restricting the development of wetlands and regulating power plants and industrial facilities in the coastal zone is already existing in our country, like in the US, the growth rates would get stabilised and demand for new beach protection projects would be reduced.

A shift in emphasis from purely technological adjustments to solutions involving a mixture of engineering and managerial techniques is the probable outcome. This, in turn, will tend to increase levels of public participation in decision making and reduce undesirable environment impacts. Like erosion and storm damage, beach pollution and overuse of beaches will gain importance in the field of beach protection and preservation. Improper and sometimes over exploitation of resources has led to degradation of coastal environment.

Water-borne transport being the cheapest mainstay of world trade, has necessitated massive waterfront constructions which have not only caused irreversible changes in the shoreline, but also irreparable damages to the coastal environment. Shipping, mineral extraction and oil exploration are the major source of release of toxic wastes. Human settlements, use of pesticides in agriculture and deforestation, discharge of sewage and industrial effluents, all have contributed to the degradation of coastal environment.

Suggestions

For a developing country like India, with a vast coastal zone, there is a need to establish a Central Coastal Zone Management Authority (CCZMA) at the national level and a suitable agency in each of the maritime states for proper coordination and implementation of the coastal zone management program of the country. These should have specified jurisdiction to deal with the problem of management of man's activity on the coast in order to see that coastal zone resources are properly utilised without adverse effects and to preserve these for the future. Each maritime state including Karnataka should re-examine their coastal protection programmes as a part of coastal zone management.

Illegal mining of sand from beaches, estuaries and upstream rivers as well as occupation of newly formed/protected beaches and spits by coastal (mainly fishing) community have to be stopped. A top priority should be given for maintenance of cleanliness and aesthetics of some selected beaches of the State for tourism development in the nearby places. This would greatly help in earning foreign exchange and creating job opportunities for the local people. Some of the locations where resistant rock cliffs are available in Uttara Kannada, may be selected for resort and tourism development. Unplanned, haphazard construction of structures, specially seawalls wherever erosion takes place is not a true solution for erosion control, but they may be constructed in those areas which are heavily developed. As the cost of beach nourishment is tremendous, it is not possible to accept this managerial tool for a developing country like India. Naturally growing beach plants and artificial plantation of casuarina, mangrove and other salt tolerant species should be encouraged in order to achieve the shoreline stabilisation.

References

Anonymous, 1990. A New Coastal Defense Policy for the Netherlands. *Rijkswaterstaat*. The Netherlands, 103.

Baba, M. 1979. Coastal Erosion in Kerala–Some Problems and Solutions, *Centre for Earth Science Studies*. Proc. Paper No. 6, 15.

Carmel, Z., Inman D.L. and Golik A. 1984. Transport of Nile Sand Along the Southeastern Mediterranean Coast. *Proceedings of the Nineteenth Coastal Engineering Conference*, Houston, Texas. 2 : 1282-1290.

Chapman D.M. 1980. Beach Nourishment as a Management Tool. *Proceedings of the Seventeenth Coastal Engineering Conference*, Sydney, Australia. 2 : 1636-1649.

Chavadi, V.C. 1996. Recent Trends in Coastal Research with Particular Reference to Uttara Kannada Coast, Karnataka. *Jour. Indian Association of Sedimentologists*, 15 : 1-27.

Dean, C. 1989. As Beach Erosion Accelerates Remedies are Costly and Few. *New York Times*, August 1, Sec. C, 1.

Dean, R.G. 1988. Eroding Shorelines Impose Costly Choices. *Geotimes*. 33-5 : 9-14.

Dixon, K. and Pilkey, O.H. 1991. Summary of Beach Replenishment on the US Gulf of Mexico Shoreline. *Jour. Coastal Research*, 7 : 249-256.

Emery, K.O. and Neev, O.H. 1960. Mediterranean Beaches of Israel. *Bulletin 26, Geological Survey*, Ministry of Development, Israel. 22.

Gilbert, S. 1986. America Washing Away. *Science Digest*, 94 : 28-35.

Ibe, A.C., Awosika, L.F., Ibe, C.E. 1991. Monitoring of the 1985/86 Beach Nourishment Project at Bar Beach, Victoria Island, Lagos, Nigeria. *Proceedings Coastal Zone.* American Society of Civil Engineers, New York, 534-552.

Lacey, P. 1991. Preserving the Coastline. *Municipal Engineer,* 8 : 301-314.

Leonard, L.A., Dixon, K.L. and Pilkey, O.H. 1990. A. Comparison of Beach Replenishment on the US Atlantic, Pacific, and Gulf Coasts. *Jour. Coastal Research,* 6 : 127-140.

Moller, J.P. and Swart, D.H. 1987. Extreme Erosion Event on an Artificially Nourished Beach. *Proceedings Coastal Sediments.* American Society of Civil Engineers, New York, 1882-1896.

Narayana, A.C., Pandarinath, K. Karbasi, A.R. and Raghavan, B.R. 1991. A Note on Silica Sands of South Kanara Coast, Karnataka, India. *Jour. Geol. Soc. Ind.,* 37 : 164-171.

Oradiwe, E.N. 1986. Sediment Budget for Montery Bay. *M.S. Thesis,* US Naval PG School, Montery, California, 101.

Sahai, B. 1994. The Use of Coastal Environmental Maps of Karnataka. *Deccan Herald,* 18th October.

US Corps of Engineers 1973. Ecological effects of Offshore Dredging and Beach Nourishment: A Review. *Miscellaneous Paper* No. 1-73. Coastal Engineering Research Centre, Washington D.C.

Vera-Cruz, D. 1972. Artificial Nourishment at Copacabana Beach. *Proceedings 13th Coastal Engineering Conference.* American Society of Civil Engineers, New York. 1451-1463.

CHAPTER 8

CASE STUDIES

I. DEFORESTATION : A CRITICAL ISSUE

D. Venkat Reddy, A.P. Sivakumar and
B.R. Manjunatha

Introduction

Of all environmental problems crumbling our country, the problem of deforestation has received maximum public attention. It may be believed that about 10,000 years ago 85 per cent of the earth surface was thickly covered with forests and during 1980 this was reduced to 33 per cent. Our culture is closely associated with nature, as we worship forests and try to find the god in all living species. Modern civilization is responsible for depletion of forests at an alarming rate (Bahuzuna, 1986). Deforestation as a cause of ecological imbalance has been noticed since a long-time. Although awareness of the problem in developing country arose only in the early 1970s, when several studies demostrated that the severity of environmental damage and shortage of firewood (Allen and Burness, 1985). The present study attempts to highlight on the causes of deforestation and its impact on immediate environment.

Forest Loss

In most of the developing countries forests are being destroyed at a rapid rate posing threat to ecotechnology. It is estimated that forests cover only about 7 per cent of the earth's surface and they house between 50 to 80 per cent of Planet's species. Such a rich biodiversity of flora and fauna are threatened by the increase of human population, urbanization and resultant deforestation. The depletion of forests in Brazil, East Africa, India, Indonesia and

(212)

many other countries represent a grave situation and the mankind seems to be at war with the biotypes. India has lost about 4.04 million sq km of forest land or about 12 per cent its geographical core forest area between 1951-52 and 1975-76 (Puri 1983). Space-based remote sensing data are being used for forestry application. In India, National Remote Sensing Agency (NRSA) and Department of Space, have confirmed that in a span of 7 years, forest cover has been reduced from 16.89 per cent to 14.10 per cent of the total geographical area. NRSA has prepared state-wise forest maps and details are presented in Table 38 and vital forest statistics are shown in Table 39.

Table 38 : State-wise forest area calculated by using satellite data

State/Union Territory	Forest area by satellite data (mha) 1972-1975	1980-82	Areas controlled by forest department, m ha 1980
Andhra Pradesh	4.90	4.04	6.41
Assam	2.11	1.98	3.07
Bihar	2.27	2.01	2.92
Gujarat	0.95	0.51	1.95
Haryana	0.08	0.04	0.16
Himachal Pradesh	1.51	0.91	2.21
Jammu and Kashmir	2.23	1.44	2.19
Karnataka	2.95	2.57	3.79
Kerala	0.86	0.74	1.11
Madhya Pradesh	10.86	9.02	15.39
Maharashtra	4.07	3.04	6.41
Manipur	1.51	1.38	1.52
Meghalaya	1.44	1.25	0.86
Nagaland	0.82	0.81	0.29
Orissa	4.84	3.94	6.77
Punjab	0.11	0.05	0.24
Rajasthan	1.13	0.60	3.49
Sikkim	0.18	0.29	0.26
Tamil Nadu	1.67	1.32	2.18
Tripura	0.63	0.51	0.59
Uttar Pradesh	2.59	2.10	5.14
West Bengal	0.83	0.65	1.18
Andaman and Nicobar	0.33	0.64	0.71
Arunachal Pradesh	5.14	5.21	5.15
Dadra and Nagar Haveli	0.02	0.01	0.02
Goa, Daman and Diu	0.12	0.11	0.11
Mizoram	1.39	1.20	0.71
Total	61.54	46.37	74.87

(mha : Million hectares)

Source : National Remote Sensing Agency (NRSA), Hyderabad.

Table 39 : Vital forest statistics (in million hectares)

	1972–73	1980–82
Forest Cover	55.52	46.35
Closed forests	46.42	36.02
Open forests	8.77	10.06
Mangrove forests	0.33	0.26
Forests Cover (% of total land area)	16.89	14.10

Source : National Remote Sensing Agency

Satellite data of 1980-82 showed that closed forests covered only 36.02 million hectares and open forests 10.06 million hectares. The study revealed that Manipur, Arunachal Pradesh, Andaman and Nicobar islands are the only areas which fulfill the stipulation of 60 per cent forest cover as it was laid down by the National forest Policy. Maximum deforestation has occurred in Madhya Pradesh, Orissa, Andhra Pradesh and Jammu and Kashmir states. While considering proportion of forest areas during the last seven years period, Punjab has lost over half the forest cover that existed in 1972-75 and Rajasthan, Haryana and Gujarat nearly half. Himachal Pradesh and Jammu and Kashmir have lost over a third of their forest cover, Maharashtra a quarter and West Bengal and Tamil Nadu over fifth.

Effects of Mining of Forests

Mining is a devastating operation that not only destroys the natural ecosystem, particularly in surface mining, but also introduces tremendous distortion into the social fabric. More than 80,000 ha of land of our country is presently under the stress of mining activities of various kinds. As there is a demand for minerals, growth of mining area would expand at a faster rate, resulting in a scarification of land scape, debris dumps, soil degradation with widening circle of deforestation and a stress to the human environment (Valdiya, 1988). Large scale deforestation is resulted due to indiscriminate open cast mining in our country. A comparison of the 1951 and 1985 data indicates, an increase of 42 per cent in the total number of mines, which can be considered substantial. Likewise production of coal is reported to have increased four times, iron ores 10.65 times, copper 11 times, bauxite 33 times and lime stone 16 times, during the said period (Narayan, 1994).

Kudremukh

It is the India's largest integrated opencast iron ore project, located at Kudremukh. Kudremukh is situated in the Gangamula hill range of the western ghats in Chickmagalur district of Karnataka. The ore body stretches to a length of about 6 km and laterally about 800 m. The deposit consists of 610 millions tonnes of weathered magnetite and hematite ore apart from the primary ore which is about 400 million tonnes. The opencast mining of Kudremukh has resulted in the removal of top soil because of deforestation (Venkat Reddy, 1988). According to Narayan (1994), mining activity has affected forest area and degradation of 'Shola' forest in the valleys, around the mining and residential areas of Kudremukh. The area under open sholas forests was reduced by 26.7 per cent, the main reason could be the submergence of land due to construction of reservoir. Mining activity has also degraded the quality of 'Shola' forests around Kaiga and where nuclear power plants are being constructed.

An attempt has been made to quantify the forest loss using satellite imagery of diferent periods by the Survey of India (maps, 1 : 50,000 scale topographical map base). The analysis was carried out by Visual and computer aided interpretation and spot checked by personal visits to ensure remote correctness. The reliability of the study was estimated to be at approximately about 90 per cent (Narayan, 1994).

Mussoorie-Dehradun Mining Area

Large-scale deforestation resulted due to indiscriminate opencast mining of various minerals in Missoouri-Dehradun area over a length of 40 km causing a great damage to vegetation, increased soil erosion and pollution of streams and springs (Valdiya, 1988). The forests around Dehradun belongs to Bhabar-Dunsal forest, while around Mussoorie it is of lower western Himalayan temperate forest type. Mining activity in Doon Valley started in 1911 and the number of mines kept on increasing till 1985, when the supreme court of India ordered for the closure of all mines, however, three mines are under public litigation files on environmental grounds. This study showed that the forest cover has reduced by 2.8 Mkm^2 during 1972 to 1988. Notably large portion of forest cover near

Hatachipaon hill has altogether been cleared due to intense limestone mining.   ·

Exploitation of soapstone and magnesite from Khirakot, Kosi valley, Almora districts has destroyed about 14 ha of forest on hill slopes, coal mining of flat terrain of Jharia, Raniganj, Singrauli areas, and strip mining and trenching has resulted in extensive deforestation. Indiscriminate mining in Goa forests since 1961 has destroyed more than 50,000 ha of forest cover.

Paper and Paperboard Industry

Large paper and paperboard mills are fully integrated to pulp production which are now based mostly on bamboo and hardwood. Most of paper mills are   located near the forests. It is estimated that large integrated mills with an installed capacity of 1.4 million tonnes use an average 60 to 65 per cent bamboo and 30 to 35 per cent hard wood. The paper industry consumes about 2 per cent of the country's annual consumption of wood and 51 per cent bamboo. The details of paperboard industries of the country and expected shortage in forest based raw materials for paper industries are listed in Table 40 and 41 (Venkat Reddy, 1993).

Table 40 : India's paper and paperboard Industry in 1984

Number of paper mills (10^2)	175
Annual capacity (million tonnes)	2.16
Production (million tonnes)	1.12
Capacity utilization (%)	58.10
Forest raw material requirement (million tonnes)	3.10
Per capita consumption (kg/year)	2.00
Estimated value of actual producion (crore)	1205.00

Source : Indian Paper Market Association 1984.

Table 41 : India's expected shortage of forest based raw materials

Year	Paperboard Production (million tonnes)	Forest raw material required (million air dry tonnes)	Newsprint production (million tonnes)	Forest raw material (million air dry tonnes)	Expected shotfall
1981	1.02	2.85	0.26	0.52	—
1991	1.57	4.40	0.43	0.87	2.06
2000	2.30	6.70	0.72	1.45	4.84

Source : Development Council for Paper, Pulp and Allied Industries.

Rural Energy Demand (Firewood Collection)

In rural areas nearly 60 to 70 per cent of the total fuel consumption is towards cooking which is mainly met with fuel wood or agricultural waste. Total non-commercial energy consumption is registered to be 179.41 million tons of coal equivalent, of which 116.62 tonnes come from firewood. The urban consumption of firewood is harmful to the country's environment. According to National Council of Applied Economic Research (NCAER), a country-wide survey of household fuel consumption reveals that three-fourth of the firewood used in the rural areas will be in the form of logs. This fact clearly illustrates that a large scale deforestation is due to use of wood logs for various purposes. Madhya Pradesh is the biggest supplier of firewood to Indian cities.

Dams and Reservoirs

Large dams are posing environmental issues. Today there are about 176 major and 447 medium irrigation projects and many are in various stage of construction. Most of them are behind the schedule. The annual report of the Department of Environment, 1983-84, revealed that the construction of the Idukki dam on Periyar river in Kerala and Kali Project near Dandeli in Karnataka state have hastened up the degradation of vegetation and forest cover. Dams and reservoirs that are under construction in our country are influencing the degradation of forests. Large dams have drowned half a million hectares of forests, about a tenth of the area that has benefited from canal irrigation.

Impacts of Deforestation

Atmosphere

Biologists have calculated that a hectare of well grown forest annually absorbs about 65 tonnes of carbondioxide and releases 3.5 to 5 tonnes of Photosynthetic oxygen which is vital for all aerobic organisms. Every year, atmosphere looses one billion cubic meters of oxygen which is replaced by carbondioxide. Technosphere consumes 15 times more oxygen than all living organisms taken together. Forests act as an excellent biological filters to separate dust and industrial wastes (Rembohrov, 1984). A team of researchers from South America are working on tropical forests have stated that the forests play an important role in weather generation. Uncontrolled deforestation may lead to fluctuations and even a severe decrease of normal rainfall, changing of weather patterns. Severe drought conditions prevailed in the states of Gujarat, Maharashtra, Rajasthan and parts of Karnataka may be the impact of deforestation. Increase of dust in arid zones through destruction of sparse vegetation and the increase of bush-fires may accelerate deforestation and loss of valuable land due to soil erosion.

Impacts on Land

Soil erosion is a global phenomenon. In India, it is seen in its worst form in the himalayan watershed, which sustains a huge population and replenishes several perennial river systems. It is also manifesting in other mountain chains such as Aravallis, Vindhyas, Western Ghats and the Eastern Ghats, at varying intensities. Green cover gives stability to the soil and is nature's protection against soil erosion. The roots of the plants bind the soil whilst leaves and stems break the velocity of the wind at ground level.

Clearing the existing forests for agricultural crops, industry, hydroelectric projects, townships, roads, bridges and plantations are cited as possible reasons for landslips as it is true for Nilgiri hills. The soil erosion is quite high in the Western Ghat hill ranges along the Karnataka state. The main reason for this is identified as due to deforestation. The alarming rate of the disappearance of trees has triggered the agitations, for instance, Appiko, which is striving to retain the vegetation cover as forests protects the precious life-sustaining soil. There are no proper estimates on soil erosion, but it is believed that the effect of soil erosion is felt in many places in the state (Venkat Reddy, 1995).

Impact on Water

Hill range forests are places for origin of rivers and streams. Deforestation in catchments has affected considerably on fertile lands adjacent to valley riverine systems. Deforestation directly increases the siltation in river channels, dams and reservoirs. High siltation rate has an adverse effect on the water bearing capacity and minimize the biodiversity which is a common problem in most of tanks, lakes and reservoirs. Because of siltation in river channels, flooding is quite frequent, paticularly in the drainage basin of the major rivers.

Biodiversity

At present, an increase of human interference has altered the natural balance in the environment by reducing biological diversity to its lowest level (Wilson, 1989). Majority of the tropical forests are rich in plant and animal species. Deforestation and greenhouse phenomena are the main causes identified for the decrease of biodiversity. The International Union for Conservation of Nature and Red Data Book of 1972 lists the Malabar civet as "possibly" extinct prone species. There are only two reports about the occurrence of this species since the past fifty years : one at Kudremukh (reported by K.U. Karanth) and the other in Tiruvella in Kerala (G.U Kurup in 1987).

The Indian subcontinent is one of the fascinating ecological and geographic regions of the world at par with Africa. Deforestation is a threat on most of the species of wild life. About 15,000 species of plants (out of 2,50,000 total species known in the world) and 75,000 animal species (out of total 1.5 million) have been identified from India (2 per cent of the total land area). India possess around five per cent of known living organisms on the earth. Information regarding the impact of deforestaion on land, water, biodiversity has been provided in this chapter. Mining and deforestation together enhance the rate of soil erosion, siltation in river valleys and reservoirs causing many problems.

References and Further Reading

Allen, J and Barnes D.F. 1985. The causes of deforestation in developing countries, *Annals of the Association of American Geographers V*. 75 pp. 163-189.

Bahuguna, S. 1986. Deforestation and its impacts, *Seminar of Indian Environmental Problems and Perspectives*, Geological Society of India, Bangalore, Memoir 5, 167-173.

Narayan, L.R.A. 1994. Impact of mining activities on environment, *The Hindu* Kasturi and Sons, Madras 9.1.1994 pp. VII.

Puri, G.S. 1983. *Forest Ecology Phytogeorgraphy and Forest Conservation*, New Delhi, Oxford and IBH Publishers), Rembohrov (1984) *Green Medicine for Biosphere Trees Vol.* 44, No. 1.

Rao, U.R. 1990. *Space technology as an Instrument for Combating Environmental problems particularly those of developing countries*, ISRDO Publications, Bangalore.

Valdiya, S.K. 1988. *Impact of mining on environment in India.* H.R.G. Publication Services, pp. 29-51.

Venkat Reddy, D. 1988. Deforestation and its impact on ecological problems with speical reference to Western Ghats, Indian *Journal of Environmental Protection Vol. 8. No. 12* 930-936.

Venkat Reddy, D. 1993. Deforestation. Its Impact on Indian Environment, *Advances in Environmental Sciences* (Ed.) Tripathi A.K., Srivastava A.K. Pandey S.N., Ashish Publishers, New Delhi.

Venkat Reddy, D. 1995. *Engineering Geology For Civil Engineers*, Oxford IBH Publishers, New Delhi.

Wilson, E.D. 1989. Threats to Biodiversity , *Scientific American*, September 1989, pp. 60-66.

II. IMPACT OF HUMAN POPULATION GROWTH ON FORESTS IN AND AROUND BHADRAVATI

A.S. Chandrasekar and B.B. Hosetti

Abstract

A demographic study was conducted in relation to forest environment of Bhadravathi, Shimoga and Tarikere Talukas covering an area of 1,256.6 km² over a period of seven decades. The literature availed on population density of the three towns and forest area in 1920 were compared with the data obtained in 1990. The study reveals that increased population enhanced the anthropogenic activities in the area resulting into the decline in forest cover. Further 56.9 per cent of the forest was degraded due to human activities during the last seven decades.

Introduction

The present day human species have an average life span of 50-60 years. It was only 30 years about 15,000 years ago. However, maximum life span of 100 years remains the same. The gradual increase in average life span may be due to the vigor of reproduction, limited environmental hazards like heat, or cold, enough food, control of diseases have triggered the human population on this globe (Culter, 1984; Rameswar Singh, 1988). As the population increased in most of the countries problems of environmental degradation was also cropped up. Increased population and various developmental activities posed a threat to ecological balance in nature. The tremendous increase in human population also envisaged the mis-management of environmental resources, large scale deforestation, unplanned discharge of residues and wastes, handling of toxic chemicals, indiscriminate construction and expansion of settlement activities. Such activities have degraded the environment by polluting the three great media of earth-water, air and soil.

Bhadravathi town was a small village about three decades ago. Now it has grown to a size of medium town. The present study is undertaken to adjudicate the changes in the quality of adjacent forest cover due to growth of Bhadravathi Village into a town (increase in population).

Study Area

The study area includes Bhadravathi as the center place having Shimoga on northern and Tarikere on southern sides. The study area covers about 1,256.6 km². This area lies between north latitude 13.4' to 14.02' and east longitude 75.31'and 75.54'. It is 585 m above the mean sea level. Rainfall varies from 391-102 mm. Temperature (17.2 to 42.2°C) and humidity (42.92 per cent) were recorded; the climate of the area is semi arid. Geologically this area consists of most ancient rocks, Dharwar Schist interspersed with younger granites and gneisses. Soil is mostly red loamy type. Most part of the study area belongs to maidan (plain). The two important rivers Tunga and Bhadra flow in the area and join to form Tungabhadra near Kudli, 15 km away from Bhadravati town.

Methods

The information on human population and forest cover were obtained from the Shimoga Gazetteer 1920 to 1981 (Anonymous, 1920-1982). The data on present conditions were obtained with the help of Census of India, 1981 and Department of Forests. Forest biomass survey was made by point center quarter method.

Discussion

The forests of almost all countries are disturbed by human activities (Krishnan, 1992). Now a days it is suspected that hardly any forest is left undisturbed due to increased human population and their activities. Having Bhadravathi, as a center, an aerial radius of 25.00 km covers the study area spread in about 1,256.66 kms². The towns, Bhadravathi, Shimoga and Tarikere being small villages in 1920 have increased their population and attained urban size by 1990 (Table 42). Bhadravathi was a small village with a population of 3,781 in 1920 has grown rapidly and became a rated industrial town in Karnataka, after the establishment of VISL (Vishweshwarayya Iron and Steel Ltd).

Table 42 : Total population of the study area

Year	Rural	Urban	Total
1920	52,029	26,411	78,440
1990	242,543	358,219	548,861

Source : Gazetteer of Karnataka (1920).

It also accommodated several other major industries (Mysore Paper Mill, MPM), Sugar industries and distillaries. Table 43 reveals that the urban growth of Bhadravathi was most conspicuous. It increased four times in a span of seven decades. Any place accommodating large human population consumes more amount of food and raw materials leading to the release of wastes and pollution of the environment. A town with a 1,000 human population requires about 180 kg food, and 240 kg natural gas, every day (Krishnan, 1992) and all the three towns of the study area hold several thousand population. Because of the increased population in the towns the human settlements have also been spread leading to the growth of urban agglomeration (Shailaja, 1993). Today Bhadravathi town has spread to an area of 5 km² at the cost of occupying the adjacent agricultural and forest lands. The construction of water related projects such as Tunga dam and Bhadra reservoir in the area have also attracted thousands of immigrants to this palce as labourers from neighbouring states and who settled here, and thus became regular citizens of Bhadravathi and Shimoga.

Table 43 : Urban population growth of the three towns, Bhadravathi, Shimoga and Tarikere

Year	Bhadravathi	Shimoga	Tarikere
1901	2.676	6,240	10,164
1911	1,810	13,118	6,618
1921	2,789	15,090	7,858
1931	9,137	20,661	8,211
1941	19,585	27,712	8,858
1951	42,451	46,524	12,343
1961	65,776	63,764	15,620
1971	101,358	102,709	20,022
1981	130,606	151,783	23,929

Source : Gazetter of Kamataka (1920–1981).

The forest cover of the areas was 18,489.04 hectare (moist deciduous), 29,497.7 hectare (dry deciduous) and 8284.7 hectare (scrub) in 1920 was reduced to 18.1 per cent (moist deciduous), 55.2 per cent (dry deciduous) and 26.74 per cent (Scrub) by 1990 respectively (Table 44). Apart from the contrasting decline in total forest cover the important wildlife is also threatened and pushed to extinction.

Table 44 : Forest cover in the study area

Year	Moist deciduous (hectare)	Dry deciduous (hectare)	Scrub (hectare)	Total (hectare)
1920	18,489.04	29,497.68	8,284.67	60,189.75
1990	6,283	19,174	9,293	35,750
Per cent reduction in forest cover (%)	18.08	55.18	26.74	
Total plantation		17,034		

Now-a-days more publicity is given to protect the environment and forests, through media. Among the three towns, Bhadravathi environment is worst affected and is attributed to industrial discharges. The two rivers of the area, Tunga and Bhadra are polluted by city wastes and MPM effluents respectively (Vijayakumar, 1993). Recently,to compensate the lost forest cover, aforestation program by the department of forests and MPM is undertaken. Under their guidance about 17,000 hectare of social forest is raised. The study of the increased population of human beings of Bhadravathi, Shimoga and Tarikere revealed that it has reduced the forest cover of Western Ghats to 56.9 per cent in the area within a period of 30 years.

References

Anonymous, 1920-1981. *Gazetteer of Karnataka State*. Govt. of Karnataka, India.

Culter, R.D. 1984, *Evolutionary biology of ageing and longevity in Mammalian species*, pp. 1-147 in J.E. Johnson Jr., editor. Ageing and cell function. Plenum Press, New York, USA.

Krishnan, N.T. 1992. *Natural resources and their control in ecology*, Ist Edition. J.J. Publ., Madurai, India.

Rameshwarsingh, 1988. Environment and ageing. pp. 55-63. *in Threatened habitats*. Jagmandir Book Agency, New Delhi, India.

Shailaja, RN. 1993. *Impace of human growth on urbanization in Shimoga*. M.Sc. Project Report, Kuvempu Univ., BR Project, Shimoga, India.

Vijayakumar, K.N. 1993. *Study of water quality of River Bhadra*. M.Sc. Project Report, Kuvempu Univ., BR Project, Shimoga, India.

III. IMPACT OF COAL MINING EFFLUENTS ON FISH FARMING AND FISH FARMERS : A CASE STUDY OF CHITRA COAL MINES, BIHAR

Arvind Kumar

Introduction

Coal mining is an ecologically devastating operation, which disturbs the pre-existing balance of the nature and upsets the prevailing ecosystems. It produces huge amounts of wastes which trigger a number of environmental problems like deforestation, land degradation, effect on flora and fauna, vibrations, noise, air and water pollution. The environmental degradation in coal mining areas commences right from the early stages of exploration. As an integral part of the mining activity, there is always a need to clear the trees and jungles in the mining areas that are to be exploited for their extraction of mining operations. Besides various adverse effects of mining, coal washery effluents cause pollution of the nearby streams and reservoirs and accelerate eutrophication. The water pollution caused by coal mining operations poses a great threat to pisciculture. A number of studies have been made on various aspects of pisciculture (Banerjee, 1967 ; Karamchandani & Pisolkar, 1967 ; David *et al.*, 1969; Lakshmanan *et al.*, 1971 ; Natarajan, 1976 ; Choudhary *et al.*, 1978 ; Sunder & Subla, 1984 ; Sudershan & Somvanshi, 1988 ; Ahmed & Singh, 1986 ; Jhingran, 1991 ; Srivastava, 1993 ; Kumar, 1994a ; Kumar, 1995a ; and Kumar & Singh, 1996). On the other hand, the physico-chemical nature of coal effluents has also been the theme of many studies (Singh, 1986; ; Sinha & Sinha, 1987 ; Sinha & Mehrotra, 1990). But till now, no attempt has been made to study the impact of coal mine drainage and coal washery effluents on pisciculture, in the tribal zone of Bihar (Santhal Paragana). Therefore, the present study is undertaken to investigate the physico-chemical characteristics of effluents and to discuss its impacts' upon fish farming in order to improve the socio-economic conditions and malnutrition status of the poor tribals of Santhal Paragana.

Study Area

The Santhal Paragana which forms the south-eastern portion of Bihar, lies between 23°40' & 25°18' NL and between 86°28 & 87°57' EL. It is an upland tract with hilly backbone running from north to south. In the north-east there is a long but narrow strip of alluvial soil which lies between the river Ganga and the Rajmahal hills. These hill systems arise abruptly from the plains, and their height varies between 1000 to 2000 feet. There lies a level of fertile tract in north-west of these ranges whereas to the south-west, this range gives rise to a series of rolling ridges and undulating uplands from which arise isolated hillocks and ridges of sharp and varied outlines.

Topography of Santhal Paragana shows that it may be divided into three distinct zones :

(1) the hilly portion, which covers about 3/8 of the entire area

(2) the rolling country covering half of it,

(3) the flat country on the north-east of the area.

The hilly part of the area stretches continuously for about 150 miles from the Ganga at Sahibganj to the southern boundary. The hills are covered with luxuriant forests and savannah of considerable sizes. The rolling country includes whole of the west and south-west of Santhal Paragana division. It consists of long ridges of rocky intervening depressions with scrub jungles. The third part consists of a fringe of low land between the Ganga and the hills, which is largely cultivated for winter and summer crops. But this part of Santhal Paragana is subject to annual inundation of flood of the river Ganga during monsoon. Therefore, there are large number of perennial water bodies for fish culture in this area. The displaced fishermen of East Pakistan, now Bangladesh, rehabilitated in this area and as those people are expert personnel in the field of pisciculture, as a result, large quantum of fish landed and transported to large cities specially to Calcutta. Thus, this area of Santhal Paragana provides a vast resource of aquaculture for the tribal and local people.

But now-a-days, due to coal mine drainage and coal washery effluents from two vast coal fields of Santhal Paragana, Lalmatia and Chitra, the fish farming has suffered a large damage, and the socio-economic condition of the tribals has also deteriorated to a great extent.

Among the coal mining centres of India, Chitra coalfield is about 50 km from the divisional headquarter (Dumka). It is one of the chief storehouses of the coal of India. This field is being mined since 1942. But in the early phase, the mining proceeded in a very unplanned and unscientific manner, only to get more and more output of the coal without even the slightest care for the environmental protection and reclamation. Understandably this has led to a severe environmental damage causing intense disturbance to the ecological balance in Santhal Pargana.

Purpose

In the state of Bihar, the main concentration of the Santhal and Pahariya tribes is in Santhal Paragana. They live in the lap of nature. Since water is used directly, it is one of the most common sources of infections and definitely a potential vehicle for the organisms causing specific human enteric diseases. Due to uncontrolled coal mining, the water quality of the natural resources has been severely affected posing various health hazards. In recent years, there have been serious outbreaks of various diseases like dysentry, diarrhoea, gastroenteritis, anemia and goiter in Santhal Pargana (Kumar, 1995b). Very high mortality rate is also reported among the population using pond/reservoir/river as the main source of drinking water (Ghosh, 1985; ICMR, 1985; Kumar & Pandey, 1995).

Though underground mining has little direct impact on surface water bodies, it may have profound long term effects on the ground water resources. By its nature, a mine tends to drain a large area and if soluble minerals are present in the coal and associated rocks, these minerals may enter into the streams draining a mining region and cause a severe degradation of water quality (UNEP, 1979). This effect is important for both where mines are drained by pumping and in mountainous area where water is drained by gravity (UNESCO, 1979). Besides coal mine drainage, coal washery effluents are also potential pollutants to the receiving aquatic ecosystems.

On the other hand, the tribals are prone to malnutrition. The fish is the main source of protein. But unplanned coal mining in this area posed a great threat to aquaculture. The fish production has declined to a very great extent. Once, this zone was able to supply the fish to the adjacent districts of Bihar, but today, there is acute shortage of fish even in rainy days. The main cause of low fish production is the presence of two big and many other small coal mines in this area. 25 per cent of the wetland in this zone turns blackish due to coal washery effluents. Therefore, the main objectives of this investigation is to explore the extent of water pollution in wetlands and their impact upon pisciculture, in order to protect the hungry and thirsty community of Santhal Pargana.

Physico-chemical Analysis of Coal Mine Effluents

The coal mine effluents were collected from January to December, 1995 in accordance with the standard methods (APHA, 1989). Table 45, 46 and 47 depict the chemical characteristics of Chitra coal mine water, coal washery effluents and CMRS annual report respectively. The Indian Standard Tolerance limits of sewage, industrial effluents and inland surface water have also been presented in 48.

Table 45 : Physico-chemical characteristics of coal mine effluents (minimum, maximum and average) during different seasons

Parameter	Summer	Monsoon	Winter
Temp. (°C)	26.2–32.5	25.8–32.2	27.2–29.4
	(28.2)	(30.3)	(27.5)
pH	7.4–9.3	7.1–8.3	7.5–8.8
	(8.4)	(7.6)	(8.1)
MO Alkalinity	170.5–275.6	220.6–349.5	150.4–189.6
	(240.48)	(289.6)	(163.4)
Phenol Alk.	20.2–35.2	31.6–70.5	26.7–50.2
	(25.5)	(50.15)	(40.45)
Total Hardness	388.4–449.6	415.2–500.4	211.4–689.4
	(415.3)	(469.4)	(463.9)
Dissolved solids	215.6–510.2	461.3–602.4	235.4–560.9
	(421.6)	(543.4)	(422.4)
TSS	85.1–209.4	91.7–241.3	89.3–289.5
	(134.2)	(169.3)	(148.4)
COD	134.4–21.6	19.4–28.4	10.5–15.9
	(19.6)	(25.4)	(16.3)
Chloride	50.1–87.4	25.3–45.3	22.4–49.3
	(63.4)	(35.9)	(40.3)
Sulfate	45.2–111.6	71.3–99.6	88.4–151.2
	(81.4)	(90.4)	(116.3)
Phosphate	109.6–189.4	81.4–111.4	86.3–153.2
	(152.4)	(97.6)	(126.8)
Iron	1.3–7.9	1.2–2.5	1.2–2.6
	(2.3)	(1.89)	(2.2)

(All values of parameters are in mg/1 except Temperature and pH)

Table 46 : Physico-chemical characteristics of coal washery effluents from January to December, 1995

Parameters	Jan	Feb	Mar	Apr	May	June	July	Aug	Sep	Oct	Nov	Dec
Temp. (°C)	27.0	28.1	27.8	28.3	29.5	30.2	31.8	31.0	29.2	29.3	28.5	27.8
pH	7.5	7.6	7.7	7.8	7.9	8.0	7.9	8.5	8.1	7.9	7.8	7.7
DO	—	—	—	—	—	—	—	—	—	—	—	—
Total Alk.	85.5	71.0	66.3	71.2	78.4	92.4	74.0	82.0	72.0	120.0	85.0	83.0
Total Hard.	316.0	350.0	289.0	348.0	349.0	448.0	365.0	360.0	362.0	348.0	335.0	330.0
TSS	1701.0	1549.0	1080.0	1550.0	1550.0	1900.00	1800.0	1950.0	1550.0	2050.0	1600.0	1650.0
TDS	1150.0	1050.0	1000.0	1650.0	1300.0	1450.0	1150.0	1700.0	1200.0	1250.0	1300.0	1100.0
TVS	825.0	830.0	780.0	1125.0	1000.0	1200.0	950.0	1150.0	1000.0	1150.0	1200.0	1230.0
BOD	310.0	240.0	200.0	280.0	390.0	380.0	440.0	400.0	390.0	440.0	405.0	400.0
COD	1150.0	1130.0	1080.0	980.0	1160.0	1150.0	1280.0	1175.0	1205.0	1280.0	1270.0	1120.0
Chloride	30.5	25.0	24.2	26.2	29.4	35.0	27.4	35.4	36.3	34.2	35.3	34.4
Sulphate	22.4	9.4	9.0	16.5	24.4	27.0	18.5	23.6	24.4	25.3	25.0	24.0
Calcium	75.5	34.4	48.3	73.2	75.2	85.4	85.2	85.1	78.2	85.3	85.8	88.4
Magnesium	55.0	42.0	38.0	45.0	48.0	54.0	50.0	55.5	56.0	48.0	55.0	48.0
Iron	0.33	0.41	0.31	0.34	0.35	0.36	0.35	0.36	0.48	0.42	0.30	0.32

All values are in mg/1 except temperature and pH —indicates absent.

Table 47 : Physico-chemical characteristics of coal minewater (Minimum, maximum and average). After CMRS Annual Report, Dhanbad.

Parameter	Minimum	Maximum	Average
Turbidity			5
pH	7.70	7.93	7.8
DO	4.86	6.20	5.43
Alkalinity	300.0	420.0	386.0
Hardness $CaCO_3$	610.0	700.0	660.0
TDS	716.0	948.0	850.0
BOD	0.60	3.66	2.45
Chloride	24.0	27.0	25.0
Sulphate	330.70	359.0	344.0
Nitrate	40.0	58.0	46.0
Calcium	92.0	108.0	97.0
Magnesium	92.0	107.5	104.0
Iron	0.02	0.03	0.0266
Cadmium	Nil	Nil	NII
Sodium	44.3	60.0	52.2
Manganese	Traces	0.06	0.04
Copper	Nil	Nil	Nil
Zinc	Nil	Nil	Nil
Lead	Traces	0.005	Traces
Arsenic	Nil	Nil	Nil
Fluoride	0.10	0.20	0.20
Chromium	Nil	Nil	Nil
Cyanide	Nil	Nil	Nil
Oil & Grease	0.03	0.04	0.033
Phenolic Compound	Abs	Abs	Abs

(All values are in mg/1 except pH)

pH value of the effluents varied from 7.1 to 9.3, showing highly alkaline nature. The effluent standard for discharging the same in inland surface water advocates the range from 5.5 to 9.0 only. Methyl orange alkalinity ranged from 150.4 to 349.5 mg/1. The phenolphthalein alkalinity ranged from 20.2 to 70.5 mg/1. Alkalinity (31.6–70.5 mg/1) was remarkably low during the rainy season reflecting high concentrations of carbonate, bicarbonate and hydroxyl ions.

The value of total hardness varied from 388.4 to 449.6 mg/1,

Table 48 : I.S. Tolerance limits for sewage (I), industrial effluents (II) and that of inland surface water (III)

Characteristics	I	II		III
	IS : 4764-1973	Inland Surface Water IS : 2490-1974	Public Sewers IS : 3306-1974	IS : 2296-1974
Temperature (°C)	—	40	45	—
pH	—	5.5-9.0	5.5-9.0	6.0-9.0
Dissolved O$_2$	—	—	—	3 mg/1
TSS	30	100	600	—
BOD (5 day 20°C)	20	30	500	3
COD	—	250	—	—
Chloride (as (Cl)	—	—	600	600
Sulfate	—	—	—	1000
Nitrate (as NO$_3$)	—	—	—	50
Sulphide (as S)	—	2.0	—	—
Flouride (as F)	—	2.0	—	1.5
Total residual chloride	—	1.0	—	—
Copper	—	3.0	3.0	—
Lead	—	0.1	1.0	0.1
Mercury	—	0.01	—	—
Nickel	—	3.0	3.0	—
Zinc	—	5.0	15.0	—
Cadium	—	2.0	—	—
Chromium	—	0.1	2.0	0.05
Arsenic	—	0.2	—	0.2
Insecticides	—	0.0	—	0.0
Phenolic Compounds	—	1.0	5.0	0.005
Oil & Grease	—	10.0	100.0	0.1

(All values of parameters are in mg/1 except Temperature and pH)

from 415.2 to 500.4 mg/1 and from 211.4 to 689.4 mg/1 in summer, rainly and winter respectively. In the present investigation very high value of hardness indicates that these waters differ from the common type of hard water due to the sulfate content but not due to bicarbonate ions. It is to be noted that ICMR permissible limits for hardness of water as calcium carbonate is upto 600 mg/1 and that of WHO (1971) is 500 mg/1.

It is also remarkable to note that the dissolved oxygen was completely absent throughout the period of investigation. The dissolved solids are within the limits (range : 215.6 to 602.4 mg/1) but the total suspended solids crossed the standard limit like phosphate. The other parameters viz. chloride, sulfate etc. are within the permissible limits but the iron content is many times more than the allowable value i.e. 1 mg/1 (Table 45).

On the other hand, the coal washery effluents are totally different in quality. There was no seasonal variations in the phys- ico-chemical characteristics of washery effluents. Temperature varied from 27.0 to 31.8°C. The pH value ranged from 7.5 to 8.5. The DO was completely absent throughout the year. The total alkalinity varied between 66.3 and 120.0 mg/1. Total suspended solid values varied betwen 1080.0 and 2050.0 mg/1 with minimum value in March and maximum in October. The total dissolved solid value ranged from 1000.0 to 1700.0 mg/1 while the total volatile solid contents varied between 780.0 and 1230.0 mg/1 with minimum and maximum values in March and December respectively.

The BOD and COD values were recorded considerably higher. The value of BOD ranged from 200.0 to 440.0 mg/1, whereas the COD values varied from 980.0 to 1280.0 mg/1 with minimum and maximum values in April and October respectively The sulfate contents ranged from 9.0 to 27.0 mg/1. On the whole the chloride contents are relatively higher than the sulfate. The calcium and magnesium concentrates ranged from 34.4 to 88.4 mg/1 and from 38.0 to 56.0 mg/1 respectively. The iron concentration was re- corded minimum (0.30 mg/1) in November and maximum (0.48 mg/1) in September (Table 46).

The above data revealed the fact that many parameters like BOD, COD, TSS were remarkably higher than the Indian Standard Limit (Table 48). Due to combined effect of BOD, COD and total solids, the effluents become very much harmful for the life support- ing systems in them. Verma & Shukla (1969) and Sinha et al. (1990) also confirmed the present findings while working on sugar factory and coal mine effluents respectively. My earlier study (1994b) also revealed that the high BOD and COD values are detrimental to aquatic biota.

During the present investigation, DO content was totally absent which justifies that the effluent does not supports any sort of life. Ghose & Basu (1968) also advocated the similar views while studying the estuarine pollution of Hooghly by effluents from a chemical factory in West Bengal. Therefore, the combined effect of various parameters increases the toxicity of the effluents which, in turn, disrupted the normal biological process and ultimately the fish production. Over addition of pollutants by excessive discharge in the receiving water may prove a slow poison and causes premature death of water bodies leading to eutrophication.

Standard Water Quality for Pisciculture

For assessment of impact of coal mine effluents on fish culture, it is essential to have a brief account of normal water quality of successful fish farming first, and then compare the impact produced by the effluents. The Indian Standard Institute has fixed certain minimum, optimum and maximum conditions for water for proper fish culture as below :

pH — 6.0 to 9.0

DO — not less than 4.0 per cent saturation

CO_2 — not more than 6.0 mg/1

Electrical conductivity–not more than 1000 mho/cm

Free NH_3N—not to exceed 1.2 mg/1

Impact of Coal Mine Effluents on Pisciculture

Although discharge of coal mine effluents is toxic to fish production, the desirable range of pH is 6.7 to 9.0 for fish culture. The range from 5.0 to 6.6 and 9.1 to 11.0 results in low productivity. The extremes are toxic to pisciculture. Good fish fauna was recorded in the pH range of 6.7 to 8.7 (Ellis, 1944) However, Banerjee (1967) reported that pH around neutral condition (6.5 to 7.5) was the most suitable for productive ponds on the basis of his survey of water quality and soil conditions of several water bodies in Assam, Andhra Pradesh, Manipur, Orissa and West Bengal.

Dissolved Oxygen below 5 ppm may be considered unfavourable for a productive water body. Under normal conditions, DO content above 7 ppm is suitable for a productive fish pond. According to Ellis (1994), good fish fauna was found in streams with DO value over 5.0 ppm. Hicking (1968) reported that the fish fed poorly and starved at low concentrations of DO i.e. 2 mg/1. Photosynthetic oxygen production is an important parameter which could be used to assess the fish production (Sreenivasan 1969).

Carbon dioxide is essential for photosynthesis. However, its high concentration adversely affect the fish. A good fishery is correlated with low CO_2 content (Bose and Sinha, 1981). Fish production has been reported to increase with the increase of alkalinity (Carlender, 1955). Alkalinity over 150 mg/1 is conducive to higher production (Ball, 1948). Wetzel (1966) reported that hard waters were more productive than the soft waters. The total dissolved solids (TDS) correlate well with plankton, bottom fauna and fish production (Northcote & Larkin, 1958). The value of TDS less than 50 mg/1 are unproductive for fish culture.

The TSS of coal effluents are the most vital parameters to influence the turbidity of receiving system and play a key role in the productivity. The high turbidity caused by the effluents lowers the penetration of sun light and reduces the photosynthetic activity which in turn, is related to the productivity of the water mass (Kumar, 1995c). The suspended particles causing turbidity may also absorb a considerable amount of nutrients, like phosphate, nitrogen and potassium in their ionic forms making them unavailable for plankton production (Kumar, 1995d). According to Nikolsky (1963), the most pronounced mechanical effect of suspended mineral particles occurs on fish when the water contains up to 4 per cent by volume of solid particles.

When the suspended solid is deposited, it fills the interstices between rocks and pebbles and also develops a separating layer between nutrient rich mud and water, thus, sterilizing the basin of the water body and ultimately eliminating much of the available surface for the growth of organisms and eliminating many benthic fauna. Cairns (1968), also got the similar results while studying on suspended solids standards for protection of aquatic organisms. The sterilization of the bottom hampers the food chain and food web and also hinders the fish production. Tarzwell & Gaufin (1967) have also opined that the water with suspended solid values may be treated as 'aquatic desert' which hampers the production of aquatic species.

The coal mine effluents with high BOD & COD are potential oxygen depleting sources for receiving water bodies. An effluent without DO content and with high requirement of oxygen utilize the DO of receiving ecosystem leading to oxygen depletion at first and then deoxygenation. Therefore, an excess organic materials can reduce the DO level to zero because the rate of consumption is far greater than its replenishment. Thus, all aerobic aquatic life is seriously threatened.

The Chitra coal field area of Bihar has experienced very tragic effect of oil and grease spill in water courses. The coal mine effluents having oil and grease exhibited a different nature of problem. The oil shows its problem by forming a thin film over the surface water. The oil film apart from being objectionable from aesthetic view point, also interferes with the normal limnological phenomenon. It stops the reoxygenation process of surface water particularly hindering at air and water interface and stopping diffusion. A visible colour of oil on water surface with a film thickness of 0.038×10^{-4} to 0.152×10^{-4} cm. A film thickness of 10^{-4} cm or higher has been reported to interfere with the reoxygenation process (Sinha et al., 1990). Only this much of thickness of oil film on water surface has also been reported to interfere in fish physiology by coating the gills.

The overall picture, thus, emerges from the analyses of physico-chemical properties of coal mine effluents is that the effluent receiving water body appears to be unfit for fish culture and needs special attention in order to improve the dilapitated socio-economic condition of the tribals in Santhal Paragana.

Suggestive Measures for Improvement of Pisciculture

1. Control of Mine Drainage

The wastewater from mines must be treated by various physico-chemical techniques, such as neutralization, aeration, reverse osmosis, drainage diversion, impoundment, reject tailing pond process, revegetation, flash distillation, surface mine reclamation etc.

2. Transfer of Water Bodies

All polluted water bodies should be transferred to the fishery department, so that no administrative problem will arise and various types of integrated fish farming may be practiced.

3. Assured Water Supply

An assured water supply should be made available through

the water pumps using ground waters so that the effect of coal
effluents can be minimized and on the other hand the fish farmers
will remain independent of monsoon water. In general, there is
acute scarcity of water and is a serious problem for Santhal Paragana in summer, even water levels in ponds/tanks/reservoirs
becomes insufficient for pisciculture.

4.   Input Supplies

Input supply particularly of fish seed of desired species is not
available in time and in required quantity due to lack of adequate
facilities in this zone. Though, fishery departments supply fish seed
to the fish farmers, they do not fulfill the total demand. During the
rainy days, the tribals have to move from pillar to post for procuring
the spawn of carps. So there should be a better arrangement.

5.   Installation of Induced Breeding Stations

There is no any pronounced river in coal mine area for spawn
collection. So natural spawn collecting ground is very few in tribal
zone, as a result of which fish seed of pure carps is really inacces-
sible. So the installation of induced breeding stations by the fishery
departments can help the problem to maximum extent.

6.   Loan and Subsidy Facilities

Banks and fishery departments should simplify the procedure
for disposal of loan applications. Although the subsidy is provided
to the fish farmers, its percentage should be increased. It should
also be extended to other items such as tubewells and to new inputs.

7.   Insurance

To protect against disease and calamities, an assurance scheme
should be introduced so that the poor tribals may not be exploited
by the money lenders.

8.   Freedom from Illegal Sharer

It is very common practice in tribal dominated division of
Bihar, Santhal Paragana, that the money lenders and middlemen
become the sharer to the lease belonging to fishermen community.

The consumers living at far off places have to pay many a time more than the actual price only because of the middlemen who devour the lion's share of the profit in the trade and the poor fishermen and tribal people who work day and night, are forced to starve as they are paid minimum possible price. So there is great need to make the fish farmers free from the clutches of such persons.

9. Research and Development Activities

Maximum attention should be paid towards various types of integrated fish farming such as fish cum poultry farming, fish cum duck farming, fish cum dairy farming, fish cum pig farming, fish cum horticulture and Singhara cum fish farming

10. Transfer of Research Results from Laboratory to Field

State fisheries departments are constantly circulating new technology models upto some extent, but there should be proper connection among the research centres and the fishery departments so that the advanced techonology can be transferred to the field level.

11. Adoption of Improved Fishery Technologies by Farmers

Importance of improved fishery technologies to increase the fish production has been discussed by several workers (Sinha, 1975; Sreenivasan & Pillai, 1976 ; Patra & Roy, 1988; Banerjee, 1994 ; Verma, 1995). In Santhal Paragana division, fish farmers are very conservative and do not adopt improved fishery technology in rural area near coal fields. It was observed that fish farmers of the distant country do not even acquainted with composite fish farming

Not only this, the fish farmers remove the weeds manually. No weedicides and insecticides are used by the fish farmers. Very low percentage of fisher folks provided supplementary feeding to the fishes. They are completely ignorant about predators of fish. In general, ponds are infested with macro-vegetations. Induced breeding was also not well known to them. So it is the need of time to organize the training programmes on these lines.

Jharkhand Movement against Coal Mines Pollution

Jharkhand means a place full of forests and in reality the geographical area of Jharkhand is full of forest and minerals. Nature has made this region the richest in the country in terms of natural resources but the man living in this region is the poorest in the country because of the other mans attitude. The coal mining has affected the local people in all spheres of life. In the name of 'National Interest', Santhal Paragana is witnessing a gigantic industrialization process for the exploitation of its natural wealth. The ideology of national interest (which is nothing but the interest of ruling capitalist class) is the guiding principle behind the process of development that is taking place in this area. The tribals and poor locals of this area feel that in the name of national interest, Santhal Paragana is witnessing no development but the rape of its peoples natural wealth through a process of capitalist exploitation.

The brutality displayed on the land and people of this zone by the process of coal mining, by the plundering of its mineral wealth, and by the decimation of its forests which provided much of the livelihood for its people, has not only reduced the majority of its inhabitants to destitution but has also brought the area to the brink of an ecological disaster.

The villagers mostly tribals, living around the tailing ponds lead a precarious life. They have become the guinea pigs of a laboratory used for experimentation to find out the effects of the coal dust in the area. The children of tribal village, about 100 yards away from the tailing pond, look like skeletons. Their bodies are covered with dust, and any wound they get takes a long time to heal. The children play and take bath in the tailing ponds. Bottles, papers, polythene bags, clothes etc. used by miners are thrown in the ponds. The poor villagers carry those to their homes and use them without knowing about the dangerous consequences. All the sources of drinking water in this area are affected by coal mining because of their closeness to the tailing ponds.

The utter disregard for the lives and the environment of the indigenous people of this area forced the people to come out unitedly under the umbrella of Jharkhand Mukti Morcha led by Mr. Sibu Soren (Gurujee). The struggle for their survival is intensifying which has brought them together. The main motto of this movement is to make our Jharkhand free from any exploitation.

The leader of the movement believes that if unplanned coal mining generates poverty, the poverty is the greatest pollution. Day by day this movement is gaining momentum.

In view of the intensity of the movement and the government favourable disposition to its demands, one can not but be optimistic and hope that this will lead to changes in policies to assure in an era when development will be achieved not at the expense of the environment, which sustains the very culture and life of the tribals. To conclude, one will have to qualify one's hope as the movement at present is caught in the cut-throat battle of votes and it will take a few months before a clear picture emerges in Santhal Paragana coal fields.

Acknowledgements

The author is grateful to Professor J.S. Datta Munshi, Emeritus Scientist (CSIR) and Dr. S.P. Roy, Professor of Zoology, Bhagalpur University, for valuable suggestions and encouragement.

References

Ahmad, S.H. and A.K. Singh 1986. Energy transformation through primay productivity and fish production in ponds and tanks in and around Patna (Bihar) *Mendel 3* (2) 142-145.

A.P.H.A., 1989. *Standard Methods for the Examination of Water and Wastewater.* 17th Ed. American Public Health Association, Washington D.C., N.Y.

Balla, R.C. 1948. A summary of experiments in Michigan lakes on elimination of fish population with rotenone, 1934–42, *Trans Am. Fish. Soc.,* 75 : 139-146.

Banerjee, S.M. 1967. Water quality and soil condition of fish ponds in some states of India in relation to fish production *Indian J Fish.,* 14 (1&2) 115-144.

Banerjee, S.R. 1994. Save natural hatcheries of Bihar. *Fishing Chimes,* 14 (2) 13.

Barrett, P.H. 1953. Relationship between alkalinity and absorption and regeneration of added phosphorous in fertilized trout lakes. *Trans. Am. Fish. Soc.,* 82 : 78-90.

Bose, K.C. and M.P. Sinha 1981. Study on effects of acute organic pollution in Ranchi lake, Ranchi (Bihar) III–Absence of definite diurnal rhythm, Abs. in Ist Congress of Env. Kota. Oct. 1-3, 1981.

Cairns, J. 1968. Suspended solids standards for the protection of aquatic organisms. *Purdue Univ. England Bull.,* 129 : 16-27.

Carlender, K.D. 1955. Standing crop of fish in lakes. *J. Fish. Res. Board Can.* 12 (4) 543-70.

Choudhary, H., N.G.S. Rao, G.N. Saha, M. Rout and D.R. Kanaujia 1978. Record fish production through intensive fish culture, *J. Inland. Fish Soc. Ind,* 10 : 19-27.

David, A., P. Ray, B.V. Govind, K.V. Rajagopal and R.K. Banerjee 1969. Limnology and fisheries of the Tungabhadra reservoir. *Bull. Cent. Inl. Fish Res. Inst.,* Barrackpore, 13 : 188 p (mimeo).

Ellis, MM. 1944. Water purity standards for freshwater fishes. *Spec. Sci. Rep.*, USFWS 1-15.

Ghose, A.B. and A.K. Basu 1968. Observation on estuarine pollution of Hooghly by effluents from a chemical factory complex at Roshra (W.B.), India, *Environ. Health.* 3 : 204-238.

Ghosh, S. 1985. Dimension of morbidity and mortality among children. Paper presented at *Workshop on Genetic Epidemiology Approach* to health care at National Institute of Health and Family Welfare, New Delhi.

Hickling, C.F. 1968. *The farming of fish.* London. Pergamon Press. 88 p.

Huet, M. 1965. Water quality criteria for fish life. In *Biological problems of water pollution.* Transaction of 3rd Seminar, 1962. Publ. Public Health Serv. U.S. (999 WP–25).

I.C.M.R. 1985. *Diarrhoeal diseases in infants and children,* Indian Council of Medical Research, New Delhi.

Jhingran, V.G 1991. *Fish and fisheries of India.* Hindustan Publishing Corporation, Delhi. 727 p.

Karamchandani, S.J. and M.D. Pisolkar 1967. Survey of the fish and fisheries of the Tapti river. *Surv. Rep. Cent. Inl. Fish. Res. Inst.* Barrackpore (4) : 29 p (mimeo).

Kumar, A. 1994a. Seasonal trends in biological and physico-chemical properties of a fish pond of Dumka (Bihar), *Acta Ecologia,* 16 (1) 50-57.

Kumar, A. 1994b. The effect of sewage pollution on the ecology of river Mayurakshi in Bihar. Abs No. 40, Annual Session of Acad. Env. Biol. held at Vellayani (Trivendrum).

Kumar, A. 1995a. Periodicity and abundance of rotifers in relation to certain physico-chemical characteristics of two ecologically different fish ponds of Santhal Paragana (Bihar), *Indian J. Ecol.,* 21 (1) 54-59.

Kumar, A. 1995b. Studies on pollution in the river Mayurakshi in South Bihar, *J. Env. Poll.,* 2 (1), 21-26.

Kumar, A. 1995c. Some limnological aspects of the freshwater tropical wetlands of Santhal Paragana, Bihar, India. *J. Env. Poll.,* 2 (3) 137-141.

Kumar, A. 1995d. Periodicity and abundance of plankton in relation to physcio-chemical characteristics of a tropical wetland of South Bihar, India. *Ecol. Env. Conts.* 1 (1-4), 47-51.

Kumar, A. and R. Pandey 1995. Comparative evaluation of potable water quality of tribal and non-tribal villages of Santhal Pargana, Bihar. *Ecol. Env. Cons.* 1 (1-4) 71-74.

Kumar, A. and A.K. Singh 1996. Fish fauna, fish spawn collection and fish trade system in Santhal Paragana (India) *J. Ag. Biol. Fish.* (in press).

Lakshmanan, M.A.V., K.K. Sukumaran, D.S. Murthy, D.P. Chakraborty and M.T. Philipose 1971. Preliminary observations intensive fish farming in freshwater ponds by the composite culture of Indian and exotic species. *J. Inland Fish. Soc. India,* 2 : 1-21.

Natarajan, A.V. 1976. Ecology and the state of fishery development in some of the manmade reservoirs in India. In : *Symposium on the Development and Utilization of Inland Fishery Resources,* Colombo, Sri Lanka (27-29 Oct.). Indo-Pacific Fisheries Council, 17th Session : IPEC/76/SYM/26/June, 1976.

Nikolsky, G.V. 1963. *The ecology of fishes.* London, New York, Academic Press.

Northcote, T.G. and P.A. Larkin 1958. Indices of productivity in British Columbia Lakes. *J. Fish. Res. Board Canada*, 13 (4) 515-540.

Patra, B.C. and A.K Roy, 1988. A preliminary study on the influence of the three organic manures on the growth of Indian major carps *J. Inland Fish Soc. India*, 20 (2) 61-63.

Singh, G. 1986. A survey of corrosivity of underground mine waters from Indian coal mines, *Int. J. Mine. Wat.* 21-32.

Sinha, V.R.P. 1975, Composite fish culture can boost up fish industry, *Ind. Farm.* 25: 17-18.

Singh, M.P. and K. Sinha 1987. Characterization of coal mine effluents from Jharia Coal fields, Bihra, India. *J. Ind. Poll. Control.* 3 (1) 11-18.

Sinha, M.P. and P.N. Mehrotra. 1990. Productivity of water bodies with reference to coal mine and allied effluents. Cited in *Impact of mining on environment*. Ashish Publishing House, Delhi.

Sinha, M.P., M.K. Singh, S.N. Singh and M.M.P. Singh 1990. Some problems of Pisciculture in particular to coal industry effluents. Cited in *Impact of mining on the environment*, Ashish Pub. House, New Delhi.

Sreenivasan, A. and K.V. Pillai 1976. Fertilization of fishery water : Experimental fertilization of small reservoir *J. Inland Fish Soc. India.*, 8 : 117p.

Sreenivasan, A. 1969. Primary production and fish yield in a tropical impoundment, Stanley Reservoir, South India. *Proc. Nat. Inst. Sci.* 35 B : 125-130.

Srivastava, M.P. 1993. Bionomics of air-breathing teleostean fishes and live fishes trade system in Koshi river basin of North Bihar, India *J. Freshwat. Biol.* 5 (1) 69-82.

Sudershan, D. and V.S. Somvanshi 1988. Fishery resources of the Indian EEZ with special reference to upper east coast. *Bull. Fish. Surv. India.* 16.

Sunder, S. and B.A. Subla 1984. Fish and fisheries of the river Jhelum, Kashmir. *Zoologica Orientalis*, 1 (II) 34-39.

Tarzwell, C.M. 1957. Water quality criteria for aquatic life Cited in *Biological problems in water pollution*. Transactions of the 1956 seminar. USD/HEW/PHS. Washington, Rabert A. Taft Sanitary Engineering Centre, pp 246-272.

Tarzwell, C.M. and A.R. Gaufin 1967. Some important biological effects of pollution often disregarded in stream surveys. Cited in: (ed. L.E. Keup, W.M. Ingram and K.M. Mackenthunan), A collection of selected papers on stream pollution. Wastewater and water treatment. Fe. Water Pollut. Contr. Admn. Cincinnati, Ohio.

UNEP, 1979. Energy Report Series. The environmental impacts of production and use of energy part I. Fossil Fuel, Pl. Nairobi.

UNESCO, 1979. Hydrobiological problem arising from the development of energy. Technical paper hydrology No. 17. UNESCO, Paris.

Verma, S.R. and G.R. Shukla 1969. Pollution in perennial stream Khala by sugar factory effluents near Laksar (Saharanpur) UP, *Environ. Health.* 1 (2) 145-263.

Verma, A.M. 1995. Inland fish culture in Koshi division. North Bihar : Problems and aspects, *J. Freshwat. Biol.* 7 (3) 207-216.

WHO, 1971. *International Standards for Drinking Water.* Geneva.

Wetzel, R.G. 1966. Productivity and nutrient relationship in Marl Lakes of Northern Indiana. *Verh. Int. Ver. Limnol.* 16 : 321-332.

IV. RESPONSE OF SOME TREE SPECIES TO IRON ORE MINE REJECTS : A CASE STUDY

A.V. Veeresh & S.G. Torne

Introduction

In order to successfully revegetate mine ore reject dumps, it is essential to know the responses of plants, to the changed conditions, on the site and which plants are suitable for revegetation. It is also important to analyse the reject for its physical & chemical characteristics which would help in understanding the cause for various responses of plants. The plants chosen for the study were, *Acacia nilotica* (Linn.) Delile, *Azadirachta indica* A. Juss., *Bombax ceiba* Linn., *Parkia biglandulosa* Wight & Arn., *Pithecellobium dulce* Benth., and *Tamarindus indica* Linn., all of which are known for their sturdy nature and drought resistant character.

Material & Methods

Garden soil was prepared by mixing one part farmyard manure with one part of sand for every five parts of top soil collected from areas undisturbed by mining activity (control). Reject soil samples were collected from various sites on the dumps bulked together and then a representative sample prepared for analysis. The soil samples were analyzed for their particle size and various chemical constituents by following the methods outlined by Allen (1974); Moore & Chapman (1976). Nitrogen was estimated following the method of Hawk *et al.* (1954) and Phosphorus was estimated according to the method of Seking *et al.* (1965).

Seeds of the plants were collected from the trees during their fruiting period, brought to the nursery, air dried thoroughly and stored. The seeds were presoaked overnight before sowing them in beds. After germination they were transplanted into small bags before being planted on the dumps. At the onset of monsoon, the saplings were transplanted on the dumps and in the nursery, control plants were grown in big bags of the size 24" × 12 " filled with garden soil. The different morphological parameters were monitored regularly for a period of one year. Analysis of chlorophyll & proline was carried out after a period of one year of growth. Chlorphyll was analysed according to the method of Arnon (1949) & Proline according to the method of Bates *et al.* (1973).

Observations

Soil Analysis

Physical analysis of the garden and dump soil are shown in Table 49. From the table it can be observed that the garden soil is mostly made up of gravel and sand with little of clay and silt, whereas in the case of the dump soil, it can be seen that there is less amount of gravel with more of sand along with silt and clay.

Table 49 : Particle size analysis of garden soil and the dump soil

Textural Class	% Amount present	
	Garden Soil	Dump reject
Gravel (>2mm)	60.33	35.30
Sand (2 –0.02 mm)	33.84	36.93
Silt + Clay (<0.0.2 mm)	05.83	27.77

The angle of the slope was found to be between 30°–40°. Chemical analysis of the soil revealed that both the soil were acidic in nature having a pH value of 5.5 (Table 50). Concentrations of the major plant nutrients and heavy metals in the soil reflect the absence of organic matter with severe deficiencies in nitrogen, phosphorus and potassium with high level of iron (46 per cent) and manganese in the dump rejects than in the garden soil. The amount of organic carbon was less in the dump reject, than in garden soil. The dump reject had organic matter content of 1.38 per cent which was poor, compared to the garden soil having 4.15 per cent. This low content of organic matter was unable to support a good vegetation cover on the dump reject. There was also deficiency of other plant nutrients like calcium, zinc and magnesium in the dump rejects.

Table 50 : Chemical characteristics of the garden soil and mine waste dump.

Sl. No.	Characteristic	Garden Soil	Dump Reject
1.	pH value	5.5	5.5
2.	Loss on ignition	15.5%	6.23%
3.	Organic carbon	2.4%	0.80%
4.	Iron	20.08%	46.00%
5.	Alumina	15.52%	7.13%
6.	Manganese	0.20%	0.78%

Contd...

Sl. No.	Characteristic	Garden Soil	Dump Reject
7.	Copper in ppm	6.0	4.0
8.	Zinc in ppm	7.6	18.6
9.	Nickel in ppm	14.0	10.0
10.	Lead	Traces	Traces
11.	Calcium	0.15%	0.08%
12.	Magnesium	0.07%	0.01%
13.	Available Nitrogen in kg/ha.	51.52	15.68
14.	Available Phosphorus in kg/ha.	>448	33.60
15.	Available potassium in kg/ha.	53.76	2.24
16.	Silica	38.17%	34.17%

Plant Responses

The different comparative morphological responses of plants are given in Table 51. Studies revealed that there is a overall decrease in the growth of plants on the dumps when compared to the control plants. Various morphological characters like size and thickness of leaf, thickness of stem, length of the root, showed that there is variation in their responses. The stem of all the plants growing on dumps were thicker than the control plants, whereas the leaf of all plants growing on dumps were thicker than the control plants, except in the case of *Bombax ceiba*. The size of the leaf was small only in *Acacia nilotica* and *Pithecellobium dulce*, which were grown on dumps when compared to control plants whereas in others the leaf was not much affected. It was observed that roots of *Acacia nilotica, Azadirachta, indica, Parkia biglandulosa* and *Pithecellobium dulce* grown on the dumps were affected. Their length and dry weight per plant was less than the plants grown in garden soil. Whereas in the case of *Bombax ceiba* and *Tamarindus indica*, the length of the root and dry weight of root per plant was almost same, showing that there was no effect of the mine reject on the growth.

Chlorophyll analysis showed that there was a decrease in the chlorophyll content of all the plants grown on reject dump, except in *Pithecellobium dulce*. The analysis for proline content, also showed that, it accumulated only in *Acacia nilotica* & *Pithecellobium dulce* grown on dumps.

Discussion

Soil Analysis

Normal soils consist an inorganic frame work of sand, silt and clay particles intimately mixed with organic material produced by the degradation of animals and plant remains. The relative amounts of each, control the physical, chemical and biological properties of the soil. Compared to natural soils, mine wastes, (1) are deficient in most plant nutrients; (2) may contain excess salts and heavy metals ; (3) are composed of unconsolidated sand that will easily erode and (4) have a mineralogy upon weathering which will affect levels and availability of plant nutrients and possibly toxic minerals (Shetron, 1983).

Investigations into vegetative stabilisation of mine wastes should include an examination of physical and chemical characteristics to assess the suitability of the material for plant growth. The physical nature of wastes can sometimes be assessed by an experienced eye, without recourse to formal tests, but the chemical content of wastes requires analytical investigation. Biological characteristics are rarely investigated, but are improved indirectly as the physical and chemical problems are overcome.

Physically, the texture, structure, stability and water content can present problems; chemically, an extreme pH, lack of nutrients and excess of toxic metals and salts are common, and biologically the materials lack typical micro-organisms and larger organisms such as worms which in a normal soil are responsible for mixing and distribution of decaying plant material (Williamson *et al.*, 1982).

Normal soils contain around 2-5 per cent organic matter. Mine materials may lack the organic matter or it may be so less that they cannot sustain vegetation on it. Besides supplying nutrients, the organic matter is loose and fibrous. It contributes considerably to the physical properties of soil. Analysis of the reject showed that it was very poor in organic matter content, which deleteriously effected the plants (Williamson *et al.*, 1982). The soil analysis revealed that there was high amount of iron and manganese with deficiency in nutrients due to acidic pH. Due to the combination of adverse physical structure and chemical quality of the mine reject an environment non-hostile to plant growth has resulted. A fertile

Table 51 : Responses of tree species to garden soil and mining dump rejects

Sr. No.	Response		Acacia nilotica		Azadirachta Indica		Bombax Ceiba		Parkia biglandulosa		Pithecellobium dulce		Tamarindus indica	
			Garden soil	Reject	Garden soil	Reject	Garden soil	Reject	Garden soil	Reject	Garden soil	Reject	Garden soil	Reject
1.	Height of the plant in cm		146.5	140.1	98.2	85.5	68.8	59.1	136.5	110.1	136.5	125..6	121.5	109.1
2.	Thickness of stem in cm		0.88	1.01	0.48	0.57	1.16	1.8	1.07	1.57	1.02	1.03	1.07	1.75
3.	Size of leaf	L	2.0	1.03	4.77	4.74	9.81	10.54	9.84	13.6	2.77	2.06	5.94	6.57
	in cm	B	0.95	0.48	1.8	1.81	1.78	3.28	1.3	1.7	1.06	0.71	1.95	2.34
4.	Thickness of leaf in mm		0.14	0.16	0.16	0.19	0.2	0.2	0.11	0.17	0.15	0.2	0.18	0.2
5.	Length of root system in cm		25.6	19.5	26.3	20.6	18.6	18.3	26.4	18.6	27.4	21.5	28.5	27.8
6.	Dry wt. of Root per plant in gm		3.5	2.3	3.5	2.8	2.1	2.2	3.1	2.6	3.2	2.4	2.9	2.87
7.	Total chlorophyll in mg/100 g fresh wt. of leaf		196.08	166.9	170.64	82.26	147.66	99.24	69.66	61.44	158.46	166.98	212.94	174.62
8.	Proline content in µg/g fresh wt. of tissue		52.5	75.0	167.5	168.5	41.25	45.0	41.25	41.25	120.0	167.5	41.25	45.0

L - Length   B = Breadth

soil contains major nutrients essential for plants, viz., nitrogen, phosphorus, potassium, calcium, magnessium and small quantities of trace elements. Mine reject is generally sterile and devoid of vegetation (Bradshaw & Chadwick, 1980) and lack plant nutrients.

Plant Responses

From the soil analytical results, it can be seen that there was a high amount of iron and manganese in the dump soil, which are known to inhibit growth of the plants when present in excess (Foy *et al.*, 1978). In the observations made here, it was seen that the plant growing on dumps were shorter than the control plants, which showed that high amount of iron and manganese in the reject affected their growth (Table 51). Sideris & Young (1949) showed that increased amount of manganese in the culture solution suppressed the growth of plants. Also they showed that with a high amount of iron in the culture solution, lead to the decreased growth of the plants. Vlamis and Williams (1964) have shown that in case of rice and barley plants, at higher levels of manganese the height of the plant reduced when compared to control plants. Kuo & Mikkelsen (1981) studied the effect of manganese on sorghum, it was observed that with the increasing amount of manganese in the solution decreased the height of the plant. Wheeler *et al.* (1985) showed that excess iron in the soil resulted into a reduced growth in the shoot of *Epibolium hirstum*. Similarly Pegtel (1986) showed that high amount of iron has reduced the growth of *Succisa pratensis*. The results obtained in the experiments carried out here, showed that there is a decrease in the growth of the plants in the presence of high percentage of iron in the reject. Our observations have confirmed the results obtained by earlier workers on different plants.

High aluminium content is known to cause stunting of the plants leading to thicker stem and leaf, with reduced size of leaves (Foy *et al.*, 1978). Morris & Pierre (1949) have shown in certain legumes, that with more of manganese in the culture solution, there was a decrease in the size of the leaves, Lohnis (1951) showed that excess manganese in the culture solution decreased the size of leaf in some vegetable crops. Singh & Sharma (1972) reported that under calcium deficiency conditions there is a decrease in the size of the leaves of potato.

The observations made here showed that the stem of the all plants growing on dumps was thicker than the control plants, whereas the leaf was thick in all plants except *Bombax ceiba*. The leaves of *Acacia nilotica* and *Pithecellobium dulce* were smaller in the plants grown on dumps. In contrast to this, in the case of *Bombax ceiba*, *Parkia bigladulosa* and *Tamarindus indica* the leaves of plants growing on the dumps were larger than the control plants, and in the case of *Azadirachta indica* it was not attacted. This particular behaviour of plants showed that these were not affected by the heavy metals present in the reject soil as far as their leaf size was concerned.

There have been frequent reports in the literature, which show that aluminium, iron and manganese have inhibitory effects on the plant growth, particularly on the growth of roots. Millikan (1949) noted that excess iron inflax produced a stunted root growth. Lohnis (1951) observed in garden crops that manganese affected the growth of the roots. Mugwira *et al.* (1978 & 1980) reported the aluminium injury to roots of triticale & wheat was characterised by decrease in roots of triticale & weight per plant. Ohki (1976) studied the effect of manganese toxicity on soybeans and found that at high levels of manganese there was a decrease in the dry weight of the roots. Shetron & Spindler (1983) studied the root pattern of *Medicago sativa* in iron tailings and natural soil and found that the dry weight of the roots grown in tailings was less than the plants grown in natural soil. Wong *et al.* (1983) studied the root growth of two grass species on iron ore tailings at elevated levels of iron, manganese and copper and found that with increasing amounts of these elements in the soil, there was a decrease in the length of the roots. Similarly, excess of iron in the soils caused a decrease in dry weight of the root in *Epibolium hirsutum* & *Juncus subnodulosus*. (Wheeler *et al.*, 1985). The studies of Pegtel (1986) on the response of *Succisa pratensis* to aluminium, manganese & iron showed that there was a reduction in the length and dry weight of root per plant with increased level of these elements. Emanuelsson (1984) was also studied the root growth in relation to calcium and found that the length & dry weight of roots were decreased at low levels of calcium concentrations.

The results of soil analysis showed that there was a higher percentage of iron and manganese and less amount of aluminium and other nutrients in the reject soil than the garden soil. The effect of aluminium was mainly due to the low content of phosphorus in the reject which helped in reducing toxicity of aluminium.

The observations made here showed that, not all plants were affected by the high amount of iron, manganese, and aluminium in the soil, but only *Acacia nilotica, Azadirachta indica, Parkia biglandulosa* and *Pithecellobium dulce* were affected. There was no effect on *Bombax ceiba* and *Tamarindus indica*.

Presence of high amount of manganese and aluminium in the absence or deficiency of nutrients are known to effect the chlorophyll content in the plants. Kelley (1912) and Johnson (1924) in the first quarter of the present century itself have shown that excess manganese at pH values above 5.5 deprive the availability of iron and thus failed to produce chlorophyll in the plants. Sideris and Young (1949) reported in pineapple plants that excess manganese in the culture solution lead to the decrease in the chlorophyll content of the leaves. Agarwal *et al.* (1964) have shown in their studies that excess manganese led to decrease in the chlorophyll content of barley leaves. Sarkunan *et al.* (1984) have shown in rice plants that aluminium affected the chlorophyll content of the plants. White *et al.* (1974) hypothesized that manganese interfered with iron uptake in leaves for chlorophyll synthesis. Fleming *et al.* (1974) have shown that, in the absence or deficiency of phosphorus and in the presence of aluminium resulted into decrease in the chlorophyll content, but after the addition of phosphorus the effect of aluminium on the chlorophyll content was lifted. The observations made here showed that there was a decrease in the amount of chlorophyll of all the plants except in *Pithecellobium dulce* grown on dumps.

Proline is known to accumulate in plants under stress conditions or due to deficiency in nutrients (Savitskaya, 1976 & Ghildiyal *et al.*, 1986). In the analysis of the plants growing on dumps for proline, it was seen that only two plants were under stress–*Acacia nilotica* and *Pithecellobium dulce*. The remaining plants have adapted to the adverse conditions.

Conclusions

The response studies of the tree species showed that, *Bombax ceiba* and *Tamarindus indica* were more resistant to the different stress conditions presented by the dump rejects as compared to other plants studied. These two being the local plant species were better adopted for the conditions found on the mine waste dumps. This shows that local plant species as mentioned above are better adapted to the conditions than exotic plant species. Also the study revealed that even though there was certain effect on other plant species, but still they can do well because of their sturdy nature.

References

Agarwala, S.C., C.P. Sharma & A. Kumar 1964. Interrelationship of iron & manganese supply in growth, Chlorophyll & iron porphyrin enzymes in barley plants. *Plant Physiol, 39* : 603-609.

Allen, S.E. 1974. *Chemical Analysis of Ecological materials*, Blackwell Scientific Publications, Oxford, London.

Arnon, D.I. (1949). Copper Enzymes in isolated chloroplats, Polyphenoloidase in *Beta vulgaris*. *Plant Physiol.* 24 : 11-15.

Bradshaw, A.D. & Chadwick M.J. 1980. *The Restoration of Land* : Blackwell Scientific Publications, Oxford, London.

Emanuelsson, J. 1984. Root growth & calcium uptake in relation to calcium concentration. *Plant & Soil, 78* : 325-334.

Fleming, A.L., JW. Schwartz & C.D. Foy 1974. Chemical factors controlling the adaption of weeping lovegrass and tall fescur to acid mine soils. *Agron. J.* 66 : 715-719.

Foy, C.D., R.L. Chaney & M.C. White 1978. The physiology of metal foxicity in plants. *Annual Review of Plant Physiol., 29* : 511-566.

Ghildiyal, M.C., Pandey M. & Sirdhi G.S. 1986, Proline content in linseed varieties as influenced by zinc nutrition. *Indian J. Plant Physiol, 29* : 368-374.

Johnson, M.O. 1924. Manganese chlorosis of pineapple-its cause and control. *Hawaiian Agr. Expt. Sta. Bull.* No. 52.

Kelly, W.P. 1912. The function & distribution of manganese in plants and soils. *Hawaiian Agr. Expt. Sta. Bull.* No. 26.

Kuo, S. & D.S. Mikkelsen 1981. Effect of P. & Mn on growth response & uptake of Fe, Mn and P by Sorghum. *Plant & Soil, 62* : 15-22.

Lohnis M.P. 1951. Manganese toxicity in field & market garden crops. *Plant and Soil,* 3 : 193-222.

Millikan, C.R. 1949. Effects on flax of a toxic concentration of boron, iron, molybdenum, aluminium, copper, zinc, manganese, cobalt or nickel in the nutrient solution : R. *Soc. Vitoria Proc., 61* : 25-42.

Moore, P.D. & S. B. Chapman 1976. *Methods in Plant Ecology* : Blackwell Scientific Publications, Oxford, London.

Morris, H.D. & W.H. Pierre 1949. Minimum concentrations of Manganese necessary for injury to various legumes in culture solutions. *Agron, J., 41* : 107-112.

Muguwira, L.M., S.M. Elgawhary & S.N. Patel 1978. Aluminium tolerance in Triticale, Wheat and Rye as measured by root growth characteristics & aluminium concentration. *Plant and Soil, 50* : 681-690.

Mugwira, L.M., S.U. Patel & A.L. Fleming 1980. Aluminium effects on growth & Al, Ca, Mg, K & P levels in Triticale, Wheat & Rye. *Plant and Soil, 57* : 467-470.

Ohki, K. 1976. Maganese deficiency & toxicity levels for Bragg Soyabeans. *Agron. J.,* 68 : 681-684.

Pegtel, D.M. 1986. Response of plants to Al, Mn & Fe with particular reference to *Succisa pratensis* Moench. *Plant and Soil, 93* : 43-55.

Sarkunan, V., C.C. Bidappa & S.K. Nayak 1984. Physiology of Aluminium toxicity in rice *Curr. Sci, 53* : 822-824.

Savitkaya N.N. 1976. Biol. Naukmi., 19 : 49.

Seking, T. 1965. *Photoelectric colorimetry in Biochemistry part II.* Nanhcod Publishing Co., Tokyo, p. 242.

Shetron S.G. 1983. Alfalfa, *Medicago sativa* L., establishment in mine mill tailings I. Plant analysis of alfalfa grown on iron & copper tailings. *Plant & Soil, 73* : 227-237.

Shetron S.G. & J.J. Spindler 1983. Alfalfa, *Medicago sativa* L., establishment in mine mill tailings. II. Root patterns of alfalfa in iron tailings and natural soils. *Plant and Soil, 73* : 239-246.

Sideris, C.P. & H.Y. Young 1949. Growth & Chemical composition of *Ananas comosus* (L), Merr., in solution cultures with different iron-manganese ratio. *Plant Physiol., 24* : 416-440.

Singh, O.S. & V.K. Sharma 1972. Alterations in growth & metabolism of potato plants by calcium deficiency. *Plant & Soil, 36* : 363-369.

Vlamis, J. & D.E. Williams 1964. Iron and Manganese relations in rice & Barley. *Plant and Soil, 20* : 221-231.

Wheeler, B.D., M.M. Al-Farraj & R.E.D. Cook 1985. Iron toxicity to plants in base rich wetlands : Comparative effects on the distribution & growth of *Epiobium hirsutum* L. & *Juncus subnodulus* Schrank. *New Phytol., 100* : 653-669.

White, M.C., Chaney R.L. & Decker A.M. 1974. Differential variental tolerance in soyabean to toxic levels of Zinc in sasafras sand loan. *Agron. Absts., 1974* : 144-45.

Williamson N.A., M.S. Johnson & A.D. Bradshaw 1982. *Mine wastes Reclamation,* Mining Journal Book Ltd., London, England.

Wong, M.H., W.M. Lau, S.W. Li & C.K. Tang 1983. Root growth of two grass species on iron ore tailings at elevated *Environ. Res., 30* : 26-33.

CHAPTER 9

WILD LIFE CONSERVATION AND MANAGEMENT IN INDIA

B.B. Hosetti and H.S. Patil

1. Preamble

In 1850's, Alexander Wilson, an ornithologist, watched a single migrating flock of passenger pigeon darkened the sky for more than 30 minutes. The flock was containing 2 billion birds spread 384 kilometer (240 miles) long and 1.6 km (1 mile) wide in the sky. By 1914, the passenger had disappeared for ever. How could the species that once the most numerous in North America became extinct in only few decades ? The major reasons for this is commercial hunting and loss of habitat. The loss of forests lead to loss of food supply and breeding places. These pigeons were easy to kill because these flew in gigantic flocks and nested in long narrow colonies. People used to capture one pigeon alive and tie it to a pearch called a stool, soon a curious flock alighted and collect around this 'stool pigeon' and were shot or trapped by nets that contain more than thousand birds.

Around 1858, massive killing of passenger pigeons became a big business. Shot guns, fire traps artillery were used in killing the birds. In 1878, a professional pigeon trapper made $ 60,000 by killing 3 million birds from their nesting ground near Petoskey Michigan. By early 1880s intensive commercial hunting was ceased, because the species has been reduced to only few thousands. At this point recovery was difficult because the bird laid only one egg per nest and were susceptible to death from infectious disease. They all died from severe storms during their annual migration from

Central to South America. By 1896, the last live passenger colony had vanished. In 1914, the last known live passenger pigeon on earth died in Cincinnati Zoo. Does it really matter that a wild species such as the passenger pigeon became extinct ? Another famous example of extinct bird is that of Dodo birds of Africa.

Why preserve plant and animal species ?

Certain wild species known as wild resources, are important because of their actual or potential economic value. Wild resources that provide sport in the form of hunting or fishing are known as game species. Wild life provides people with a wide variety of direct economic benefits as sources of food, species, flavouring agents, scents, soap, cooking oil, lubricating oil, waxes, dyes, natural insecticides, paper, fuel, fibres, leathers, fur, natural rubber, medicine, fire wood etc.

Most of the important crops that supply 90 per cent of world food today were once wild plants in the tropics. Other wild species may be needed for agriculture scientists to develop new crop grains to get higher yields and increased resistance to diseases, pests, heat and drought. Pollination by insects is essential for many food and non-food food plant species. Predatory insects, parasite and microbial pathogens are increasingly used for biological control of various weeds and pests, thus keeping the loss of crops and trees under control.

About 40 per cent of drug used throughout the World have active ingredients extracted from plants and animals. Many animals are used to test drugs and vaccines and to increase our understanding of human health and diseases. The nine banded Armadillo is used to study leprosy and prepare a vaccine for the disease. An estimated 10 per cent of World's marine species contain anti-cancer chemicals. Very little is known about the earth's 1.4 million identified species and nothing about <30 million undiscovered species. Less than 1 per cent earth's discovered species have been thoroughly studied to determine their possible usefulness. Loss of this biological and genetic diversity reduces our ability to respond to new problems and opportunities.

Aesthetic and Recreational Significance

Many wild species are a source of beauty, wonder, joy and recreational pleasure for a large number of people. The observation of leaves changing colour in autumn, smelling aroma of wild flowers, watching an eagle soar, porpoise glide through water, Jackal call in the night, feathers of wild fowl, peacock dance, beautiful coloured spots of curculinid beetles in jawar field, jumping style of a green coloured grasshopper in paddy fields are pleasurable experiences that cannot be measured in terms of rupees or dollars.

Ecological Significance

The important contribution of wild species may be their role in maintaining the health and integrity of the World's ecosystems. Ecosystem services of wild plant and animals on land and in the sea include, production and maintenance of oxygen, other gases in the atmosphere and in water, filtration and detoxification of poisonous substances, moderation of earths climates, water cycles, decomposition of wastes, nutrient recycling, soil fertility, crop pests, diseases, photosynthesis, fossil fuel etc.

Ethical Significance

It is believed by most human beings that to hasten the extinction of any species is ethically and morally wrong. Some ethical theorists are of the opinion that every wild creature has an inherent right to survive without human interferences just as human being has the inherent right to survive.

How species become endangered and extinct ?

At least 90 per cent of different species lived on earth have either became extinct or has evolved into a form, sufficiently different to be identified as a new species. Over 3.6 billion years ago life is believed to have begun. The rate of evolution of new species is higher than the rate of extinction. Hence it is believed that presently more than 30 billion different unknown species might be existing in nature. Since the agriculture began about 10,000 years ago, then onwards extinction of species increased at an alarming rate, especially since 1990, as human settlements have expanded world wide. The rate is accelerating rapidly. In 1975, about 100 species became extinct i.e. one species per every three days. The extinction rate increased significantly now-a-days. According to

Edward C. Wilson, extinction by 1985, had increased to ten folds i.e. 10000 species a year at an average of 3 species/day.

Wilson and several other biologists warn that if deforestation (especially tropical moist forests), desertification and destruction of wetlands and coral reefs continue at their present rates, at least 500,000 and perhaps one million species will become extinct as a result of human activity between 1975-2000. An average of 2000 species in a year or 1 species every 30 minutes may disappear. A 200 fold increase in the extinction rate in only 25 years may be evidenced. Most of these species constitute plants and insects. Animal extinction receive most publicity than plants because most animal species depend directly or indirectly on plants for food.

Threatened and Endangered Species

An endangered species is the one having few individual survivors that the species could soon become extinct in all or disappear from most of its natural range. eg. In USA, whooping crane and whirled pangonia and in India slender loris (*Loris tardigrandes*), lion tailed monkey, pangolin, wolf, jackal, hyaena and other mammals are the good examples for threatened species. Threatened species such as grizzly bear and bald eagle in USA are still abundant in their range but are declining in numbers and likely to become endangered in foreseeable future.

2.   Causes for endangering the species

1.   Habitat disturbance and loss

Greatest threat to wild animal and plant species is destruction or alteration of habitat, the area where species seek food, find shelter and breed. As the human population increased, settlements started draining and filling of wetlands, cleaning of forests and production of food, minerals, energy and other resources, destroy or disrupt habitats for many wild species. Disturbance and loss of habitat has been a major factor in the process of extinction. Eg. Ivory billed wood pecker, whooping crane, california condum (vulture) and Asiatic cheetah.

2. Hunting : There are three major types of hunting.

1.   **Commercial hunting :** Animals are killed for profit from sale of their fur and other parts.

2. **Subsistence hunting** : The killing of animals to provide enough food for survival in accordance with the carrying capacity of the habitat.
3. **Sport hunting** : This is killing of animals for recreation. Although subsistence hunting was one of the major cause of extinction of some species. Sport hunting is closed and regulated in most of the countries. Game species are endangered only when protective regulations do not exist or not properly enforced.

On a worldwide basis commercial hunting threatens a large number of animal species. The tiger, snow leopard and cheetah are hunted for their fur. Alligators are hunted commercially for their skin, elephants for their ivory tusks, and rhinoceros for their horns. Every year about 90,000 elephants were slaughtered for their ivory tusks.

Single Rhino horn a mass of compact hair is worth as much as $ 2400 (Rs. 72,000) in black market. It is used to make handles for ornamental knives in North Yemen. Although about 60 countries agreed for not to import or export these horns. However illegal traffic is going on, because of high market values. Between 1970-1986 the number of black Rhinos in Africa was dropped from 65000 to 5000 and only 100 white Rhinos were left by 1986. If poaching continues at the present rate, all species of Rhino will be extinct within a decade.

3. Predator and Pest Control

Extinction or near extinction can also occur because of attempts to exterminate pest and predator species that compete with people and other livestock for food and also with game species. The parakeet was exterminated from Carolina, in the United States around 1914, because it fed on fruit crops. It's disappearance was hastened by the fact that when one member of a flock was killed the rest of the birds moved over its body, making themeselves easy target. Eg. Common Crow is facing similar problems in India. Ranches, hunters were used by common public and government employees to control predators.

USA—Timber wolf, montane Lion, and Grizly bear.

India—Tiger, Lion, wolf, bear, elephant etc.

4. Pets Medical Research

More than 6 million wild birds are sold every year which end their lives as pets in the countries like USA, UK,West Germany etc. Large number of animals die during shipment.

Some species of exotic plants especially orchids and cacti are endangered in India and USA because they are gathered often illegally and sold to collectors and used to decorate houses, offices and land scapes. A single rare orchid may be sold to a collector for $5000.00. One third of cacti of USA (Texas and Arizona) are thought to be endangered because they are collected and sold as potted plants. About 71 million animals, rats, mice, dogs, cats, primates, birds, frogs, guinea pigs, rabbits and hamsters are used each year, throughout the world for toxicity testing, biomedical and behavioral research and drug development. Medical research coupled with habitat loss is a serious threat to endangered wild primates such as *Chimpanzee* and the *Orangutan*.

Under pressure from animal-right groups, scientists are trying to find alternative hunting methods that do not subject animals to suffering or better yet do not use animals at all. Promising alternatives include the use of cell and tissue culture, stimulated tissues and body fluids, bacteria and computer generated mathematical models that enable scientists to estimate the toxicity of new compounds from knowledge of chemical structures and properties. Public zoo, botanical gardens and aquaria are under constant pressure to exhibit rare and unusual animals. Eg. Orangutan. The dolls or wild dogs once abundant in Western Ghats and castle rock forests of Goa are now exterminated. For each exotic animal or plant they reach a zoo or botanical garden alive, many die during transport. Since 1967, many reputable zoo and aquaria agreed not to purchase endangered species although some abuses still occur.

5. Pollution

Chemical pollution is a new threat to wild life. Industrial wastes, mine acids deposition, excess heat from electric power plant have wiped out some species of fish. Eg. Hump backed chub. Slowly degradable pesticides, DDT and dieldrin have been bio-magnified in food chain and caused reproductive failures and egg shell thinning in some birds of prey (*Pere falcon*) (California brown Pelican) and bald eagle. Now-a-days persistent pesticides are banned in USA and Europe. However these are still in use in India.

6. Introduction of Alien Species

When Alien species are introduced into a new geographical area, these may be able to establish themselves with seriously affecting the local population size or natives or they may decrease or even cause extinction of one or more species by preying on them, beating them in the competition for food, or destroying their habitat. They may also cause extinction of population of some species by killing their predators. In 1859, a farmer in South Australia imported two dozen of wild european rabbits. A pair was escaped from the captivity. By 1930, they were increased to more than 750 million and started competing with others for grass and affected the food crops, sheep population, water holes, young trees etc. In the early 1950s, about 90 per cent of rabbit population was killed by deliberte human introduction of viral disease. Similarly in South India, seeds of a plant, 'Eupetorium' (understory weed) was sprayed in the buffer zones of forests to increase the green cover. Now it is a nuisance weed growing everywhere and competing with other grass plants for nutrients and space.

7. Treatise and Laws

To maintain a satisfactory biodiversity, the role of animal group organizations and the law is essential. Organizations such as International Union for the Conservation of Nature and Natural Resources (IUCN), International Council for Bird Preservation (ICBP) and World Wild Life Fund (WWLF) have identified endangered and threatened species and publish them in the volumes of 'Red Data Book'.

Several international treaties and conventions now offer protection to wild species. A treaty passed by UN Environmental Programme in 1975 is - Convention on International Trade in Endangered Species (CITES) was developed after 60 years of work. This treaty was signed by 93 countries, banned hunting or capturing of 700 endangered and threatened species. Although a number of countries officially offer protection to endangered or threatened species, the most strictly enforced protection is provided by the

United States, Soviet countries and Canada. Several laws are passed to protect wild species throughout the World. Development of wildlife refuge is also encouraged. Establishment of botanical gardens and aquaria, national parks and protected forests are essential to save and increase the population of endangered species.

3. General Importance

In India, wild life resources are so rich and diverse, that a few places on the globe can be compared with it. Indian subcontinent is having sheer natural beauty and wealth of diverse species of plant and animal life. Our wild life have spread from rain forests of Assam to barren deserts (Thar) of Rajasthan and hostile mountains of high himalayas. There are 400 species of mammals, 1200 species of birds, 2350 species of reptiles and 29,70,000 or more species of insects reported from India (Khajuria,1957). Some animals like black buck, The golden langoor, lion tailed macaque and pigmy hog are unique species in India.

Man is interested in the animals and plants since the dawn of the civilization. He has hunted wild animals for joy, food and clothing. According to the Red Data Book, about 600 species are threatened with extinction. India has already lost 200 animals and bird species since the beginning of Christian era. Now about 250 species of animals in India, are considered as extinct prone species.

4. Historical Developments

The idea of preservation of wild life is not new to India. It is the integral part of Indian culture. The Vedas contain hymns in praise of animals. Kautilya's Artha Shastra provides several penalties for killing, entrapping or molesting deer, bison, birds and fish in protected areas. In third B.C. Ashoka mentioned about the protection law of game fishes. During medieval period, Mughal emperors showed deep interest in protecting animal life of the country. Game was protected and preserved in hunting resources called 'Sikar Guah'. This trend continued for Ist 100 years (with accelerated pace) during british rule. There was a mad destruction of animals during this period. As a result many animals specially lions, tigers and cheetahs were almost wiped out.

In Assam, Colonel Pullock, a military engineer shot a rhino or buffalo for every breakfast almost everyday. A former Raja of a state in Madhya Pradesh, probably killed highest number of tigers (i.e. 1170) in the world. During this period, more wild life was killed than through centuries of hunting. Animals such as nilgai, black buck, four horned antelope, Indian gazelle, single horned rhinoceros and many species of birds were near extinction. India was probably the first country to enact a wild life protection act. The wild birds and animals protection act was passed in 1887 and was repeated in 1912. The forest act XVI of 1972 deals with the game protection in the states. Various states have established their own wildlife protection laws. In 1952, Indian Board for Wild Life (IBWL) was established. This was followed, by setting up of Wildlife Boards in individual states. Separate wild life preservation or organization were also created in some states (Wildlife circle). In 1972, new wild life protection act was passed. Under this act, possession, trapping, shooting of wild life, alive or dead are prohibited.

5. Important Wild Animal Species in India

India is the seventh largest and second most populous nation in the world. In the north it is bound by Himalayas and stretches towards south and tapers off into Indian Ocean between Bay of Bengal on the east and Arabian sea on the west. It has an area of 12, 61, 597 sq. miles. The entire area can be divided into three well defined regions:-

(1) The great mountain zone of himalayas.

(2) The Indo-gangetic plain.

(3) The southern peninsula.

These regions have varied climatic conditions and support different wild life. The important Indian wild life comprises of : bison, spotted deer or cheetah, swamp deer, musk deer, sambhar, elephants, wild pigs, rhino, tiger, bonnet monkey, black buck, leopard, Indian lions, white tigers, black bear, wild sheep, wild goat, golden cat, pelicans, egrets, purple heron, painted stork and a vast variety of flora.

Deer

These are the only few wild animals at par with the beauty of most beautiful animals. The deer belongs to the family Cervidae.

India possess totally 9 different species of deers such as (¹)
Kastura (Musk deer) (2) Kotra (Barking deer) (3) Cheetal (Spotted
deer) (4) Hoghiran (Hog deer) (5) Sambhar (Dancing deer) (6)
Barasingha (Swamp deer) (7) Thamin (8) Hanglu (9) Sangai. These
are herbivores, control the overgrowth of vegetation on grassland
and in the open forests. These become food for carnivores. They are
found in large numbers and are most commonly hunted for meat
and fur. Due to limited shelter and food supply, mass hunting and
habitat deterioration have pushed most of these species to near
extinction. Eg. Sangai of Manipur or Brow antelered deer occurs
only in dozens around Logtak lake in Manipur.

Unlike other deers musk deer, *Moschus moschiferous* has no
antlers but has a gall bladder. It is found in higher himalayas,
Kashmir, Himachal Pradesh, Northern U.P., Nepal and Sikkim.
Mostly it is found at the height of 2500 to 4000 m MSL. The musk
deer is shy and lead solitary life. The males (bucks) have a gland in
abdomen, which contains musk. The musk is an important ingre-
dient of several Ayurvedic and homeopathic drugs and is also used
as a perfume fixative. In Asia, Europe and America there is a
demand of 2500 kg musk every year. The musk extract being sold
at the rate of 40,000 to 60,000 $/kg in the International market. To
meet this demand deers are slaughtered in large numbers. The
Government of Himachal Pradesh established a national park of
1,000 sq km at Manali in Kulu valley and is the principal musk deer
region in the country.

Four species of barking deers are found in India also called
Kotra or Ribfaced deer. These live in lower hills at the height of 2000
to 3000 m. The largest of deer tribe is sambhar, also called as
dancing deer. It is very rare and extremely beautiful variety. It is the
largest in size and is 150-160 cms in height. It is found in the forests
all over the country. Its antlers are very long sometime upto 100 cm.
It is a browsing animal found in the herds of 4-12, shy and comes
to the open during nights.

Swamp deer *(Cervous unicolor)* may have tiny antlers and live
in the tall grasses with plenty of water around. It is a gregarious
animal in season they move in large herds. There are two species in
India live in himalayas from Northern U.P. to Assam, one more
species lives in M.P. It is estimated that 5,000 swamp deers are alive
today in India. Out of these only 200 are found in Kanha national

park in M.P. Spotted deer is one of the most beautiful deers found in open type of forests. It is a good grazers along flowing streams. It is spotted with white on a brown coat. The antlers are long and slender. Spotted deers are gregarious, during all the seasons pairing, breeding and shedding antlers around the year was noticed. The hog deer *(Axis arcinus)* is smaller and stouter in built. It has hog like appearance. Usually solitary occasionally in pairs in Punjab to Assam, live in grassy patches bordering forests and graze during cool hours of the day.

Dhole (Asiatic White Dog)

It is a rust-sand coloured weighing around 18 kg and about 60 cm in height. The tail is black and hairy. It is found in entire Asia, also occur in the islands Sumatra and Java. There are totally nine subspecies and three of them occur in India.

(1) *Kuon alpinus laniger* - in Kashmir.

(2) *Kuon alpinus primaevus* - in Kumaon.

(3) *Kuon alpinus dukhuenis* - in South gangetic plains.

It is a carnivore and consumes large amount of meat. It eats even bones and skin of young deer. In India, this animal is very common in Bandipur Wildlife sanctuary.

Gaur (Bison)

It is commonly called as bull gaur *(Bos gaurus)* and in the north east India is called as mithun. It is a relative of ox, sometimes it is as big as 6-4" and weighing 1000 kg. Now it is confined to only Bandipur and nearly 50 per cent of them found in Bandipur, Karnataka state. The gaur population was abundant in Bandipur and Madumalai of Tamil Nadu was reduced by disease called Reindeer pest disease. The yalk is another close relative of gaur found in the snowy hills of himalayas. This gaur is extinct in Sri Lanka, north India and a few animals exist in A.P., M.P. and Bihar.

Wild Buffalo

It is widely distributed over the grass jungles and riverine systems in the gangetic plains, in the terai of Assam and in eastern peninsular India. This animal number was reduced by hunting. It always preferred marshy grass grown jungles in the vicinity of rivers and lakes. It also enters cultivated fields in search growing grain hence it was killed.

The wild buffalo, *Bubulus bubalis* is a huge animal. It is much larger than the domestic variety more robust and it has more streamlined body. It is an animal of uncertain temper. Today it is noticed only in Assam (Manas and Kaziranga sanctuaries) the Nepal terai and Bastar district of M.P. In 1965, their number was 500 in Bastar, in 1975 and now it may be less than only 100 in that area.

Antelope

In India, four species of antelopes are present. These include, (1) The black buck (2) Chowsingha (3) Nilgai and (4) Gazelle or Chinkara. The black buck is one of the most graceful antelopes. It is one of the fast running animal of the world. Similar to cheetah. The young ones are light brown in colour. On attaining maturity the males turn black. The males possess hallow spiral horns which fully develop at the end of third year and are about 46 cm long. They are polygamous, as many as 50 females are appreciated by a single male. The black buck is a gregarious animal. It is exclusively Indian species occurs in the plains and avoid hilly and moun-tainous terrain and forests. Once it was found in large herds in Gujarat but now found in small herds of 20-30.

The four horned antelope or Chowsinga, *Tetracerus quadricornis* has two pairs of horns. The front pair being shorter than the latter, found in the foot hills of himalayas and Bundel-khand region. It mostly leads solitary life. The Indian gazella or chinkara found in southern part of U.P. in small thin bushes. It is a slender animal resembles the black buck in form. The peculiarity of the animal is that both sexes have horns.

Wild sheep and goat

Wild sheep and goat of India are restricted to northern parts of Tamil Nadu. The Nelagiri Thar (*Hemitragus hylocrius*) is the only to be found in south. This is also greatly endangered and only about 500 animals are alive today. They move in herds and when a herd is settled down to rest a sentinel soldier is posted to keep watch.

Big Cats

Among carnivores cats stand supreme in equipment of tooth and claw and in the combination of strength and ability. The important wild cats include,

(1) Tiger *(Panthera tigris tigris)*

(2) Leopard *(Panthera pardus)*

(3) Snow leopard *(Panthera uncia)*

(4) Leopard cat *(Felis bengalensis)*

The tiger is well distributed in U.P. from Himalayas to Vindhya forests in the south. It lives in a variety of habitats from thorny forests to dense terai of Assam. It is found in himalayas at an elevation of about 2,000 m above sea level. If it is left undisturbed then it live in one locality. It is a nocturnal and can walk 20-30 km during its night prawl for prey. It generally live solitary life. In 1949, the tiger population was 20,000-25,000. In 1958, it was reduced to 3000. In 1970, the number reduced further. To save the tiger from extinction 'Project Tiger' has been launched in 1972. Since 1972, tiger has become the national animal. The WWLF (World Wide Life Fund) and IUCN (The International Union for Conservation of Nature and Natural Resources) have contributed a sum of Rs. 8 million for equipments, expertise and study abroad.

Leopard, *Panthera pardus* is smaller and spotted cousin of tiger. It lives in all types of forests. It is more dreadful as it lives anywhere and eat anything, all animals, birds, reptiles, crabs, cattle, deer, monkey and even man. India possesses only one species of leopard. It breeds throughout the year. The cubs born 2-4 at a time born after 13 weeks of gestation. Once it was found all over India, Burma and Sri Lanka. Now no exact figures are available. They are still allowed to shoot under license from Indian Wild Life Act.

Snow leopard, *Panthera unica* is most beautiful creature found in high himalaya from Kashmir to Sikkim near snow line. Its foods are wild sheep, goat and domestic livestock. It has a creamy gray coat with large black rings. It has a shorter face and long fur. The fur of this is attractive and costs 5,000 each. This is a greatly endangered animal. Another endangered species, clouded leopard *(Neofelis nebulosa)* also hunted for skin found in Sikkim and Assam.

Bear

The brown bear is found in mountains over 3000 m above MSL. It hibernates during winter. Black himalayan bear is found at the height of 1500 to 3000 m MSL. Sloth bear found in the foot hills of forests and plains of himalayas and also in the forests of South India.

Elephant

Indian elephants, *Elephas maximus*, have been widely used as a means of transport and riding during peace and war times in the past. They are the largest mammals on land. They are confined to terai and other foot hills, feed on succulent grass, bamboo and drink plenty of water. They cannot tolerate hot sun and live in groups. Now-a-days Periyar sanctuary in Kerala, Sakrebail and Heggadedevana Kote in Karnataka possess elephants. Wild elephants are menace, as they destroy our crops, attack even human beings when they are furious. The age is about 70 years and less intelligent than Chimpanzee. They can be tamed and used for transport of logs and other materials.

Lion

The Asiatic lion, *Panthera leo persica* is distinguishable by its yellowish brown colour and tuffed tail in both sexes. About 150 years ago it ranged over most north India. Now it is present in gir forests in Kathiawar, the only home in India.

Indian Rhino (*Rhinoceros unicornis*)

Rhinos existed during Mohenjaodaro era about 5000 years ago in the plains of Indus river, which is now a part of Pakistan. Due to capturing and killing and clearing their habitat rhino gradually disappeared in the west. It possesses a horn on the snout sharper and longer in the males and blunt and short in females. A full grown rhino may attain 3.9 m long and 1.8 m in height with a horn about 30 cm in length. It is confined to grasslands of foot hills of himalayas, plains of West Bengal and Assam. It prefers swamps, savanna along river valleys etc. It is a slow solitary animal and is strictly territorial. It breeds throughout the year. Single calf is born at one time lives for 50-70 years. In the whole world only five species are

existing. Two are in Africa and three in Asia. Three live in India are
near extinction. In Assam, Kaziranga sanctuary possess significant
number of live animals.

Birds

In India, there are 1200 species of birds. The Pheasants, (*Para
cristatus, Ophrysai superulosa, Gallus gallus*), Ducks (*Anser indicus,
Sarkidiornis melanotos, Anas poecilopyncha*), Pigeons and Sandgroves
(*Columba livia, Steptopelia decocta*), Storks and Egrets (*Leptotilos
dubuis Egretta garetta*), Spotted billed pelican (*Pelacanus philippen-
sis*), Golden eagle (*Aguilla chrysaetos*), Sarus crane (*Grus antigone*).
Indian great horned owl (*Bubo bubo*), and great Indian hornbill
(*Buceros bicornes*). All these form an important bird wild life of
India. The osprey, great Indian bustard, cranes, Jerdons courses,
mountain quail, horned owl, red stork, avocet, Pink headed duck
are the threatened bird species of our country.

The great Indian bustard *Choroitis nigriceps* has been of good
interest in India and abroad recently. An international conference
was held in Jaipur in 1980 to assess its present day position and to
suggest steps to save it from extinction. It is a large Indian game bird
reminisecent of young ostrich having heavy body long neck and
long bare running legs. The plumage is dull brown above, and
white below. It has a height little over a meter and wing span 2.5 m.
It is omnivorus, feeds on all types of animal food specially
arthropods, lizards and vegetable, wild berries, grass, seeds, crops
etc. It breeds in July to September in grass fields or near shrubs or
grass cover. Usually single olive-brown egg is laid, females take
care of egg and young chiks. It is a inhabitat of wide open dry
scrubby plains of Rajasthan, West Punjab and Gujarat. Formerly it
occurred in the entire southern peninsula of India. As it is a larger
and spectacular bird and as its flesh was greatly relished and is
killed to near extinction.

Crocodiles

India possesses three types of crocodiles. All the species are
present in Orissa. The esturine crocodile *Crocodylus porosus* is
widely distributed in Orissa and Sunderbans of West Bengal. It
grows to a length of 7 m. The 'Save Crocodiles' project begun in
1974 under the guidance of Dr. H.R. Bustard. Protection of these
was undertaken in A.P., Bihar, Gujarat, Kerala, Orissa, Rajasthan,
Tamil Nadu, U.P. states.

The freshwater crocodile is known as mugger or marsh crocodile, (Crocodylus palustris) is widely distributed in India. It grows to a length of 5 m. The gharial species (Gavialis gangeticus) makes a lot of noise during mating season. It grows to a length of 8 m. It is shy, prefers larger rivers with sites for basking in the sun. It breeds during May-June and lay as cluster of 50 eggs in sand. These eggs hatch before the out break of the monsoon rains and migrate to the longwater system. Now it is present in sparce numbers in River Ganga. Its presence in Brahmaputra and Mahanadi is doubtful. Dr. H.R. Bustard of F.A.O. conducted a survey in May-July, 1974 to study all the three Indian crocodiles and stated that gharials are on the verge of extinction and could be saved by active management. He referred mugger as depleting species,not yet endangered and still present in most of the states.

6. Protected Species of India

Since man has learnt to live in one place, his demand on land has been increasing. The cleaning of forests is continued uninterruptedly for farming, industries, estates, multi-purpose valley projects, power lines etc. The continuous shrinkage of natural habitat of wild life caused havoc in wild life population throughout the world. Due to human interference most of the species could not adjust themselves to changed conditions. Some become extinct, while others are struggling to survive. In order to afford full protection the following animals have been declared as protected.

White eyed duck	Snow leopard
Indian antelope or black buck	Bustard
Four horned antelope	Peafowl
Bharel	Monal pheasant
Swamp deer or gond	Koklas pheasant
Eastern Pangolin	Chir pheasant
Elephant	Monitor lizard
Indian gazeelle or chinkara	Water lizard
Musk deer	Gharial
Serow	Marsh crocodile
Golden cat	Lethery python turtle

7. Management Packages

The Indian Board for Wild Life (IBWL) established in 1952 and the Ist phase of wild life preservation was confined to the protection of wild life from poachers and unscrupulous hunters. During second phase the importance was given to the development and creation of national parks and wild life sanctuaries, where threatened species could be preserved or protected. During this period detailed study of wild life is also made to collect necessary data required for future work.

The third phase covers most important aspects i.e. census operation and effective protection to threatened species, speedier development of wild life sanctuaries, detailed study and improvement in the existing food and water resources. The important steps suggested to preserve wild life in India are :

(1) **Study of Habitat:** The wild life management staff should have a correct idea about the exact habitat, which the species needs. A species can be preserved and managed only if its habit and habitat is known. To acquire such knowledge takes time, effort and scientific training. The information obtained on such aspects may be some time incorrect and incomplete in India.

(2) **Habitat Protection:** The habitat destruction is the main reason why India has lost most of its wild life. Some species like swamp deer have been threatened severely when their habitat was changed into fields.

(3) **Improvement of Habitat:** This is usually achieved by creation of parks, sanctuaries and construction of water holes, salt, licks, raising plantation of foliage yielding type and providing fodder, grass and trees.

(4) **Statistical data:** There should be a clear cut information on how many species and what is their number. For example, a thousand tigers in a small forest are not safe. A thousand hangul (deer) in the hills of Kashmir valley would be reasonably safe if fully protected. To control the population by hunting or game the knowledge on the number, how fast they reproduce and what percentage of young survive to adult stage etc. is required. Such statistics may vary from area to area , to each population. India has still

a long way to go for knowing exact number of individuals present in its forests.

(5) **Strengthening legal provisions:** In order to punish heavily to the poachers, the wild life conservation bill has been proposed for enactment. Under the provisions of existing laws there is hardly any scope for effective steps against poachers.

(6) **Closure of shooting:** Protection of wild life species threatened to extinction has to be reviewed from time to time and protective measures are afforded when desired. Various migratory water birds, vultures, kites, owls etc. have been recently closed for shooting.

(7) **Research on wild life:** Setting up a research cell to study the biological requirement of different species.

(8) **Organization of veternary unit:** To take care of wild life in case of epidemics and suggest precautionary measures establishment of veternary units in the fields are needed. Such units will help in preventing the spread of diseases.

(9) **High altitude zoo:** To preserve and propagate wild life of high altitude zoo in the Himalayas under controlled conditions in natural surroundings are suggested.

(10) **Census of different species:** Census of pea fowl and crocodiles is on hand, more species to be undertaken for census every year.

(11) **Introduction of exotic species:** It is hoped that carefully planned experiments of this kind will not only enrich the fauna but also serve an alternative home to those species.

(12) **Breeding farms:** It is proposed to establish breeding farms especially for threatened species so that these can be reared in semi natural conditions and then introduced into parks and sanctuaries and in areas where formed once their natural home.

(13) **National park and sanctuaries:** Some existing sanctuaries are to be raised to the status of national parks, while some more sanctuaries are to be established. Particularly for musk deer, black buck and other protected species. Similarly more water bird sanctuaries also proposed to be established for migratory as well as resident species of birds.

(14) **Legal measures** : Protection of wild life by law is a major step. Though many laws are enacted previously by Wild Life Act, 1972, under this act, possession, conservation, trapping, shooting of wild animals alive or dead, serving their meat in houses, their transport and export are prohibited. The offenders are fined Rs. 2000 and six years imprisonment. Wildlife Protection Amendment Act, 1986 prohibit ivory trade and materials made out of wild animals.

References

Agarwal, V.P. and Chaturvedi, L.D. 1995. Threatened Habitats, Society of Biosciences, Muzaffarnager.

Kaul, M.K. and Singh V. 1985. Conserve Himalayan Hogwood. Himalaya Plant Journal 3: 42-44.

Koshoo, T.N. 1995. Environmental Concerns and Strategy, Ashish Publishing House, New Delhi-26.

Miller, G.T. 1990. Environmental· Science: Conservation and Management. Wordsworth Publishers. Belmount, U.S.A. II Ed.

Ambio-Volume 22 (2) March, 1994. A Journal of Human Environment. The Royal Swedish Academy of Science. Stockholm, Sweden.

Ambio-Volume 24 (1) Feb., 1993. A Journal of Human Environment. The Royal Swedish Academy of Science. Stockholm, Sweden.

Ambio-Volume 22 (7) Nov., 1993. A Journal of Human Environment. The Royal Swedish Academy of Science. Stockholm, Sweden.

Captivating antlers. Science Reporter. Feb., 1995. CSIR Publications, New Delhi-12.

CHAPTER 10

TERMITE INFESTATION IN THE B.R. PROJECT AREA : A CASE STUDY

A. Naveed and E.T. Puttiah

1. Introduction

Termites are soft bodied insects inhabit in large numbers, mostly in the tropical and semitropical areas. A few species also found in the cooler parts of the world. These are social insects live in colonies with a definite division of labour. All functions of the colony are shared by different caste members like workers, soldiers and reproductives. The morphology of each member caste are modified according the special function required to perform by the group. Each caste is specialised to carry out functions like swarming, dissemination, construction, fungus growing, foraging, feeding and maintenance of the colony. The secondary reproductives are called as alates, swarm out of their nest at certain months (pre-monsoon) of the year, shed their wings, pair and attempt to establish a new colony.

A few genera of termites also found in the cooler parts of the world. These are of two types (1) one group inhabit the earth and are referred as subterranean species. The insects belonging to genera, Reticulitermes live on the trees and the mounds will have no connection with the ground and are called as nonsubterranian species. Mastotermitidae is the most primitive group, closely related to cockroaches as indicated by the anal lobe in the hindlimb. The members of Kalotermitidae construct nest in dry wood without soil. They are commonly called as dry wood termites. The important genera of this family kalotermitidae are eucryphotermes

(270)

and Glyptotermes. The Hodotermitidae are known as harvester termites.

2. Methods

To assess the termite infestation in the B.R. Project area , two different places were selected. They are staff quarters area of Irrigation department and Staff quarters of Karnataka Powe. Corporation. These two places are situated near the Bhadra Reservoir Project. The termite attack was acute in these regions. A house to house survey was conducted in these areas and also in agricultural fields. Mounds were enumerated in particular selected areas for height and dimension of each mound. The soldiers, workers, queen and king termite casts were collected and identified. In addition, the termite population density and caste ratio were estimated by using hand sorting methods.Then the mound soil from 6 different spots was collected and analyzed for chemical properties.

3. Classification of Termites

At present about 2000 living and fossil species of termites have been described. According to Syndes (1949) and Emerson (1955) termites are classified as follows.

Family I	Mastotermitidae	
Family II	Kalotermitidae	
Family III	Hodotermitidae	
	Sub family	1 Termosinae
	Sub family	2 Stolotermitinae
	Sub family	3 Porotermitinae
	Sub family	4 Creatatermitinae
	Sub family	5 Hodotermitinae
Family IV	Rhinotermitidae	
	Sub family	1 Psammotermitinae
	Sub family	2 Heterotermitinae
	Sub family	3 Coptotermitinae
	Sub family	4 Termitogetonae
	Sub family	5 Rhinotermitinae

Family V   Serritermitidae
Family VI   Termitidae
        Sub family   1 Amitermitinae
        Sub family   2 Termitinae
        Sub family   3 Macrotermitinae
        Sub family   4 Nasutitermitinae

The first five families are known to be primitive termites and possess symbiotic protozoans in the intestine which help in the process of digestion of the cellulose. The family Termitidae includes approximately 75 per cent of the known species of termites. Termitidae is the largest family of termites and these build nest below the soil surface and many species construct large earthen mounds of varying shapes and sizes above the soil surface. The important genera are, Specalitermes, Microcerotermes, Macrotermes, Odontotermes and Trinervitermes. Out of 2000 species in the world there are 300 species known from Indian region. Of the several genera found in India the genus Odontotermes is dominant and so far 40 species have been described under this genera (Sen-Sarma *et al.*, 1975).

4.   Biology of Termites

(i)   *Egg:* Eggs cylindrical, capsule like, translucent white, 0.86 mm in length and 0.53 mm in width, usually found in the main depository or incubation cavities in groups. Eggs were also noticed in the fungus garden of the main depository throughout the year.

(ii) *Alates:* These are different from the other nymphs in having wing buds. They attain 5 to 6 mm size after some days. As the size increased, the color changes from white to light yellow to pale brown followed by scleritisation of the body. During later stages of the development, they found in separate flat chambers with tightly packed fresh combs. They move freely between the flat chambers and broad chambers. The alate nymphs were nourished by the workers till their swarming. The young mounds up to 0.75 mm circumference were without alates.

(iii) *Nymphs :* Nymphs are milky white, 1 to 5 mm in size. Newly hatched nymphs concentrated in the main depository, later on distributed to other fungus chamber. They were the most numerous casts in the colony. Some peripheral and top fungus comb were without Nymps. The young Nymphs do not take part in any activity of the colony.

5. **Caste system**

A termite colony is mainly differentiated in 2 types

1. Sterile caste
2. Fertile Caste.

Workers and soldiers constitute the sterile caste and alates, king and queen constitute the fertile caste.

Workers

The workers are the most numerous caste in the colony. Usually workers are 4 to 5 mm in length and 1.5 to 2.0 mm in width. They take part in carrying eggs and feeding. Other castes in the colony participate in construction, expansion of the mound, cultivation of food for colony etc. During egg laying the workers carry these eggs to main depository. Some workers were found carrying the food materials and damaged nymphs for feeding the queen.

Soldiers

The soldiers are distinguished by their modified mouth parts and by large strongly chitinised often pigmented heads. The main function of the soliders is to defend the colony in the nest and also foraging workers, which remain at distance from nest. These are also sterile, 5-6 mm in length, 1.5 to 2.0 mm in width, they spray a brownish yellow liquid from the interior region of the head and bite the enemies when disturbed. Soliders always guard the workers and become very aggressive when disturbed.

The soliders always concentrate on more numbers around the royal chamber in the colony. Both the workers and soldiers are very sensitive to disturbance. The first sign of disturbance was given by the major workers in the foraging area by running around in all directions and touching other individual by their antennae. When royal pair was exposed, both castes found to be immediately congregating around the queen and repair the royal chamber. Observation on *O. obesus* mound revealed that if any part of the mound was demolished, first soldiers appear immediately after 3-4 minutes and workers follow them. The mound construction material is the soil particle mixed with saliva.

Royal pair

The reproductive caste consists of male and female. Usually king is small in size but queen length ranges with species to species depends on the stage of physogastricity. It is usually 7-8 mm in length. They spend their lives in Royal chamber. The queen is attended by several workers and soldiers. It reaches its peak of production in about 4 to 6 years, laying hundreds to a few thousand eggs/day depending upon the species.

6.   Infestation in the B.R. Project Area

Two places namely Irrigation department quarters and karnataka power corporation colony were severely damaged by termite attack. The staff quarters windows, doors, and roofing were made by different type of wood materials in 1955-56. In these areas most abundantly available termite species were *O. obesus and O. wallonensis*. Out of 300 species known from Indian region, hardly 40 species have so far been found to be injurious to economically important crop plants. Literature survey reveals that species of *O. obesus* and *O. wallonensis* affect many species of economically important plant. There are a number of reports of damage to maize crops by *O. obesus*. The attack is at the ground level and the damage occurs throughout the growing season from germination onwards (Patil and Basalingappa, 1994). Termites consume large portion of roots and continue their excavations to the stem, packing with soil to heights varying from 1-2 cm to 1 m in severe attacks.

Other crops subjected to the attack of *Odontotermes* species include sorghum, and wheat. *Odontotermes obesus* is known to attack groundnut foliage, seeds, seedling, stems, pods of developing and matured plants, harvested plants and seeds in storage. *O. obesus* is also known to attack the roots and collar region of cotton, Eucalyptus and chillies (Besson, 1941).

Table 52 : Chart showing the infestation by *O. wallonensis* and *O. obesus* on different plants.

Species	Hosts	Parts damaged
O.obesus	Coconut, Wheat, Barley, Bajra, Maize, Jawar, Groundnut, Cotton, Sugar cane, Vegetables, Pulses, Jute, Castor, Rubber, Mango, Guava, Citrus, Grape, Chilli, Eucalyptus.	Seedling roots and stems
O. wallonensis	Casurina, Mango, Silver oak, Groundnut, Maize, Ragi, Castor, Red gram, Coconut, Jack, Cashew Grasses.	Roots and stem.

Among the above mentioned list of plants in the B.R. Project, ground nut and coconut plants were infested from the underground (Table 52). First they attack roots of these plants due to which the plant wilts and dies. Dead wood is the stable food of the large number of termites. Some timber plants are known for their resistance to pests and the organisms of decay. Others may be made resistant by treatment with wood preservatives and some timber are very sensitive to termite attack. Termite resistance in timber is not a simple property but depends on the kinds of termites to which they are exposed as well as the conditions under which the timbers are employed.

For doors and windows, the construction authority used jungle wood for roofing purpose, teak and jungle wood for rafting. Hence in both the regions mainly windows, doors and rafting were affected. In the KPC colony most of the houses have been fenced by bamboos. All the bamboo parts are severely affected by termites.

Timber

Most of the plants species have natural resistance to termite attack. These wood plants contain specific chemical substances which drive away termites. Termites attack only on cellulose containing materials. Lignin in plants is neither harmful nor repellent to termites. After consumption of wood lignin is excreted out through fecum. The following are some of the commercial resistant timber plants found in India.

1. *Adina corchifolio* (Hardy)
2. *Cederala toona* Roxb. (Burma ceder)
3. *Choloroxylon swietenia* (Ceylon salim wood)
4. *Hopea odoralta* Roxb. (Thingan)
5. *Tectona grandis* (Teak)

In the study areas the wood materials used for construction of roofing, rafting and windows are of different species of trees. It was interesting to note that in the B.R. Project area the famous natural insecticide resistant plant, neem was also affected by termites. In the KPC colony out of 131 houses surveyed about 90 houses were infested by termites. The infestation was more in the garden region than inside areas. The houses in the irrigation quarters were infested both on outside and inside as well. Out of 572 houses 472 houses were affected of which 280 houses were severely damaged.

An interview with the inmates both in KPC colony as well as in irrigation department, revealed that the subterranean termites in the soil destroy the roots of garden plants, due to which many garden plants such as roses, crotons and other orchard plants like Mango and Guava were affected. In the KPC colony most of the house owners use DDT to prevent termites. Use of DDT reduced the termite infestation to a little bit but the inmates of those houses were unaware of any knowledge regarding the adverse effects of DDT and also about other insecticides suitable for control of termites. In some of the fields in and around B.R. Project BHC and Gamaxin were also used to control termite menace to crops.

7. Abatement

To keep control over the extensive damage caused by termites two private organizations viz., Rakesh Pest Control, Bangalore and City Pest Control, Hubli, were assigned the task. These two firms charge Rs. 15 per every 1 meter treatment. The Rakesh Pest Control and City Pest Control agencies, use different chemicals and different methods for antitermites task. The quantity of chemicals used by these firms was prescribed by Indian Standard Institute.

1. Pre construction antitermite activity : Soil treatment

Treating the soil beneath the building and around the foundations with a soil insecticide is a preventive measure. The purpose of treatment is to create a chemical barrier between the ground and the basement from where termites enter and damage wood and other cellulosic materials in the building. Anyone of the following chemicals (confirming to Indian standard) in water emulsion are effective when applied uniformly over the area to be treated.

1. Dialdrin concentrates   0.5% by weight
   (IS: 1054 - 1952)

2. Aldrin concentrates
   (IS: 1307 - 1956)   0.5% by weight

3. Heptachlor Emulsifiable concentrates
   (IS: 6439 - 1972)   0.5% by weight

4. Chlordane Emulsifiable concentrates
   (IS: 8693 - 1966)   1.0% by weight

The bottom, surface and sides (up to height of about 300 mm) of the excavations made for column pits, walls or arches and basement were treated with the chemical at the rate of 5 L/m^{-2} surface area.

Post Construction Antitermite Activity : Soil treatment

The object of soil treatment is to establish chemical barrier between the termites in the soil and the building for protection. Basically it consists of treating the soil adjacent to or under the building with a chemical toxicant which kills or repels termites. Water emulsions of one of the chemicals given above, shall be used in soil treatment and applied uniformaly at the prescribed rate.

Treatment Outside the Foundations

The soil in contact with the external wall of the building shall be treated with chemical emulsion at the rate of 7.5 liters per m^2 on the vertical surface of the sub structure to a depth of 30 cms. To facilitate this treatment a shallow channel was excavated along and close to the wall face. The chemical emulsion was directed towards the wall at 1·75 liters per running meter of the channel. Rodding with 12 mm diameter mild steel rods at 150 mm apart shall be done in the channel if necessary for uniform dispersal of the chemical to 300 mm depth from the ground level. The balance chemical of 0.5 litre per running meter was used to treat the backfill earth as it is returned to the channel directing the spray towards the wall surface. If there is a concrete or masonry apron around the building, approximately 12 mm diameter holes were drilled close to the plinth wall at 300 mm apart, deep enough to reach the soil below and the chemical emulsion pumped into these holes to soak the soil below, at a rate of 2.25 liters per linear meter.

The treatment described above is applied to masonry foundations. For R.C.C. foundations, the soil (back fill earth) in contact with the column sides and plinth beams along the external perimeter of the building was treated with chemical emulsion at the rate of 7.5 liters per square meter of the vertical surface of the structure. To facilitate this treatment, trenches were excavated equal to the width of the shovel exposing the sides of the column and plinth beams if this level is less than 300 mm. The chemical emulsion shall then be sprayed on the backfill earth as it is returned into the trench, directing the spray against the concrete surface of the beam or column as the case may be. If there is a concrete or masonry apron around the building, approximately 12 mm diameter holes were drilled close to the plinth wall, about 300 mm apart deep enough to reach the soil below and the chemical emulsion pumped into these to soak the soil below at a rate of 2.25 liters per linear meter.

Treatment of Soil in the Floor

The points where the termites are likely to seek entry through the floor are the cracks at the following locations.

(a) At the junction of the floor and walls as a result of shrinkage of concrete.

(b) On the floor surface owing to constructional defects.

(c) At construction joints in a concrete floor.

Chemical treatment was provided within the plinth area on the ground floor of the structure wherever such cracks are noticed, by drilling vertically 12 mm holes at the junction of door and walls, constructional and expansion joints. Chemical emulsion was squirted into these holes using a hand operated pump at the rate of 1 litre per hole. Later the holes were sealed.

Treatment to Voids in Masonry

Termites are known to seek entry into masonry foundations and work their way through voids in the masonry and enter the building at the ground and upper regions. The movement of termites through masonry wall at plinth level may be checked by squirting the chemical emulsion into the holes to soak the masonry. The holes shall be drilled at a downward angle of about 45 degree, preferably from both sides of the plinth wall at 300 mm intervals and emulsion squirted through these holes. This treatment shall also extend to internal walls having foundations of the soil. Holes shall be drilled at critical areas such as wall corners, and where the door and window frames are embedded in the masonry or floor at ground.

Treatment at the Points of Contact of Woodwork

All existing woodwork in the building which are in contact with the floor or walls, shall be treated by spraying at the points of contact and the adjoining masonry with the chemical emulsion by drilling 6 mm holes at a downward angle at about 45 degree at the junction of wood work and masonry and squirting chemical emulsion into them till refusal or to a maximum of half liter per hole. The treated holes shall then be sealed.

Treatment of Woodwork

For the purpose of treatment, wood work may be classified as follows:

(a) Wood damaged by termites and does not need replacement.

(b) Wood damaged slightly by termites and need replacement.

The woodwork damaged beyond repair should be replaced. All damaged wood work which does not need replacement may be treated. Infested woodwork in chaukats, shelves, joints, purlins in contact with the floor or walls may be provided protective treatment by drilling holes of 3 mm diameter with a downward slant to the core of the woodwork on the inconspicuous surface of the frames.

Treatment of Electrical Fixtures

If infestation is in electrical fixtures (Like switch boxes in the wall) or cover of the switch box, such cases may be treated with 5 per cent Chlordane powder. The covers of the switch boxes may be refixed after dusting. Periodical inspection and vigilance are necessary after carrying out the preventive treatment. It is essential that follow up action is maintained during subsequent humid seasons if termites re-appear.

The following are some of the controlling measures practiced in agricultural fields to control termites.

1. The mounds within the fields and its surrounding should be dug and the queen termite may be killed. Aluminum phospide tablets are placed inside the mound and the mouth is air closed.

2. Crude oil emulsion mixed with water should be applied to the crop land. In case, if the crop is attacked in different areas, then 30 per cent aldrin or 20 per cent chlordane insecticide dissolved in water and 0.1 per cent solution prepared and poured to the bottom of the crops.

3. For termites which build nests inside the earth and destroys the crops, in such cases before sowing 5 per cent aldrin and chlordane or 6 per cent heptachlor powder at the rate of 12 kg/Acre may be applied by mixing soil and then sowing may be carried out.

4. 2 per cent Chlorophyriphor or 20 per cent chlordane or 30 per cent Aldrin should be prepared in terms of 0.1 per cent and 18 liters of this solution may be poured into each mound. By doing this queen termite and others are killed and then the mound is destroyed.

5. Prepare 400 ml solution of Aldrin and then apply to 100 kg seeds and then sow which prevents termite attack.

6. In cultivated lands, making water to stand time to time in proper periods prevents termite attack.

Tabl 53 : Inscticides used for the control of termites

	Insecticides	Available form
1.	Aldrin	Solution
2.	Benzene Hexachloride (BHC)	Wetable powder
3.	Chlordane	Solution
4.	D.D.T.	Wetable powder
5.	Dialdrin	Solution
6.	Heptachlor	Solution
7.	Copper-chrome-Arsenic	Solution
8.	Methyl Bromide	Solution
9.	Napthenates	Solution
10.	Aluminium phosphide	Tablet
11.	Chloropyriphos	Solution
12.	Crude oil emulsion	Solution

8. Termite Mounds

The mound structure of commonly available species such as *O. obesus* and *O. wallonensis* was studied in the University campus.

1. Nesting pattern: Termites are commonly called as the engineers of insect world, because they construct their mounds in such a fantastic manner that even man cannot imagine. The termites build many strikingly discrete nest or mounds with a very complicated architecture. Their nests are constructed for the comfort of the colony inmates and protect them not only from climatic hazards such as heat, cold, drought and rain but also served a barrier to natural enemies. Termites construct 3 types of nests.

1. Nest below the ground level (Subterranean)
2. Nest constructed above the ground level and are called as mounds.
3. Nest constructed on trees or poles (Arboreal type)

Commonly, the termites construct their nest by cementing soil particles with the saliva. The simplest type of nest are just excavation in wood or ground and are normally built by primitive termites. Members of the family Kalotermitidae, live by burrowing in dry wood. The members of Hodotermitidae, have been adopted to a life in dry grass land, construct nests below the ground, in a series of large chambers with interconnecting galleries. Rhinotermitidae the moist wood termites show more developments in organization of nests. Species of Reticulitermes and Heterotermes have fixed nest sites in old tree roots, buried logs or timber.

In higher termites, the nest systems are more complicated within which micro-climate is controlled with a great accuracy. The food is also stored in the nests. The variety of nests constructed by different groups of termites throughout the tropical areas has been reviewed earlier (Emerson, 1956; Noirot 1970; Patil and Basalingappa, 1994). In the present investigation emphasis is given to commonly available species, *Odontotermes obesus* and *Odontotermes wallonensis* and their nesting pattern is explained as follows :

(a) The Structure of young nests of O. obesus

The shape of the mound of *O. obesus* was conical with one or more hollow conical turrets erected on the surface of the ground.

The thickness of the wall of the turrets in the young mounds was 0.7 - 1.5 cm. The turrets wall consists of aeration pits. The workers of this species collect soil particles and were found placing particle by particle along with salvia and thus raise the turrets. The inner surface of the turrets and the mound wall were finely plastered by their mouth parts. The active construction of mound was found during rainy season than in summer.

In this species the royal chamber was situated beneath the fungus garden and it was half moon shaped. On the floor of royal chamber king and queen were placed. The royal chamber is about 2.3-2.5 cm in size. The mound was having many cavities in which fungus gardens were placed and their colour was yellowish brown and conical in shape.

(b) The Structure of established nests of O. obesus

The mound structure of the established nests of this species was complicated with the addition of various structures such as vaults, runways etc. The wall of the mound was quite thick. The inner wall of the mound was similar to that of young mounds but differ with many pores termed as aeration pits. The turrets were opened at the tip by a hole called chimneys. The different chimneys of the mound are interconnected by galleries. The fungus garden of the established nest was quite similar in shape to that of young mounds but differ in shape and blackish brown in colour. The place of royal chamber in the established mound varied, a little bit bigger than that of young mound and encircled by the fungus garden.

(c) The Structure of young nests in O. wallonensis

The mound of young O. wallonensis was dome shaped, with one or two turrets small and erected. The wall of the turret walls were 0.8 to 2.7 cm thick and smooth inside and have aeration pits in the turrets. The mound internally contain a simple cavity and an aggregated fungus garden. The shape of the fungus garden varies. It may be plate, dome or conical in shape. The colour of the fungus garden was yellowish brown. The shape of the fungus garden was similar to that of O. obesus i.e. half moon shaped the upper surface was arch like and length is ranging from 3.9 to 6.3 cm and width 1.8 to 2.3 cm respectively.

(d) The Structure of established nests of O. wallonensis

The shape of the turrets of the established mound nest of *O. wallonensis* was conical and their number was more. The turrets of *O. wallonensis* were scattered on the ground. In some nests turrets were opened. These mounds contain numerous vaults arranged horizontally or vertically. The vaults are connected by galleries. The fungus gardens are placed in vaults. The shape of the fungus garden was conical and it was blackish brown in colour. The place of royal chamber was eccentric and contained many holes. The length and width of royal chamber varied from 6 to 14 and 2.6 to 3.8 cm respectively.

9.   Fungus Growers and Nonfungus Growers

(a) Fungus growers: All species of subfamily Macrotermitinae cultivate fungi in the nests. The important genera are Macrotermes, Odontotermes and Microcerotermes. Several species of Macrotermes and Odontotermes construct nests above the ground but microcerotermes species construct their nests below the ground surface.

In India, *O. obesus O. redemani, O. wallonensis* and *O. microdentatus* are mound building termites. The mounds of *O. redimani, O. microdentatus* are dome shaped, low in height with sub-conical out growth without any buttress. Internally the fungus combs are situated in separate chambers. In *O. obesus* the mound was built above the ground level. A large central hollow cavity contain a number of fungus comb arranged unilocularly.

In *O. wallonensis,* fungus chambers vary in number and size. They are round, oval or globular in shape separated by earthern walls and enclosed but connected by pathways for the movements of workers and soldiers. Fungus chambers were distributed in the entire mound both below and above the ground level. All the chambers were filled with fungus combs. These fungus garden help in maintaining humidity as well as food for termites, as depository for eggs, nymphs and other population of the colony.

In the mound of *Odontotermes obesus* each chamber is filled with a fungus comb, whose size varies according to the size of the chamber. The combs are convex and dark brown dorsally and concave and grayish brown ventrally, with a series of horizontal platforms one above the other. There are rows of openings from 3 to 10 mm wide leading into a network of narrow tunnel through the body of the comb. The combs are made of vegetable matter and wood mixed with the workers excreta. Masses of white spherical spores of the fungus are found inside the tunnels. The combs are filled with sterile casts and nymphs. The large combs in the chambers surrounding the royal cell act as nurseries and contain masses of eggs, with small red or scarlet red, spherical plant bodies among them which serve as food.

(b) Non fungus growers: Many species of termites do not maintain fungus garden in their nests. Microcerotermes and Nasuti-termes and associated genera feed on wood and dry vegetation. Their excreta contain mainly lignin, cellulose and other vegetable residue. These nests are compact with transverse holes throughout their structures and are normally found on trees, fence, poles etc.

10. Swarming Behaviour

1. Swarming : Emergence of alates take place soon after the onset of early monsoon showers during April-May, which is in conformity with the observation of Roonwal (1990). In case of *O. obesus* swarming occurred after dusk between 18.45 to 22.00 hours and lasted for 25 to 30 minutes depending on the density of alates present in the mound. Normally alates emerge out only once a year from a mound but the day of emergence varies from mound to mound and probably depends on the stage of development and maturity of the reproductives.

Swarming was observed to be slow in the beginning and became more vigorous later. Alates practically pushed out from many holes on the mound surface. These were found to crawl 4-6 cm on the surface and then flew away in the air. If alates found unable to fly, the workers immediately carry them back to the nest through the same hole. During this time the soldiers guard them from predatory ants.

2. Pre Swarming : After wing development the alates were observed to congregate in a chamber situated in the peripheral region of the mound. They were attended by the workers and soldiers. The workers make several transverse, irregular emer-gence holes just before the emergence of alates at both ends of the chambers. These holes were 6-7 cm in length and 1-1.5 cm in width. The number of emergence holes varied 3-12 depending on mound size. Workers and soldiers guard the exit hole during alate emergence.

3. Post Swarming : After the completion of emergence of alates, workers and soldiers move back to the mound through the emergence holes and seal the openings. The alates were usually attracted to the light after the emergence, if any light source was available. Otherwise they fly in the air for a certain distance and drop to the ground. Male and female join together and start searching for their future abode after shedding their wings. The males were seen moving erratically in all direction in search of females. When once a male finds the female and touched the tip of the abdomen with its mouthparts the female reciprocates by touching the male. Both male and female select a place and start constructing royal cell. The swarming activity varied in different species of termites. Basalingappa (1972) has observed swarming in case of *O. assmuti* varied from April to December.

4. Establishment of Colony : Establishment of new colony follows a general pattern in all mound building termites. Males and females, after shedding their wings, form a pair and go in search of a place where they mate and lay their first batch of eggs from which the workers are developed. Subsequently the workers take care of the mound. A minimum period of 2 years is needed to see a mound to grow above the ground level in most of the mound building termites (Pomeroy, 1976). A well established colony include all the stages, eggs, nymphs, workers, soldiers, king, queen and reproductive alates in addition to other associates of termites. There is a definite division of labour in termite colony where the production of eggs and defense of the colony are the duties of the queen and soldiers respectively and rest of functions are carried out by workers.

5. Foraging : Food supplies for the termite colony are made by the workers who feed the other dependent castes, i.e., nymphs, soldiers and primary and secondary reproductives in the colony. All termites construct a system of galleries or rurnways either covered or uncovered, along which they travel in search of requisites such as food. For foraging several hundred individuals emerge out from their subterranean galleries via holes at the soil surface. They climb up standing grass blades and cut the grasses into pieces of 2-20 mm in length. Some of the cut grasses is left on the ground but most of it was transported to the mounds via the subterranean galleries.

Species of *Odontotermes wallonensis* make their galleries at 10-30 cm below the round level. The diameter of such galleries range from 2-3 cm. Normally foraging occurs at temperature of around 27°C (Veeranna and Basalingappa 1989). In the colony of *O. wallonensis* only workers take part in foraging while soldiers guard them. Foraging was mainly carried out in cool morning or late evening hours of the day. Two peak foraging periods recorded one from April to June and the other October-December. Higher temperatures and more sunshine hours reduced the foraging activity.

11. Population Estimation

The individuals present in the fungus combs were carefully transferred to the plastic container by using handgloves. The mobile population was collected by scraping along with the soil and places of fungus combs that were fallen with the help of hand spade and transferred directly to the buckets filled with water. The animals from the buckets were collected along with the fungus combs in a separate container. Both fungus and mobile population were transferred to the laboratory for the study. Since standardized methods are not available for enumeration of the population from fungus combs in mound building termltes, the following three methods were tried.

(a) Berlse Funnel Technique

This consists of three metallic funnels of 25 cm diameter placed inside of each of the funnels. Three funnels were fixed on a 38 cm height, 120cm long wooden stand on which another stand was fitted to hold 100 watt electric bulbs just above the funnels. Fungus combs of 250 gm weight were placed in the funnels. Underneath each funnels, a 250 ml beaker with 70 per cent alcohol half filled was kept for the collection of termites. The bulbs were switched on for 24 hours. After 24 hours, the volume and number of castes or termite collected in the beakers were recorded. At the same time, the fungus comb kept in the funnel were examined for the remaining termites.

(b) Hand Sorting Method

In this method, the entire fungus combs were placed in a 60 sq cm rectangular metal chamber. The termites were anaesthesized using chloroform dipped in cotton and the chamber was closed for 30 minutes. After this, the fungus were broken into small fragments. Five samples of 100 gm each were drawn and the termites sorted out using a camel hair brush.

(c) Flotation Method

This method involve immersing the crumbled fungus combs in water to make the termites float. All the termites including those that were present in galleries (mobile) and fungus combs were collected by the methods already explained, brought to the laboratory, anaesthesized, crumbled, weighted and 50 gm samples, 3 from mound soil (mobile or running galleries) and 5 from fungus combs were drawn and placed separately in 32 cm by 15 cm wide mouth cylindrical jars. Then the water was added gently to fill up three fourth of the jar and allowed the material to settle for 5 minutes. The floating population was slowly poured out to a 0.5 cm sieve. The termites collected in the sieve were placed on a blue blotting paper for moisture absorption. This was repeated once again to extract all the termites, nymphs, workers and soldiers and counted by using digital counter.

The population density was measured in both the species of O. obesus and O. wallonensis by hand sorting method.

Table 54 shows the percentage of different castes. To estimate the population density the fungus gardens were selected from the pheripheral region. In both the species the percentage of nymph and workers were more. The percentage of workers in the peripheral region was high, it might be due to the foraging construction and repairing of the mound. The nymphal population at pheripheral region was more because the fungus garden for population estimation collected was in summer season. The population in mound fluctuates depending on seasons (Darlington, 1977).

Table 54 : Percentage of different castes

Caste	O. obesus	O. Wallonensis
Nymph	82.83%	78.29%
Workers	11.64%	13.82%
Soldiers	5.64%	7.84%
Total population	26.20×10^2	19.67×10^2

Table 55 : Characteristics of mound soil

Name of Place	pH	E.c.	O.C.	P_2O_5	k2.02
KPC Colony	6.6	0.2	0.62	10	70
Control	6.1	0.26	0.75	8	150
Rangamantapa Termite I	6.7	0.16	0.44	10	160
Control	6.1	0.21	1.12	8	180
Rangamantapa Termite II	6.3	0.21	0.58	7	110
Control	7.2	0.18	0.27	4	· 160
Ramamandir	7.8	0.27	0.87	6	90

12. Characteristics of Mound Soil

Collection of Soil Sample: Mound sample from 6 places were selected the sample representing 3 zones first upper part of mound, second from the middle part and third from the base of mound and mixed thoroughly before analyzing. The surrounding soil sample was also collected in the same manner from three differed spots was mixed and analyzed.

The chemical characteristics of soil samples of different mounds are shown in Table 55. Usually the pH of the mound soil was more alkaline compared to the adjacent areas as reported by Wild (1952). According to Sen (1944) increase in pH often associated with the accumulation of calcium carbonates. He has also stated that decreased pH may be due to high organic matter content in the mounds of Odontotermes species. The organic carbon level flactuated in different mounds. This may be due to the addition of organic materials like salivary secretions to cement the soil particles by the termites.

Soil samples from Rangamantapa termite mound soil, Kudreshed termite soil and in Narasimhaswamy termite soil showed higher levels of organic carbon compared with surrounding soils. In KPC colony the organic carbon was low compared with surrounding soil whereas Ramamandir soil shows equal percentage of organic carbon both in termite mound as well as in surrounding soil. The available P_2O_5 content in the mound soil was considerably greater than the surrounding soil. The higher phosphorus content in the mound soil was attributed to the incorporation of organic matter by the worker termites.

Conclusion

Termite are popularly known for the destructive activities such as destruction of wooden furniture, doors, frames, pillars, books and trees of commercial importance. Beside this, termites also cause damage to crop and reduce the yield. Out of 300 species discovered in India only 50 species cause heavy damage to crop and thereby annual loss of about 280 crores of Rupees as per expert reports in India.

There are also some useful activities of termites such as:

(a) Curing of Asthama patients by obtaining the smoke of burning curtain mounds of Nasutitermes sub family, which build their nest with their own excretory matter. These nests are called curtain mounds.

(b) Termites are also called as farmers friend, because they increase the aeration and fertility of the soil by making galleries and adding dead alates and other insects in soil. Analysis of mound soil revealed the enrichment of nutrients like fertilizers required for the growth of plants, hence this soil can be used as manure in fields to increase the agricultural produce.

(c) The mound soil is also useful in the construction of walls, making pots and to repair fields, because it is hard enough due to the presence of insect saliva.

(d) Termites contain vitamin-A and vitamin-B, particularly in queen and Alates. They form food nutrients for man and animals. The consumption of one queen provides energy equivalent to 4 kgs meat. Finally the social life of termites is model for us, as these small insects build separate chamber for its originator and provide security in particular, and sacrifices its life for the welfare of its colony.

References and Further Reading

Sydnes, T. E. 1949. *Catalogue of Termites (Isoptera) of the World*, Smithson Misc Coll. 112, 1-140.

Noirot, C. 1970. The nests of termites, in *Biology of termites* 2 : 73-125. Eds. Krishna, K and Wessner F. M., Academic Press, London, NY.

Besson, C.F.C. 1941. *The ecology and control of forest termites of India and neighbouring countries*, 10008 pages, Vasanth Press, Dehradun.

Darlington, J.P.E.C. 1977. Structure and distribution of population within the nest of the termite *Macrotermes subhyalinus* Proc. *VIII International Congress of the International Union for the study of social Insects*, Wegenigen, Netherlands, 245-248.

Darlington, J.P.E.C. 1985. The structure of mature mound of the termite *Macrotermes michaelsoni* in Kenya, *Insect Sci Appl 6*, : 149-156.

Sen-Sharma, P.K., Thakur, M.L. Mishra, S.C., and Gupta, B.K. 1975. *Wood destroying termites of India*, FRI Institute and College of Forestry, Dehradun.

Patil, D.S. and Basalingappa, S. 1994. Occurrence, nature and structure of the mounds of the termite, *Odontotermes brunneus* Hagen (Isoptera: Termitidae) from Belgaum, *J. Ecobiol. 6*: 17-26.

Pomeroy, D.E. 1976. Some effects of mound building termites on soil in Uganda, *J. Soil. Sci., 27*, 377-394.

Basalingappa, S. 1972. *Termites from North Karnataka*. Ph.D. Thesis, Karnatak University, Dharwad.

Veeranna, G. and Basalingappa S. 1989. Nesting pattern of the termite *O. obesus (Rambur)* and *O. Wallonensis* (Is soptera : Termitidae) *(Insect. Sci. Appl., 10.* : 69-180.

Bose, G. and Das B.C. 1982. Termite fauna of Orissa state, Eastern India, *Record Zool. Surv. India, 80*: 197-213.

Gautam, P.C. 1980. Growth ratio of termite mounds with reference to vegetation cover south Rajasthan, *Acta Ecol. 2* : 149-156.

Holmgren, H.A. 1912. Termites from British India, *J. Nat. Hist. Soc. 21*: 774-793.

Roonwal, M.L. 1973. Mound structure, fungus combs, and primary reproductives in the termite, *O. bruness* in India, *Proc. Indian Nat. Sci. Acad. (b) 39*: 63-76.

Roonwal, M.L. 1990. Termites of the oriental region, in *Biology of Termites*, 2:315-391. Krishna, K. and Wessner F.M. Academic press, London.

Verma, S.C. and Thakur, R.K. 1977. New records of termites (Isoptera) from Bihar and Orissa, *Zool. Survey of India, 3*: 361-365.

Emerson, A.E. 1956. Regenerative behaviour and social homeostatsis of termites, *Ecology. 37*, 465-522.

Wild, H. 1952. The vegetation of southern Rhodesian termitoria, *Rhodesia Agric. J. 49*, 280-296.

CHAPTER 11

USE OF AQUATIC PLANTS AS BIO-LOGICAL FILTERS FOR WASTEWATER TREATMENT: AN OVERVIEW

B.S. Mohan, N.C. Tharavathi and B.B. Hosetti

Water the essential component of life, without which living beings can not survive. It is required for various purposes in community life. But as the community developed and with the concentration of the population and products in cities, the demand for water is also increased. The rapid industrialisation also induced pressure on demand for water. For all these needs it is no longer possible to depend on natural sources of supply.

The population of India is likely to be around one billion by the end of the year 2010. The urban population would be around 400 million by that time, which would be almost double of the present urban population. The annual requirement of water is expected to increase in the same ratio by 2010. India is likely to use 61900 thousand M ha\m which will be more than half of the total water available annually.

Water pollution due to the domestic and industrial discharge has already become a grave problem in certain regions of the country and the degree of the problem increased constantly. The current practices adopted for the discharge of these wastes include direct discharge into water courses without any treatment due to the cost benefit ratio for purification.

Current awareness of the long-term danger to the environment and degradation of many natural aquatic resources has focused attention on alternative methods of processing wastewater so as to decrease adverse impacts and promote reuse of water and production of useful by-products (Hosetti & Frost, 1995). Among various processes available for pollution abatement, the use of algae in secondary and tertiary treatment of domestic sewage and industrial wastewater is well established. Aquatic macrophytes were suggested as a means of removing excess nutrients from eutrophic waters several years ago.

Those aquatic macrophytes which possess several desirable features and enhance their potential for use in the treatment of wastewaters should be recognized. Certain plants are capable of metabolizing complex organic substances such as phenols and in turn produce antibacterial substances. Many aquatic plants absorb and accumulate heavy metals and nutrients from waste waters in a ratio more than their requirement.

Over many areas of the ecosphere water has become the limiting factor to plant and animal biodiversity and therefore to primary production. The high population growth of human beings has triggered the accumulation of wastes which posed disposal problems and also a great demand for resources including water. The factors that need to be examined in order to initiate this approach in India are discussed in this paper.

Classification

Aquatic vascular plants can be broadly classified as nonrooted plants, rooted vascular plants, submerged plants and emergent plants. Hydrophytes mainly comprise of Dicotyledons, Monocotyledons and a few representatives of Pteridophytes.

Pteridophytes are tracheidal seedless plants with definite roots, usually herbs. Gametophyte and sporophyte generations are independent at maturity. Sporophyte, the dominant generation is differentiated into root, stem and leaf. Sporophylls bear sporangia. The following are the representatives of this group.

Family	Example
(1) Isoetaceae	*Isoetes sp.*
(2) Equisetaceae	*Equisetum sp.*
(3) Marsileaceae	*Marsilea minuta*
(4) Salviniaceae	*Salvinia natans*

S. molesta

Azolla pinnata

A. imbricata

Among angiosperms species of hydrophytes belonging to both dicotyledons and monocotyledons are encountered.

Dicotyledons comprises different species of aquatic macrophytes belonging to different families as follows.

Order: Polypetalae

Family	Example
(1) Nymphacea	*Nymphaea stellata*
	Nelumbo nucifera
	Euryale feron
(2) Brassicaceae	*Nasturtium officinale*
(3) Elantinaceae	*Elatine triandra*
	Bergia capensis
(4) Haloragaceae	*Myriophyllum indicum*
(5) Onagraceae	*Jussiaea repens*

Order : Gamopetalae

(1) Trapaceae	*Trapa bispinosa*
(2) Gentianaceae	*Nymphoides indicum*
(3) Convolvulaceae	*Ipomoea aquatica*
(4) Scrophulariaceae	*Limnophila aromatica*
	L. heterophylla
(5) Lentibulariaceae	*Utricularia stellaris*
(6) Acanthaceae	*Hydrophylla auriculata*

Order: Monochlamydae

(1) Ceretophyllaceae	*Ceretophyllum demersum*

Among monocotyledons different families comprises different species of hydrophytes as follows.

(1) Hydrocharitaceae	*Hydrilla verticillata*
	Hydrocharis dubia
	Vallisneria spiralis
	Ottelia alismoides

(2)	Pontederiaceae	*Eichhornia crassipes*
		Pistia stratiotes
(3)	Lemnaceae	*Spirodella polyrhiza*
		Lemna minor
		Wolffia arrhiza
(4)	Potamogetanaceae	*Potamogeton pectinatus*
		Ruppia maritima
(5)	Najadaceae	*Najas indica*
(6)	Poaceae	*Hygroryza aristata*
		Pseudoraphis spinescence
(7)	Araceae	*Cryptocoryne retrospiralis*
(8)	Alismataceae	*Sagittaria sagittifolia*
		Alisma plantagoaquatica
(8)	Butomaceae	*Butomopsis lanceolata*
(9)	Aponogetonanceae	*Aponogeton natans*
(10)	Eriocaulaceae	*Eriocaulan setaceum*
(11)	Cyperaceae	*Cyperus ariculates*
		Scirpus articulatus

Economical Aspects

Use of hydrophytes in pollution abatement include both beneficial as well as harmful aspects.

Beneficial Effects

(1) Heavy Metal Uptake

The problem of heavy metal pollution has been attracting an increasing attention of scientists during the last few years throughout the world. Various heavy metals such as Pb, Cd, Ni, Al, Hg, Zn, Cu, etc are released into the environment through various industrial processes like mining, manufacturing, agriculture and waste disposal practices. These ions are absorbed by the plants through their root system or by foliar absorption. These ions are translocated to different plant parts and interfere in normal metabolic processes and thus lead to reduction in growth. Although many heavy metals are not known as essential minerals for plants, they are readily absorbed by the root system of many plant species leading to the expression of toxic symptoms.

Aquatic plants are employed in the evaluation of water quality studies to monitor heavy metals and other pollutants of water and submerged soils. Their selective absorption of certain ions combined with their sedentary nature is a reason for using hydrophytes as biological monitors (Ronlet 1975; Kenaga and Moolenar, 1979; Clarke et al., 1982; Adema and Dezwart, 1984; Jenner and Jansenmomen, 1993; Sawidis et al., 1995).

The hazards of non-ferrous metallurgy and the investigations of environment pollution with Pb, Cu, Zn, revealed that different types of grasses can be used as monitors and rather rarely lichens, mosses and trees as indicators.

Vascular plants in comparison to bryophytes and lichens have a well developed root and transport system. Therefore they have a mixed mechanism of accepting nutrients and pollutants from their surroundings which increases the processes of localization, transport and relocalization of substances in the plant tissue.

The degree of exposure, the direction and the level of transportation and localization, depends on the stages of the development of plant which brings about individual difference. Individual difference in accumulation of heavy metals like chromium and nickel have been reported in Salvinia (35-83%) and Spirodella (10-53%) (Srivasthav,1994). It is better to define exactly the sampling parameters with respect to age, height, position and exposure to overcome the individual variation (Djingova et al., 1986).

In an estimation of heavy metal accumulation in plant species collected from same hydrotype, significant difference in metal concentration was observed among different species. (Sawidish et al., 1995)

Sawidish et al. (1995) also observed that the amount of metal accumulated in different tissues of plant body varied in different individuals with the time factor. In general roots revealed a greater metal accumulation than leaves, while stems had the lowest concentration with the exception of Mn and Cd in potamogeton crispus. These results agree with the reports of Taylor and Crowder (1981), Hutchinson et al. (1975); Wong et al. (1978) ; Pip (1990). However, Welsh and Denny (1976) and Beham et al. (1979) have suggested that some of the difference in accumulation behaviour may be due to adherence of some sediment particles to roots.

Pip (1990) conducted an experiment on *Phragmites australis* and *Potamogeton gramineus,* and concluded that the difference in metal distribution could be attributed to the relationship between metal accumulation and metabolic status of plant tissue. The heavy metal content in the flower tissue was lower than leaves of the comparing plants. Stem tissue exhibits lower physiological activity than the leaves. Seeds were less contaminated than other tissues as a result of low level bulk flow and redistribution along the phloem from leaves. (Sawidish *et al.,* 1991). Boyd (1970) has also reported that considerable translocation of elements from the leaves to fruits occurred in *Typha latifolia.* So he concluded that the accumulation of metal is strongly dependent on the kind of metal species and specific attention is necessary for the selection of the species to be used as bioindicators.

Hydrophytes as Sources of Food

There are two principal types of organs in hydrophytes: the seeds and swollen vegetative perannating organs which by virtue of their accumulated food reserves serve as potential nutritional source to man.

Various fruits and seeds rich in the oil, starch or protein may be eaten raw, or dried and ground to flour which can be baked with water or milk to give quite palatable bread. Many rhizomes and tubers are similarly rich in carbohydrate, especially starch, sugar and mucilage and are perfectly edible when raw or cooked. The foliage of a few hydrophytes provides acceptable salad ingredients or cooked vegetable dishes.

On the banks of river Nile *Nelumbo nucifera* is widely cultivated and its fruits and rhizomes are used in a variety of cooked and fresh dishes for many centuries. The rhizome of *Sagittaria trifolia,* the farinaceous seeds of *Euryale ferox* and the starch and fat laden horned fruits of *Trapa* spp. are used as food. The Chinese, probably introduced *S. trifolia* to Hawaii, Indonesia and the Philippines on account of its food value. The fruits of *Trapa bicornis,* T.*natans* and *T. incisa* form a staple food component in the continental parts of Asia, Malaysia and India.

The starch rich seeds and vegetative organs of those hydrophytes that are used as food in tropics of America, Africa, India and Malaysia are the species of *Nymphaea caerulea, N. capensis, N. lotus, Enhalus acoroides, Victoria amazonica, Typha latifolia* and *T angustifolia.*

In India young leaves, stems and roots of *Ipomoea aquatica* are used as food. Several species of *Ceratopteris* are cultivated as a green salad crop in parts of Africa and tropical Asia.

Wastewater Treatment

Conventional primary and secondary treatement methods are capable of producing clear odourless waters with acceptable oxygen levels, but disinfection and some form of expensive tertiary treatment is required to remove nutrients and eliminate disease bearing organisms (Mitchell, 1978).

Any system employing aquatic macrophytes would have to meet some criteria to warrant selection of plants for investigation. The important prerequisite conditions, for the hydrophytes to be used in the pollution abatement are explained on following lines.

(1) Rapid growth

Many floating aquatic weeds exhibit very high rate of growth under conditions, where space is available for colonization and abundant nutrients are available over a period of time when temperature and light values are high. For example Water hyacinth doubles in every 62 days in sewage oxidation ponds and Salvinia in every 36 hours.

(2) Easy propagation

Many aquatic plants can be propagated readily from a piece of vegetative tissue containing a growing point. Some of these plants are incapable of sexual propagation (salvinia) or have spread extensively without even setting seeds.

(3) Relatively constant growth

Many aquatic macrophytes exhibit a seasonal growth pattern, characterized by periods of rapid growth and periods of slow growth. Plants of tropical origin which are predominantly vegetative in growth, such as water hyacinth are however capable of sustained growth, provided temperature are maintained at high levels. It is possible that other species which do not respond to changes in day length, would behave similarly but this requires further investigation.

(4) Tolerance of high eutrophic condition

Physicochemical conditions of the water are improved when they are treated in the presence of aquatic plants. Under these conditions plants for different concentration of O_2, Organic matter, nutrients and turbidity are tested. This would make it possible to locate different species effective in treatment.

Use of aquatic macrophytes in stabilization ponds is one of the most economic and efficient system of treatment, especially in the elimination of sanitary risks (Gloyna, 1972; Hespanhol, 1990).

The water hyacinth ponds in mediterranean arid climate removed COD and TSS to the extent of 78 per cent and 90 per cent respectively (Mandi et al ., 1993).

In India, consideration should be given to the use of existing swamp systems to filter the run-off and effluents, before these enter into water bodies, which are liable to eutrophication. For this reason, the present tendency to reclaim swamps for agricultural use should be reconsidered. Swamp systems are valuable components of Indian ecosphere and should not be destroyed.

Biogas and Compost Production

At the rate of 20-40 tonnes fresh hyacinth can be harvested / hectare/day from the nitrogen waste of 2000 people and phosphorus wastes of 800 people. The decay of these aquatic weeds in water, results further into eutrophication which imparts the foul odour and make unhealthy situation for other aquatic life. Therefore for obtaining better return from these nutrient enriched aquatic systems an early harvest and utilizing the flora for composting in developing countries.

Mitchell (1978) stated that 100 Kg of fully fermented water hyacinth compost may contain nitrogen 3.75 Kg, phosphorus 0.86 Kg potassium 2.30 Kg that is the total NPK is 6.91Kg/quintal. Along with these, trace elements such as sulfur, calcium, magnesium, sodium, iron, copper, manganese, zinc, boron, molybdenum, and cobalt are present in the fully grown water hyacinth plants, which are essential for healthy growth of crop plants and vegetables and also for fish.

At 35^0C to 40^0 C temperature, 7-8 PH, C N ratio of 10:30 water hyacinth composes within 120 to 150 days resulting the ratio of 2.51:1 from fresh biomass to compost manure, for agriculture fields or fish ponds (Naskr *et al.*, 1985). The compost pits contain aerobic and anaerobic bacteria, blue green algae, and number of fungal flora. Some weed species of higher plants, some sedges, annelides, mollusks, mites and insect larvae can also be seen in those compost chambers.

Besides the high compost value of water hyacinth, it may also produce biogas, chiefly methane, like several other organic matter. In the gobar gas plant this weed has recorded to produce 374 Lt. organic gas from 10-12 kg dried *E. crassipes* of which about 60-80 per cent is methane and that gas can convert about 21000 BUT/m³. In water hyacinth composting the labour involvement is more though implementation is simple. Therefore where the commercial mineral fertilizer are more expensive and where cheaper labour is available, composting are more useful and encouraging.

Antibacterial Property

Certain natural swamp communities are capable of handling wastewaters so as to reduce bacterial load either by sedimentation or by producing exudates, which are toxic to bacteria.

Scirpus lacustris, an emergent plant is capable of metabolizing complex organic substances such as phenols and of producing exudates, which are toxic to bacteria (Mitchell, 1978). In an experiment with water hyacinth in stabilization ponds, the percentage of removal of faecal coliforms and fecal streptococci was 96.2 per cent and 94.7 per cent respectively with better decontamination efficiency if the retention time is considered (Mandi *et al.*, 1993).

Another aspect of the use of aquatic plants for the treatment of wastewaters in India which requires investigation is the relationship between the aquatic plants growing in swampy lands and human disease organisms. Indigenous plants should be tested to establish whether these plants have the same capacity as *Scirpus lacustris* to exudate antibacterial agents. The passage and survival of bacteria and viruses through such systems will have to be worked out so that it will minimize the breeding of mosquitoes and other vectors of human disease in proximity to human habitation.

Nutrient Removal

Many aquatic plants routinely absorb more nutrients than they require from nutrient rich waters and it will be accumulated in the body. Natural swamp communities are capable to reduce nutrient and sediment load in both freshwater and estuarine condition. However, their capacity to do so depends on the species of plants present. For example widespread strands of sawgrass (*Cladium Jamaicans*) in the Florida were rapidly saturated with phosphorus and were adversely affected by continuous addition of surplus nutrients. Artificial systems in which fast growing water weeds are used, probably offer more promise. Recently considerable attention has been paid to water hyacinth (*Eichhornia crassipes*) in this regard (Mitchell, 1978).

The environmental factor most likely to limit the growth of aquatic plants during the growing season is the availability of nutrients. Many of these plants appear to have compensate for this by developing a capacity to absorb large quantities of such nutrients over and above their immediate requirements during periods when these are readily available. This characteristic provides one of the main bases for employing such plants to "strip" nutrients from wastewaters. However, unless the nutrients are ultimately utilised in growth or removed by harvesting the plants, a proportion of them will be returned to the system, either through leaching or following death and decay of the plants.

A number of studies have been made on the mineral composition of aquatic vascular plants in various fresh waters but, generally, the concentration of minerals present in the plants are not compared with the ambient concentration of the same minerals in the water or show no correlation with them. Under nutrient poor conditions, plants respond principally to limiting nutrients by increase in growth rather than by increase in nutrient concentration in their tissues. When the nutrients become nonlimiting, luxury uptake may occur and tissue concentration increase without significant biomass. McNabb *et al.* (1970) showed that there is a linear increase in the phosphorus content of hornwort with the increasing concentration of soluble phosphorus in the water with an uptake levels of 3mg-1. Similar response was found in other submerged species. However, the response to inorganic nitrogen was different there being no increase in tissue concentration with increase in ambient concentration up to 14mg/1. Musil & Bren, (1977) used growth kinetics to calculate the quantities of water hyacinth that would be required initially, to harvest and the frequency of harvesting in order to remove specified amount of nutrients from a system under different temperature regimes.

Aquatic plants can play the major role in nitrogen removal from aquatic systems. The results from their study suggested that harvesting need not be frequent to prevent the return of nutrients during decaying process. Waste water treatment by stabilisation ponds with water hyacinth plants can furnish B category effluents (WHO, 1989) which can be reused to irrigate cereals, industrial crops, fooder crops, pasture and trees (Mandi, *et al.*, 1993). Many rooted emergent species absorb nutrient from the sediment rather than from the water (Boyd, 1970). For example Cumbungi obtains its phosphorus primarily from the sediment. Such plants are less suitable for nutrient absorption from effluent than those such as hornwort (MC Nabb *et al.*, 1970). However, this factor could be disregarded if it could be shown that nutrients were absorbed from a whole system such as a natural swamps and bound in sediments without detriment to the swamp itself or to anyone using its waters.

Easily Harvested and Preferably Useful

For a system employing aquatic plants to be effective, it will be necessary to maintain the growth of plants to entail regular harvesting without damaging the potential of the plants to continue to function in the system. Emergent species such as Cumbungi and the common reed, (*Phragmites australis*) may initially appear to be attractive in this regard as they are morphologically similar to many crop plants for which effective mechanical harvesters are available. However, care has to be taken not to damage the rhizome which are the basis for plant regrowth and most mechanical harvesters could not penetrate far into swamp systems. Free floating aquatic plants are the easiest to harvest without damaging the plants potential for immediate regrowth and do not require specialized machinery, although various quite complex devices have been came into market. Submerged species provide more problems but these can be minimised by designing the system to accommodate the harvesting device. If suitable machinery is available then careful consideration has to be given to the stage of growth of the species concerned. Various uses for harvested material have been investigated. It is desirable to concentrate on those uses which make minimum demands on expensive labour inputs.

The basis of production of useful materials by the aquatic plants are the substances that initially formed during the process of photosynthesis. This is conventionally expressed as dry matter production expressed in terms of tonnes/ha/year or $g/m^2/day$. Aquatic plants are capable of some of the highest yields known for example; Cumbungi (emergent) can have a yield of $52g/m^2/day$ (compared with $26g/m^2/day$ for sugarcane in Indonesia) (Westlake, 1963). Water hyacinth (free floating), a yield of $24g/m^2/day$ and hornwort (submerged) a yield of about $10gm^2/day$ (Mc Nabb et al., 1972) have been reported.

No much attempt has been made to develop methods of utilizing aquatic plants in India although these are present in significant quantities.

HARMFUL EFFECTS

Water Clogging

In most cases, an aquatic weed problem is simply an excessively large growth of one or more aquatic macrophyte species which causes blockage or amenity problems within the water body or watercourse. The root cause of such problems are attributable to the human interference with the system by nutrient enrichment.

Aquatic weeds may interfere with water movement and navigation, so the problems caused are particularly serious when they occur in navigable waterways and irrigation or drainage channel networks. In many cases the worst problems are caused by invasion of allien macrophyte species. These are transported by water currents or wind to produce piles, up to 5 m thick extending over many hectares of lakes or water course. The blockage problems produced by such enormus weed accumulation may cause flooding, break bridge, and obstruct or completely halt boat traffic movement.

The submerged weeds *Hydrilla verticillata* and filamentous algae, *Vaucheria dichotoma* are also known to cause problems in fresh water systems. In flowing waters these plants increase the frictional resistance to water flow which may increase the risk of flooding. In navigable canal systems excessive submerged weed growth can obstruct boat movement. Most submerged weed problems are found in low land often eutrophic and oligotrophic waters.

Oxygen Depletion and Eutrofication

Eutrophication can be visualised as the result of a system subjected to stress caused by excess input of nutrients. The eutrophic systems represent the enhanced primary producers growth, following the increase in the concentration of phosphorus (P) and nitrogen (N), which control the growth. On the other hand there are processes in the eutrophic system directed to remove these elements : P to the sediment and N to the atmosphere. If aquatic plants are to be used to purify the wastewater, it will be necessary to establish the limits of tolerance of these plants for different concentration of oxygen, organic matter, nutrients and turbidity (Mitchell, 1978).

Caffrey (1990) proved that *Potamogeton pectinatus* is a more compitative taxon than *Ranunculus penicillatus*. It is likely that if there is a situation of increased nutrient supply without the stress caused by competition from filamentous algae, the Potamogeton would forage nutrients more efficiently than the Ranunculus, show a greater plasticity in its growth response and so out compete the Ranunculus. Indeed there is evidence (ibid, 1990) to suggest that this has already happened in some organically polluted rivers in the British Isles.

Conclusions and Recommendations

Certain studies and concepts are particularly relevant to Indian condition. Much of the recent work that has been carried out to explore the potential use of aquatic plants for the treatment of wastewaters in other parts of the world has concentrated on weed species such as water hyacinth and alligator weed. Investigation should also concentrate on some species which exhibit rapid growth rates and are tolerant to wide ranges of environmental conditions. *Eleocharis sphacelata* is a particularly high phosphorus uptaking plant under natural conditions. More work is required to establish its maximum levels of uptake under hyper eutrophic conditions. One more aspect of the use of aquatic plants for the treatment of wastewaters which require investigation is the relationship between swamp systems and human disease organisms. Native plants should be evaluated to establish their capacity to exude the antibacterial exudates and minimise the breeding of mosquitoes and other vectors of human disease. Possible methods of harvesting and potential uses for harvested material also have to be studied.

References and Further Reading

Adema, D.M.M., and de Zwart, D. 1984. On derzoek naar een toxiciteit van milieugevaarlijke stoffen. *Bijlage 2. R/VM/TNO, Rapportnr* CL 81 /00b (RIVM 668114003).

Beham, M.J., Kinraid, T.B. , and Selver, W.I. 1979. Lead accumulation in aquatic plants from metallic sources. *J. Wildl. Mgmt.* 43, 240-244.

Boyd, C.E. 1970. Production, mineral accumulation and pigment concentrations in *Typha latifolia* and *Scirpus americanus. Ecology* 51, 285-290.

Caffrey, J. 1990, Problems relating to the management of *Potamogeton Pectinatus* L. in Irish rivers. proc. EWRS 8th Symp. on aquatic weeds 8 : 61-68.

Clarke, J.R. , Vanhassel, J.N., Nicholson, R.B., Sherry, D.S., and Cairns, J., Jr. 1982. Accumulation and depuration of metals by duckweed (*Lemna perpussila*). *Ecotoxicol. Environ. Saf.* 5, 87.

Djingova, R., Kuleff, I., Penev, I., Sansoni, B. 1986. *Sci. Total Environ.* 50, 197-208.

Gloyna, E.F. 1972. Bassins de stabilisation des eaux usees. *O.M.S. Genova.* 187p.

Gutteridge, Haskins and Davey Pty Ltd. 1977. Planning for the use sewage. *Aust. Govt. Publ. Sery.* Canberra.

Hespanhol, I. 1990. Guidelines and integrated measures for public health protection in agricultural reuse systems. *J. wat. SRT Aqua*, 39, 237-249.

Hosetti, B.B. and Stanley Frost. 1995. A review of the sustainable values of effluents and sludges from wastewater stabilization ponds. *Ecological engineering* 5. 421-431.

Hutchinson, T.C., Fedorenko, a., Fitchko, J., Kuja, A., Van Loon, J., and lichwas, j. 1975. Movement and compartmentation of nickel and copper in an aquatic ecosystem. *Trace substances in environmental Health* (D.D. Hemphill, Ed.), Vol. 9, pp. 89-105. Univ. of Missouri Press, Columbia.

Jenner, H.A., and Jahnsen - Mommen, J.P.H. 1993. Duckweed *Lemna minor* as a tool for testing toxicity of coal residues and polluted sediments. *Arch. Environ. contam. toxicol.* 25, 3-11.

Kenaga, E.E., and Moolenar, R.J. 1979. Fish and Daphnia toxicity as surrogates for aquatic vascular plants and algae. *Environ. Sci. Technol.* 13, 1479-1480.

Mandi, L., darley, J., Barbe, J. and Balexu, B. 1993. Essais dipuration des eaux usees de Marrakech par la jacinthe d, eau (charges organique, bacterienne et parasitologique.) *Revue des Science de e, Eau*, 5 (3) 313-333.

McNabb, C.D., Tierney, D.P. & Kosek, S.R. 1970. The uptake of phosphorus by *Ceratophyllum demersum* from wastewater. Report of project A-031. *Mich, Inst. Wat. Res.* mich. State Univ. (Mimeo.)

Mitchell, D.S. 1978. The potential for wastewater treatment by aquatic plants in Australia. *Water* 5 (3).

Musil, C.F. & Bren, C.M. 1977. The application of growth kinetics to the control of *Eichhornia crassipes* (Mart.) Solms through nutrient removal by mechanical harvesting. *Hydrobiologia* 53: 165-171.

Naskar, M., Saha, S.K. and Saha, P.K. 1985. Compost of water hyacinth for agriculture and aquaculture *Environ and Ecol.* 3 pp. 249-253.

Pip, E. 1990. Cadmium, copper and lead in aquatic macrophytes in shoal lake (Manitobe, Ontorio). *Hydrobiologia* 208, 253-260.

Ronlet, M.G. 1975. Wasserlinse (*Lemna minor* L.) als Testplflanzen Zur pruefung Von Krschlamm. *Schweiz Landwirtsch. Forsch.* 14, 79-82.

Sawidish, T., Stratis, J., and Zachariadis, G. 1991. Distribution of heavy metals in sediments and aquatic plants of the river Pinios (Central Greece). *Sci. Total. Environ.* 102, 261-266.

Sawidis, T., Chettri, M.K., Zachariadis, G.A., and Stratis, J.A. 1995. Heavy metals in aquatic plants and sediments from water Systems in Macedonia, Greece. *Toxicol and Environ safety* 32, 73-80.

Srivastav, K. 1994. Accumulation of heavy metals from industrial effluents in some plants, *Him. J. Environ zool.* 6 pp. 164-166.

Taylor, G.J., and Crowder, A.A. 1981. Uptake and accumulation of heavy metals by *Typha latiofolia* in wetlands of the Sudburry, Ontario regions. *Can. j. Bot.* 61, 63-73.

Techobanoglosus, G. 1987. Aquatic plant systems for water treatment: Engineering considerations. In: Reddy, K.R. and Smith, W.H.(Eds.). *Aquatic plants for water treatment and resource recovery*. Magnolia Publishing Inc., Orlando, Florida, pp. 27-48.

Welsh, R.P.H., and Denny, P. 1980. The uptake of lead and copper by submerged aquatic macrophytes in two English lakes. *J. Ecol.* 68, 443-455.

Wong, P.T.S., Silverberg, B.A., Chau, Y.K., and Hodson, P.V. 1978. Lead and the aquatic biota. In the *biogeochemistry of lead in the Environment* (J.O. Nriagu, Ed.), pp. 279-343, Elsevier, Amsterdam.

CHAPTER 12

SURFACE AND GROUND WATER BUDGETS FOR DAKSHINA KANNADA

B.M. Ravindra, B.R. Manjunath and N.A. Harry

Introduction

The geomorphi- environment of Dakshina Kannada district in the southwestern part of Karnataka State extends from the coastal shores of the Arabian Sea upto the foothill of Western Ghats. The district is endowed with heavy rainfall amounting to average of 400 cm/year. Civic and industrial development activities are relatively pronounced in this part of Karnataka. Though the terrain is traversed by a number of west-flowing rivers which run in spate during monsoons but go dry during summer. Consequently, people increasingly are dependent on ground water resources and the ground water consumption is steadily on the rise. Apart from growing dependence of farmers on borewells and openwells for irrigational source, most of the rural and several urban water supply programmes and community irrigation projects are designed or based on exploitation of ground water resources.

Department of Mines and Geology (DMG), Government of Karnataka in collaboration with NABARD has carried out the census of irrigation wells (Karnataka) during 1993-94. It revealed about the over exploitation of ground water resources in several taluks of the Dakshina Kannada District. As per these estimations, the Bantwal and Sullia taluks are regarded as 'dark' blocks and Beltangadi taluk as 'grey' block in respect of the status of ground water development and exploitation. This alarming state of affairs calls for better restraint and increased conservation efforts on the part of people and the administration. An attempt is made in this paper to discuss the pros and cons of the problem and possible preventive measures.

Geologic Setting

The terrain under discussion is founded on a basement of Early Precambrian gneisses dating back to 3000 to 3400 million years. In the southeastern sector, covering parts of Puttur and Sullia taluks, minor Ancient Supracrustal rocks of Sargur type represented by high grade garnetiferous, kyanite sillimanite schists and allied schists occur. Younger Dharwar schists (about 2600 to 2800 million years) border the eastern boundaries of Kundapur, Karkal and Beltangadi taluks. The southern part of the district has undergone granulitization, possibly in the 2600-2500 million years period. Contemporaneously, gray granites have invaded in part of Udupi, Karkal and Kundapur taluks (Manjunatha and Harry, 1994). Precambrian tectonomorphic activity came to an end with the emplacement of several sets of dolerite dykes into the stabilized craton during the interval of 2100 to 1700 million years. Pervasive lateritization has affected the district in the Neogene period. The Western Ghats were uplifted around 20 million years ago (Miocene epoch), with the downghat coastal terrain floundered into a down thrown block.

Several sets of neotectonic fractures especially the coast parallel and transverse strike slip faults have affected the district, the development of these faults being connected with plate tectonic movements in the region (Ravindra and Krishna Rao, 1987).

GEOHYDROLOGIC SETTING

Geohydrological section of the area in general consists of laterites followed immediately by weathered gneisses and/or schist and further by fresh gneiss. The structural discontinuation in them decreasing with depth. The fractures joints and shear zones in the hard rocks from ground water repositories which get recharged annually through infiltrating water derived from the surface water and then rains.

The terrain receives copious rainfall in the order of 400 to 600 cm/year, distributed mainly in the months of June to September. A major part of the rainfall flow on the soil surface, down the gradient as 'run-off' making its way to the nearby streams and rivers and eventually merging with the Arabian Sea. A portion of the run-off infiltrates into the ground percolating through soil particles and pore spaces in the laterite and clay horizons. Laterite is a formation with high porosity and permeability. Thickness of the laterite capping in various parts of the district ranges from less than 10 m to more than 60m. It is normally underlain by a clayey horizon which has high porosity and low permeability. Consequently the laterite capping with the underlying clay bed together constitute an unconfined aquifer system that is characterized by rapid recharge and discharge characteristics. This can be designated as the 'upper aquifer'. Open wells can be constructed to tap this upper aquifer in laterite. Advantages of the upper lateritic aquifer is rapid water recharge connected with continuous spells of intensive rainfall. Similarly, the disadvantage lies in rapid declination of water levels in the summer. Obviously, a part of the ground water stored in the upper aquifer, flows out of the system as lateral flow and another part trickles down to formations underlying lateritic clay, into structural discontinuities and pore spaces in the hard rocks. Hard rocks encountered in the district are mostly grey gneisses, granites, granulites, quartzites or schists. The grey gneiss, schists and quartzites are mostly well deformed and are repeat with structural discontinuities like joints, cracks and fractures and consequently constitute better aquifers. The granite and ground water are relatively less deformed and are normally poor aquifers in this region. The aquifers in hard rocks underlying lateritic section can be designated as 'lower' or 'deep aquifer'. The bore wells in this region normally are sunk into the hard rock using casing pipes to block the laterite and clay horizons, hence the borewells essentially tap the lower or the deep aquifers. Casing pipes are inserted into hardrock so as to prevent friable particles of lateritic clay horizons from falling into the borewell.

Thus the ground water in general is an annually replenishing natural resource, the decline of recharge is mainly dependents on rainfall intensity as well as the percolation characteristics of the weathered mantle. The upper aquifer is recharged rapidly with rainfall, whereas the lower aquifer gets recharged variably in different areas depending upon the structural/topographic setup of the area.

Groundwater Development

The ground water in the terrain is extracted through openwells, borewells and combination of both (ie., dug cum borewells). Openwells in general vary in diameter ranging from 2 to 8 m. Normally, they are dug from 6 m upto 25 m depth. Hence they are essentially dug in the superficial geohydrological zone, consisting of soil + laterite / clay + weathered zone, depending upon the depth dug into. The thickness of soil horizon ranges from less than a meter to about three meters. The lateritic + clay horizon ranges from a meter to about 60 meters and rarely upto to 80 meters. In granite areas, the lateritic cap may be totally absent. In several areas, the reddish clay horizon may be present without any porous laterite, resulting in very poor recharge into wells. The degree of weathering in the underlying rocks also varies. In lateritized areas, fairly well weathered rocky sections are encountered. However, in areas with granitic outcrops, the degree of weathering and consequently the ground water recharge in the rocks may be very low. Good accumulation of ground water in openwells is dependent on the thickness of porous zone (soil + laterite + clay)and the degree of weathering and deformation in the underlying rock. Thus in areas with thick lateritic cap or with thick weathered zone, good accumulation of ground water has been noticed.

Borewells are currently drilled up to depths ranging from 40 to 100 m, rarely going upto 150 m. Several exploratory borewells were drilled in the range of 100-300m by the Central Ground Water Board in the district, the deepest being 303.3 m drilled at Belman, during the year 1987-88. Borewells are drilled into the hard rocks, after inserting casing pipes to cover the overlying soil plus lateritic clay zone, consisting of loose and friable particles. Thus overlying porous friable zone is completely blocked and consequently the borewell taps essentially the ground water accumulated in inter connecting fractures in rocks at various depth levels. Deeper borewells encounter more number of water bearing fracture zones / joints and generally are liable to produce more water on pumping.

Ground Water Assessment

The component of hydrological cycle entering into the weathered ground and forming 'ground water' reserves available for safe exploitation annually, can be estimated by several methods. (Subhashchandra, 1994). It may be estimated as percentage of total rainfall over a measured territory. Or it may be estimated based on the average annual fluctuation of ground water table in representative observation wells. The State Department of Mines and Geology (Ground Water Wing) in consultation with the Central Ground Water Board has estimated the talukwise annual ground water recharge for Dakshina Kannada and are given in Table 56.

Table 56 : Annual ground water recharge for Dakshina Kannada District

Taluk	Recharge (Ham) Gross	Groundwater Net
Bantwal	11595	9856
Beltangadi	15883	13500
Karkal	12945	11003
Kundapur	25244	21457
Mangalore	12263	10424
Puttur	15385	13077
Sullia	6897	5863
Udupi	24296	20652
D. K total	124508	105832

Source : Department of Mines and Geology, Karnataka 1995.

Gross recharge refers to total annual groundwater recharge and the net recharge refers to the ground water component available for annual utilization. It was taken as 85 per cent of the gross annual recharge (the Ham ie., Hectares meters) can be converted into Mcm (million cubic meter), the relationship being 1 Mcm = 100 Ham. According to the estimation, 1058.32 Mcm of ground water is available annually for use, in Dakshina Kannada district.

Ground Water Utilization

The annual ground water utilization in the district was estimated by considering the number of effective ground water structures (open wells, borewells and open wells with borewell). The Department of Mines and Geology with the help of village accountants, conducted census during 1993. Based on the survey, annual ground water consumption in different taluks were calculated (Table 57). For these calculations, unit annual ground water draft for a pumping borewell was taken as 1.7 Ham and 1.4 Ham and 0.90 Ham for pumping dug cum borewell and dugwell respectively. For other types of use 0.20 Ham was taken. The total annual ground the water draft computed by the DMG is given inTable 57.

Rable 57 : The total annual groundwater available in the district

Taluk	Number of wells					Annual Draft	
	BW	DCB	OW	Others	Total	Gross	Net*
Bantwal	1016	38	10768	—	11822	11471	8030
Beltangadi	357	—	11337	—	11694	10810	7567
Karkala	72	1	4769	67	4909	4428	3100
Kundapur	49	—	4250	—	4299	3908	2736
Mangalore	263	—	4105	—	4368	4142	2899
Puttur	1314	—	7853	—	9167	9302	6511
Sullia	325	—	7196	—	7521	7029	4920
Udupi	12	—	8281	—	8293	7473	5231
D.K. Total	3408	39	58559	67	62073	58563	40944

BW: Borewell, DCB: Dug cum borewell, OW Openwell.

*Net draft is taken as 70% of total draft in the above table.

The Danida-DEE team in their Environmental Master Plan Study (EMPS, 1994) of Dakshina Kannada district estimated annual consumption of ground water mainly based on data from Central Ground Water Board. There is slight variation between the annual consumption estimation of D.M.G. and EMPS. The figures are provided in Table 58 for comparison.

Table 58 : Comparison of data obtained by EMPS and DMG

Taluk	Net ground water resource (Mcm)	EMPS estimate		D.M.G. estimate	
		Mcm	% of net resource	(Mcm)	% of net resource
Bantwal	95.56	73	74	80.83	81.0
Beltangadi	135.00	62	46	75.67	56.0
Karkala	110.00	55	50	31.00	28.0
Kundapur	214.57	45	21	27.36	13.0
Mangalore	104.24	67	64	28.99	28.0
Puttur	130.77	49	37	65.11	50
Sullia	58.63	40	68	49.29	84.0
Udupi	206.52	80	39	52.31	25.0

Total estimated ground water (gw) consumption (in Mcm)

It can be seen that there is a marked variation in the estimated consumption figures arrived at by EMPS and DMG. The unit annual draft assessed by the DMG are rather on higher side of realistic figures for Dakshina Kannada. However, the figures comprehensively suggest that the people are increasingly dependent on ground water for their irrigational and domestic purpose. The relative degree of consumption is reaching an alarming proportion especially in Bantwal and Sullia taluks.

NABARD has formulated the concept of categorization of areas based on the degree of ground water consumption relative to the net available resources. These areas or blocks where in the annual consumption is less than 65 per cent of the net annual rechargeable reserves in the next five years is considered as the 'white' area. Areas where the annual ground water consumption in the fifth year is in the range of 65 to 85 per cent is declared as 'grey' and the area with consumption is excess of 85 per cent in the fifth year is declared as 'dark' (Karanth, 1994). Obviously this welcome concept of white - grey - dark blocks not only indicate the relative degree of ground water consumption and the colours, grey and dark realistically suggest the gloominess of the alarming situation.

Surface Water

Dakshina Kannada district has a number of west-flowing rivers which have highly contrasting seasonal patterns during the year, with full floods during monsoon and almost nil flow during summer. There is only one medium river - Netravati (drainage area = 2000 - 20,000sq. km) and more than ten minor rivers (drainage area = <2000 sq. km). Average annual flows for several rivers have been computed based on gauging station records. The following data are from a DANIDA report (1994-95).

Table 59 : Average annual discharge of DK rivers

River	Average Annual discharge (Mcm)
Netravati	7581
Kumaradhara	325
Gurupur	1276
Swarna	1470
Sita	916
Haladi	1999
Total	13567

Gauging records are not available for a number of rivers like Mulki, Pavange, Udyavara, Chakra, Kolluru, Yedamavinahole, Bindur and Shirur. On the whole, it can be estimated that average

annual flow from coastal rivers in the district would amount to
about 15000 Mcm or even more.

Discussion

Consider that large quantity of water 15,000 Mcm is being
discharged annually into the sea from Dakshina Kannada. This is
a huge sum of water, in view of the fact that only about 1000Mcm
of ground water is available annually for utilization. On the whole,
it suggests that the ratio of available proportion of surface and
ground water is about 15 : 1. However, according to EMPS estimation the ratio of consumption of surface water and ground water in
Dakshina Kannada is 1 : 4. This is an important aspect that should
be pondered over by the policy makers.

The main problem with surface water is that it is seasonal and
is available mainly during June through December period. Surface
water in many of these streams can be stored for prolonged
consumptive uses by constructing appropriate vented dams across
the streams/rivers. Currently, the civic water supply to the city of
Mangalore is from the water stored in vented dam built across the
Netravati river near Thumbe a place about 15 km east of Mangalore. Similarly, the Mangalore Refineries and Petrochemical
Limited have built another vented dam across the upstream of
Netravati river, near Sarpadi.

A large amount of water approximately 7000 Mcm or more
flows through Netravati and it should be utilized through proper
management. There are a number of urban water supply projects
utilising the resources of river water. This programme of utilization
of river water should be encouraged and be extended to all urban
areas. This would ensure proper utilization of surface water resources and beside it would reduce the burden on ground water
exploitation.

Construction of vented dams would have an additional benefit
such as increasing the recharge to the groundwater aquifers in the
proximity of the impounded structure, the degree of recharge being
dependent on the degree of weathering and structural discontinuities in the underlying rocks in the area. After the construction of
Thumbe vanted dam, marked ground water recharge has been
reported from a number of proximal areas in open wells as well
as borewells connected to the recharge zones by way of fractures.

Similarly, increase of water levels and shortage characteristics have been observed in many dugwells after the construction of vented dams in the district. Another issue of importance is that in zones where the upper aquifer is endowed with reasonably good ground water resources, it is advisable to exploit it rather than going for exploitation of deeper aquifer. This suggestion is based on the simple observation that the upper aquifer gets recharged with ground water more rapidly than the deeper aquifer because of the higher porosity and permeability of the former.

The deeper aquifer apparently, has been recharged over a long duration of geological time possibly during the last twelve to twenty million years, since the upliftment of Western Ghats and initiation of southwest monsoon over the Indian subcontinent. The annual rate of ground water recharge to the deeper aquifer is much slower compared to the rate of exploitation in many areas. This has been evidenced by the depletion of ground water yields in many borewell. Earlier borewell drilling technology, permitted drilling only upto 100-150 feet depth. These borewells then mostly yielded copious amount of water. Currently borewells are being drilled to the depth of 200 to 300 feet and occasionally upto 400 feet or more. It is found in many of the average yielding areas that the shallow joints in the depth range of 40-90 feet almost have gone dry or yield generally less than 100 gallons per hour (gph) unlike the earlier days. This observation led to the suggestion that due to increase of ground water development over the years the upper most joints in the hard rocks have been depleted in ground water yield as a result of general lowering of ground water recharge and storage levels in the lower aquifer. In the light of this discussion, it is suggested that the borewells be used judicially in this region and only in seasons where the ground water resource in the upper aquifer has diminished considerably.

Summary

A comparison of the annually available water resources in Dakshina Kannada suggests that ratio of surface water and ground water is about 15 : 1. The cumulative average annual flow of rivers in the district being approximately 15,000 Mcm—as against annually available net ground water recharge being about 1000 Mcm. However, as per EMPS estimates, the ratio of surface water and ground water usage in the district is 1 : 4. This aspect suggests that the current pattern of water utilization in the district is against the principles of nature and needs to be refined and restrained in view of the alarming tendencies of ground water over expliotation assessed for some taluks of the district. In view of this, the following suggestions are made:

1. More emphasis should be given for the planned utilization of annually available surface water resources for civic, industrial and irrigational necessities.

2. Properly designed vented dams can be built across surface water streams and rivers to harness the precious water resources that mostly find its way to the Arabian Sea. Incidentally, impounding structures like vented dams increase the profit of ground water recharge.

3. When utilizing ground water resources, primacy should be given to open wells to tap the rapidly recharging upper ground water aquifers. Large diameter wells (upto 8 m diameter) with deeper depths (10-15 m), constrained by peak summer deeper static water levels are advisable.

4. Drilling of borewells may be restricted to contingency areas where it is impractical to tap water from surface and ground water resources.

References

DANIDA-DEE 1994. Environmental masterplan study for Dakshin Kannada District. Dept. of Forest, Ecology and Environment, Karnataka State, India and Ministry of Foreign Affairs, Danida, Danmark.

DMG 1995. Ground water resource status of Karnataka (as on 31.12.94). Ground water studies status No. 286, Department of Mines and Geology Bangalore, 30p.

Karanth, K.R. 1994. Groundwater and Evaluation and development in Karnataka (cited in GeoKarnataka, Eds., Ravindra, B.M. and N. Ranganathan), Mysore Geological Department Centenary Volume, pp. 337-359.

Manjunatha, B.R. and Harry, N.A., 1994. Geology of Western Karnataka, (cited in GeoKarnataka, Eds., Ravindra, B.M. and N. Ranganathan), Mysore Geological Department Centenary Volume, pp. 109-116.

Ravindra, B.M. and Krishna Rao, B. 1987. Relation of coastal faults and river morphology to sea erosion in Dakshina Kannada, Karnataka. Journal of Geology Society of India, 29: 424-432.

Subhaschandra, K.C. 1994. Occurrence of ground water aquifer characteristics of hard rocks in Karnataka., (cited in GeoKarnataka, Eds., Ravindra, B.M. and N. Ranganathan), Mysore Geological Department Centenary Volume, pp. 337-359.

CHAPTER 13

APPLICATION OF REMOTE SENSING TECHNIQUES FOR EIA

B.R. Manjunatha and D. Venkat Reddy

Introduction

There is a growing concern about the consequence of population explosion, industrial growth, deforestation, sea level rise,
drought, flood, ground water depletion and environmental pollution. Environmental management is relatively one of the young
interdisciplinary subjects that concerns about the impact of anthropogenic activities on the natural environment. There is a conflict
between forest policy and industrialisation of backward areas. The
industrial policy offers incentives to develop the industrially backward areas, while the forest policy aims to conserve forests to
preserve biodiversity to reduce ecological imbalances. To resolve
such conflicts, Environmental Impact Assessment (EIA) is a prima
facie that provides potential range of alternatives to achieve the
desired goals (Khanna and Kondwar, 1992).

Environmental Impact Assessment (EIA) is an interdisciplinary decision making tool for the alternative routes for the development, process technologies and project sites (Hosetti, 1997).
Though the policy of EIA seems to be complicated because of many
potential alternatives to achieve sustainable development, this
may be resolved through the hierarchical approach and thus EIA
acts as a potential formula for ensuring ecological sustainable development. This requires systematic identification and management of cumulative trend of all environmental variables on a
regional basis (Khanna and Kondwar, 1992). The usual steps

involved for working out EIA are identification, prediction and evaluation.

1. Identification

It includes the description of existing environmental components, the project and the project site/area, and potential impacts are presented to identify impacts.

2. Prediction

It means that the impact of the project activities predicted through mathematical modelling and simulation for all the components, after careful calibration, verification of the model are for prediction.

3. Evaluation

It involves the conversion of impacts into environmental quality predictors with resource value to function graphs for evaluation, of various environmental pollution problems, aesthetics and human interests.

The most appropriate stage for implementing EIA could be at the district planning because at this level, there are a number of alternatives for the development, this would greatly reduces the complexities faced at the regional/national level.

Purpose and Method of Study

There are a number of techniques available for EIA that have been given in the earlier chapters. This article explains the application of remote sensing techniques for EIA. Space-based artificial orbiting satellites like IRS-1a and 1B, Landsat, NOAA, SPOT, etc., are useful in collecting information on changes in the earth's environment through remote sensors. Remote Sensing is a multidisciplinary branch of science of obtaining information of an object without actual contact, through electromagnetic radiation (Deekshatulu and Joseph, 1992).

The remote sensing system comprises of a platform which may be a satellite, aircraft, baloon, automobile or a ground based stand having electromagnetic sensors. Of the wide range of electromagnetic spectrum (10^{-10} m - 10^6 m wave length), the visible (0.4 - 0.7 microns), the reflected infrared (0.7 - 3 microns), the thermal infrared (3-5 microns and 8-14 microns) and the microwave (0.3 to 300 cm) regions are useful for remote sensing studies. However, for EIA assessment through remote sensing, the visible, near infrared and the middle infra red regions of the electromagnetic spectrum are useful. The information gathered by the sensors is manipulated and transported back to the earth's monitoring station through films and magnetic tapes.

The data so obtained can be reformatted and processed in the laboratory to produce either images/photographs or computer compatible magnetic tapes, further, interpreted visually/digitally to produce thematic maps for obtaining the earth's environments information (Deekshatulu and Joseph, 1992). The electromagnetic remote sensors that are mounted on a satellite platform assist to get not only a synoptic view of fairly a large area, but also provide multispectoral and multi temporal data which are useful for monitoring the earth's environmental changes and natural resources (Chandrasekhar, 1992). Earth's environmental changes are categorised (ibid, 1994) into (i) gradual processes and (ii) episodic events. The former includes vegetation dynamics, land degradation, urban growth, industrialisation, environmental pollution, sedimentation, coastal morphological changes, whereas the latter includes episodic events like cyclones, floods, landslides, earth quakes, volcanoes, forest fire, etc.

Application of Remote Sensing Techniques for EIA

Remote sensors help in data collection on various fields, such as agricultural crop coverage and yield estimation, drought warning and assessment, landuse and landcover mapping for agroclimate planning, wasteland management, water resources management, marine resources survey and management, urban development, mineral prospecting, forest resources survey and management, natural disaster assessment, impact of mining activities etc., (Rao, 1992). Because of these wide applications, both central and state governments have several plans for harnessing the remote sensing techniques in solving problems of relevance to our country. Some of the applications of remote sensing for EIA are explained in this chapter.

Desertification and Forest Management

Desertification is a phenomenon of the transformation of fertile land into desert which is a severe problem faced by many countries today. It is a slow process by which arable land is converted into sandy soil through many intermediate steps. It is a non-linear process that may be restricted to some areas or increased by a sequence of drought and dry seasons. The factors that increase desertification include, over grazing, deforestation, improper agriculture, water logging etc. A total of 3.5 billion hactares of the world's fertile land is prone to desert formation. The data generated through remote sensing techniques have helped in suggesting remedial measures to prevent to expansion of desert lands (Rao, 1992).

Among several environmental problems, desertification has received much attention, because it disturbs bio-diversity and · ecological balance. It is believed that about 85 % of the earth's surface was covered by forest at about 10,000 years ago and it came down to 33 per cent in 1980 (Bahuguna, 1986).

The hot desert regions of the northwestern India are Thar, Maru, Jangala, Bagar, Shekhawati, Thai, which are together called as Marudesa or Marusthali covers about 10% of the total area of our country. Space-based images are particularly useful for monitoring the spread of deserts. For instance, eastward shift of sand dunes in the Aravalli hill ranges through wind gaps has been identified on the basis of remote sensed data. Efforts are on to prevent desertification of the border areas. In this context, multitemporal remote sensing data are particularly useful in combating desertification in the northwestern regions of India.

Remote sensing plays an important role in the survey and monitoring of forests through maps generated by using the remotely sensed data. Although the present forests cover is only 7% of the earth's surface, it account for 50 to 80 % of the species existing today. Forest ecosystems are destroyed by the growing human population for agriculture, industry and settlements. The loss of forest is noticed in most of the countries, but it is more pronounced in developing countries like Brazil, East Africa, India, Indonesia, etc. It is estimated that about 100 species becoming extint everyday because of deforestation and predicted a loss of 40 % of the forest resources by 2000 A.D.

India has about 12 per cent of forest land during the survey conducted between 1951-52 and 1975-76 (Puri, 1983). Forest maps prepared by using remote sensing data indicate that the forest area has fallen from 14 to 11 per cent within a duration of 8 years since 1974 (Rao, 1990). Space-based remote sensing data are now being used for forest survey and deforestation. National Remote Sensing Agency (NRSA), Department of Space, Government of India have confirmed that the forest area has reduced from 16.89 per cent to 14.1 per cent of the total geographic area of our country. NRSA has prepared state-wise forest map to account for deforestation.

Flood Monitoring and Management

Flood monitoring and damage assessment can be precisely known by using data collected by the Synthetic Aperture Radar (SAR) of an European remote sensing satellite, as this technique can be applied even during the cloudy days. Remote Sensing techniques are useful in flood mapping and damage assessment within a week of the disaster. NRSA is monitoring the floods caused by the major rivers of India over a decade. Predictions regarding this is also available through news media like news papers, radio and television as well.

India is blessed with an average annual rainfall of about 119cm, which is ironically the largest anywhere in the world for a country of comparable size (Rao, 1975). However, India is cursed by the distinction among many tropical countries of being severely affected by floods and drought. Floods can cause a huge loss of life and property. According to the National Commission of Floods, the total-flood prone area in India is about 40-45 million hectares of which 32 million hectares area is protectable from disaster. Among the major, and medium rivers of peninsular and extra peninsular India eighteen rivers can cause floods during the peak periods of discharge which drain an area of 150 million hectares (Venkat Reddy, 1995). Floods are frequently occur due to shallowing of rivers caused by increased siltation. This may further reduce the water bearing capacity of reservoirs, which is a common problem noticed in most of dam sites. For instance, the Tungabhadra reservoir in Karnataka state has high siltation rate which caused loss 1.23 hectares of the reservoir's land.

Drought Management

Most of the countries are frequently suffering from drought. Statistical data indicate that severe drought may occurs once in every five years in most of the equatorial countries. Drought may not be restricted to certain regions, it may cause untold misery of humans and livestock. The spread of drought has been extensive in recent years, particularly in developing countries. Satellite based remote sensing data provide an accurate information about the extent and severity of agricultural drought conditions (Rao, 1992). For this purpose landsat images are most useful to know the influence of drought on the ecosystem. Drought affected areas in India are demarcated by using NDVI data collected from NOAA satellite under the National Drought Monitoring and Assessment project (Chandrasekhar, 1994). This is done by normalizing the difference of vegetation data to identify drought affected areas. Information will be published as "Drought Assessment Bulletin" and circulated to all district authorities, planners and agriculturists.

Drought is a major factor of uncertainty that continues to haunt agricultural production. It often results in economic and demo-graphic shift and disturbs the natural ecosystem. Though drought has been defined in many ways, generally it is related to shortage of water in an area over a long period. However, several factors control the extent and severity of deserts (Balakrishnan, 1986). They include variations in rainfall, number of days without rainfall, evaporation, temperature, humidity and dew point depression, soil moisture and ground water conditions, changes in surface water bodies, crop and vegetation conditions, etc. Severity of drought can be quantified through "Drought Intensity Index" by combining the above factors. Remote sensing by virtue of its synoptic and temporal attributes has the capability to provide the timely information about drought conditions. Remotely sensed data are used for detection and monitoring of agricultural drought under the programme of Large Area Crop Inventory Programme Experiment (LACIE).

Recently, an integrated land-water resources management scheme has been launched in India to provide short and long term solutions for soil conservation and water resources management towards mitigating drought.

Water Quality Assessment

Pollution of inland and coastal waters is a severe problem in densely populated areas. In order to understand the dispersal patterns of pollutants, monitoring of colour, structure, in lake, rivers, estuaries and coasts are essential, which can rapidly be measured by using remote sensing technique. For this purpose, data obtained through Coastal Zone Colour Scanner (CZCS), Landsat, IRS, NOAA are useful for the identification, density, distribution of pollutants, chlorophyll, plankton, oil spill and burning of oil wells. In this context, remotely sensed data are useful for rapid assessment of water pollution. Rapid assessment of water quality is needed because, eight per cent of the diseases in India are caused by water pollution. The W.H.O. reports that many waterborne diseases in India are epidemic and appear during flood and drought seasons.

The major rivers drain as large as 73 per cent of the total runoff in the country. In spite of several pollution controlling agencies like Central and State Pollution Control Boards, there is a legislation on water pollution control of rivers, estuaries and coastal ocean, but still pollution is increasing due to untreated sewage discharged into rivers (Chaturvedi, 1983). Among the 3119 towns of India, about two hundred have a partial/full sewage treatment facility. Still there are as many as 1700-2700 water polluting industries of which about 160 have wastewater treatment facility and the rest do not treat the effluents. Run-off from agricultural lands is rich in pesticides, components of chemical fertilizers which may cause eutrophication of rivers. Pollutants liberated from thermal power plants can greatly worsen the quality of natural waters.

Multiband photography by virtue of its ability to record differences in spectral reflectance provides a better tool for detecting sources of pollutants and their dispersal patterns in the environment (Meyer and Welch, 1975). Applications of remote sensors for water quality assessment are given in Table 60.

Satellite remote sensing is particularly useful for pollution monitoring of the large area, while aerial remote sensing technique with high resolution sensors is useful for detecting pollution in a smaller geographic area. Remote sensing studies are being carried out in India with regard to polluting industries in the region of major rivers as well as selecting distilleries to prevent ground water pollution.

Table 60 : Application of remote sensors for water quality assessment (after Bala-krishnan et al., 1986)

Water Quality Parameter	Sensors (wavelength) used
1. Salt water intrusion	thermal infrared, and colour infrared
2. Suspended solids	colour and thermal infrared
3. Biological contaminants	colour, colour infrared and thermal infrared
4. Oil spill	ultraviolet, thermal infrared and microwave
5. Waste effluents	colour infrared, thermal infrared and microwave
6. Pollutants from agriculture, mining, & land development activities	colour and colour infrared

Wasteland Management

Land degradation phenomenon is noticed all over the world. It happens due to excess of soil erosion as well as due to drought conditions. About 20 per cent of the land area of our country is considered as wasteland which is formed due to salinity of soil that caused by excess use of fertilizers, improper irrigation, use of slash and burn techniques to clear the forest land for agriculture and in spread of deserts. About 150 million hectares of land is affected by soil erosion by the activity of water and wind, and another 25 million hectares by gullies and ravines, shifting cultivation, increased salinity, alkalinity, water logging, etc. (National Commission of Agriculture, 1976). Of these, salt affected soil and water logged lands account for 6 -8 million hectares respectively. The percapita land available in our country has declined from 0.9 hectares in 1951 to 0.5 in 1980-81. Similarly, the percapita of cultivable land has reduced from 0.45 in 1951 to 0.25 ha in 1981, and likely to decline further to 0.15 hectares by 2000 AD (Narayanan, 1995).

Wasteland can results from inherent imposed disabilities such as location of environment, chemicals as well as physical properties of the soil. It is globally estimated that two billion hectares of land has been destroyed through soil erosion and another 952 million hectares of land through increased salt content. National Wasteland Development Board, Ministry of Environment and Forest,

Government of India have described the wasteland as degraded land which can be reclaimed by adopting remedial measures suggested through the remote sensing study. The National Wasteland Development Board has took up a project to map the wasteland of 145 critical districts. This has helped in identifying 13 categories of wasteland and almost half of land is for agriculture (Rao, 1992).

References

Bahuguna, S. 1986. Deforestation and its impact. Proc. of Symp. on *Indian Environmental Problems and Perspectives*. Geological Society of India, Mem 5, pp.167-173.

Balakrishnan, P. 1986. Issues in Water Resources Development and Management and the role of remote sensing. *ISRO-Tech. Rep.* 67-86.

Centre for Science and Environment 1982. The citizen's report - New Delhi.

Chandrasekhara, M.G. 1994. Remote Sensing in Environmental Impact Assessment - ISTE *Lecture Notes on Methodologies for Environmental Impact Assessment Management Plan KREC, Suratkal.*

Chaturvedi, R.S. 1983. *Natl. Symp. on Remote Sensing Development and Management of Water Resources.* Int. Photo. Inter., Dehradun, pp 267-276.

Deekshatulu, B.L. and Joseph, G. 1992. Science of Remote Sensing, *Current Sci.*, 61 (3 & 4), 129-135.

Hosetti, B.B. 1997. Environment Impact Assessment. Cited in *Concepts in Wild life Management*. Daya Publs., Delhi.

Khanna, P. and Kondawar, V.K. 1992. Application of Remote Sensing techniques for Environmental Impact Assessment, *Current Sci.*, 61 (3 & 4), 252-256.

Meyor, W. and Welch, R.I. 1975. Manual of Remote Sensing, *Water Resource Bulletin*. American Society, v. 21, Falls Church - Virgenia.

Narayanan, L.R.A., 1995. The impact of multipurpose river project - The Hindu, Kasturi & Sons, Madras.

Puri, G.S., 1983. *Forest Ecology, Photogeography and Forest Conservation*, Oxford I.B.H. Publishers, New Delhi.

Rao, K.L. 1975. *India's Water Wealth : Its Assessment. Uses and Projections.* Orient Longman Ltd., New Delhi.

Rao, U.R. 1990. Space technology for combating Environmental problems, particularly those of developing countries. ISRO- SP, 49-90.

Rao, U,.R. 1992. Remote Sensing for National Development, *Current Sci.*, 61 (3 & 4), 121-128.

Venkat Reddy, D. 1995. *Engineering Geology for Civil Engineers.* Oxford I.B.H. Publs. New Delhi.

CHAPTER-14

BASELINE BACKGROUND RADIATIONS: A PRIMAFACIE STUDY AS A PART OF EIA FOR KAIGA NUCLEAR POWER PLANT

H.M. Somashekarappa

1. Introduction

Studies on environmental management have acquired a great significance, today due to rapid industrialisation and urbanization. Industrialisation and the associated technological endeavours such as power generation using various sources have resulted in increased pollution of biosphere. With regard to safeguarding the environment, probably no other issue has drawn so much attention and raised so many controversies as the radioactive pollution. It is here, that the studies on baseline radiation background and distribution of radionuclides of natural and artificial origin in the environment play an important role.

The radiation exposure in excess of certain lower limits due to both natural and artificial radionuclides can produce unacceptable deleterious effects in man, it is necessary to keep a watch on the levels of radioactivity in the human environment.Therefore, in dealing with the problem of radiation pollution, the concern should be whether our technological endeavour in general and nuclear power production in particular have altered the total radiation level and consequent exposure to human beings in any significant degree. This requires a precise knowledge of the environmental radiation levels, their variations due to natural sources and a thorough understanding of environmental processes and their dynamics which modify radiation levels.

Although a large number of studies have been carried out to address this aspect of environmental radioactivity these studies have not covered the whole environment which, of course, is a stupendous task. The present work constitutes an effort to supplement such studies by providing precise information on background radiation level, natural and artificial radionuclide distribution, transportation and the uptake behaviour of radionuclides and accumulation in the plant tissues in the environment of Kaiga.

2. Sources of background radiation

All living beings have evolved in an environment of natural background radiation. Natural background radiation has two components: one originating from extraterrestrial sources such as cosmic rays and the other having a terrestrial origin i.e. radioactive nuclides that exist in the earth's crust. In most places on the earth, natural radiation levels vary only within relatively narrow limits. The average annual dose due to all natural sources in normal background area is 2.4 mSv (UNSCEAR, 1988). However, in some localities wide fluctuations from the normal level have been observed. Apart from the exceptional cases of wide variations, the natural radiation level may increase slightly due to scientific, industrial and technological activities.

2.1 Extraterrestrial sources

The extraterrestrial high energy particles which originate from the cosmos are known as primary cosmic rays. These rays impinge continuously on earth's atmosphere. These particles consist of 87% protons, 11% alpha particles and 1% nulcei of atomic number between 4 and 26 and about 1% of electrons of very high energy. The interaction of the primary cosmic radiation with the atmospheric nuclei of nitrogen, oxygen and argon produce electrons, gamma rays, neutrons, mesons and many other radionuclides (Table 61). At sea level, the mesons account for 80% of cosmic radiation while electrons account for about 20% ^{3}H, ^{7}Be, ^{14}C and ^{22}Na are other important cosmogenic radionuclides to which mankind is exposed. Among these, ^{3}H and ^{14}C are the isotopes of major elements found in the human body tissue.

The dose rate in the air due to cosmic rays varies a little with the latitude but significantly with altitude, doubling aproximately every once in 1,500 m. As a result, the passengers and crew of the high flying air crafts get an additional exposure due to cosmic rays. Dose rates in air are about 30 nGy h^{-1} at sea level for any latitude and increase to about 4 μGy h^{-1} at an altitude of 12 km.

2.2 Terrestrial sources

Terrestrial sources of radiation are primordial radionuclides, produced during the birth of the universe. The half lives of these radionuclides are sufficiently long as to be active even today. Secondary radionuclides are derived from the radioactive decay of primordial radionuclides. These radionuclides can be divided into those that occur singly (Table 62) and those that are components of the three radioactive series. These are Uranium (^{238}U) series (Table 63), Actinium (^{235}U) series and Thorium (^{232}Th) series (Table 64).

Among the singly occuring radionuclides ^{40}K is the most important one. It has a half-life of 1.3 x 10^9 years and decays by beta emission to ^{40}Ca followed by K-capture to an excited state of ^{40}Ar. ^{40}Ar returns to the ground state by gamma ray emission. The abundance of ^{40}K is 0.0118 per cent and its specific activity is about 29600 Bq kg^{-1} of potassium (Eisenbud, 1987). Since potassium is widely distributed in the environmental matrix, ^{40}K constitutes a major source of environmental radiation. However, potassium is an essential nutrient which is under close homeostatic control in the body. The average mass concentration for an adult male is about 2 g of potassium per kg of body weight. The isotopic ratio of ^{40}K is 1.18 x 10^{-4} and the average activity mass concentration of ^{40}K in the body is about 60 Bq kg^{-1}. The highest annual absorbed dose (270 μGy) is received in red bone marrow and the lowest in the thyroid (100 μGy). It delivers an annual effective dose equivalent to about 180 μSv (UNSCEAR,1988).

The families of radionuclides belonging to ^{238}U, ^{235}U and ^{232}Th series account for much of the background radiation in the environment. Among these, the ^{235}U series is less important since it makes up only 0.73 per cent by weight of natural uranium compared to 99.27 per cent for ^{238}U. In addition, ^{223}Ra and subsequent members of the series are relatively short-lived and do not appear in the environment in significant concentrations (UNSCEAR, 1988).

The ^{238}U is the head of a series of 14 principal nuclides. This series can be divided into five subseries; (i) $^{238}U \rightarrow {}^{234}U$; (ii) ^{230}Th; (iii) ^{226}Ra; (iv) $^{222}Rn \rightarrow {}^{214}Po$; and (v) $^{210}Pb \rightarrow {}^{210}Po$. In each subseries the activity of the precursor controls to a large degree the activities of the decay products. These subseries deliver an annual dose of 5 µSv, 7 µSv, 7 µSv, 850 µSv and 120 µSv respectively due to intake by inhalation and ingestion of nuclides into the body (NCRP-45, 1975; UNSCEAR, 1988). ^{238}U is the most abundant (99.27 per cent) isotope of natural uranium and is found in all type of rocks and soil with varied concentration. Further, because of its presence in soil and phosphate based fertilisers, ^{238}U finds its way into human tissue through food chain. The daily intake of uranium from all dietary source is about 0.011-0.018 Bq (5 Bq annually). Measured values of the activity mass concentration of ^{238}U in the bone of adults who have lived in areas with normal dietary levels lie in the range of 5-150 mBq kg^{-1} of ^{238}U in dry bone with an average concentration of 50 mBq kg^{-1} (UNSCEAR, 1988).

^{226}Ra, ^{222}Rn and ^{210}Pb (which head three subseries of ^{238}U series) and ^{210}Po which belongs to ^{210}Pb subseries are other important radio nuclides in ^{238}U series. The ubiquitous presence of these primordial radionuclides and their decay products in the environment and human bodies results in external and internal radiation doses. ^{226}Ra has a half-life of 1620 years and becomes ^{222}Rn through alpha decay. The chemical properties of radium are similar to the calcium and, therefore, enters the human body through the food chain easily and gets concentrated in bones. The biological half-life of ^{226}Ra is also quite a long (2.5 years). Radium has a metabolic behavior similar to that of calcium and an appreciable fraction is deposited in bone. More than 70 per cent of the radium in the body is contained in bone, the remaining fraction being distributed rather uniformly in soft tissues. The average annual dietary intake of ^{226}Ra in areas of normal radiation background is 15 Bq (UNSCEAR, 1988).

The ^{222}Rn is a radioactive inert gas with a half-life of 3.8 days. The radon gas exhaled from the earth's surface into the atmosphere is rapidly dispersed and diluted by vertical convection and turbulence. ^{222}Rn and its daughter products enter the human body mainly through inhalation. The average annual intake in normal background areas is about 2,00,000 Bq through inhalation and 300 Bq through ingestion (UNSCEAR, 1988). ^{222}Rn and its daughter products are estimated to contribute about three quarter of annual effective dose received by individuals from terrestrial source and about half of the dose from all natural sources put together (UNEP, 1985).

^{210}Pb occurs in nature in partial radioactive equilibrium with its immediate daughters ^{210}Bi and ^{210}Po. The complete radioactive equilibrium of ^{210}Pb- ^{210}Bi- ^{210}Po in the biosphere is usually not reached due to the difference in the influence of biological, meteorological, chemical and other factors on each of these nuclides (Jaworowski, 1967). ^{210}Pb has a half-life of 22 years and decays to ^{210}Bi by beta emission. ^{210}Bi is also a beta emitter with a half-life of 5 days and decays to ^{210}Po.

The ^{210}Po has a half-life of 138 days and is an alpha emitter (5.3 MeV). Consumption of food is usually the most important route by which ^{210}Pb and ^{210}Po enter the human body. The absorbed dose from the ^{210}Pb subseries depends mainly on the highly energetic alpha particles of ^{210}Po, as the contribution from the beta emissions of ^{210}Pb and ^{210}Bi amounts to just about 10 per cent of the total absorbed dose (UNSCEAR 1988). Therefore ^{210}Po causes considerably greater damage compared to ^{210}Pb which is a beta emitter. It is estimated that the equivalent dose resulting from a single distintegration of ^{210}Po is a thousand times greater than that of ^{210}Pb decay (Parfenov 1974). Therefore Morgan et al. (1964) have rightly included ^{210}Po in the group of the most toxic radioisotopes.

Another major terrestrial source of natural background radiation is decay series of thorium. ^{232}Th is the head of a series of 11 radionuclides and can be divided into three subseries: (i) ^{232}Th itself; (ii) ^{228}Ra$\rightarrow$ ^{224}Ra; and (iii) ^{220}Rn $\rightarrow$ ^{208}Pb. These subseries deliver an annual dose of 3 μSv, 13 μSv, and 160 μSv respectively due to intake by inhalation and ingestion (NCRP-45 1975, UNSCEAR, 1988). ^{232}Th has a half-life of 1.4x10^{10} years and decays to ^{228}Ra by alpha emission. Although, ^{232}Th is not as widely distributed as uranium in rocks and soils, certain rocks such as igneous are found to contain ^{232}Th four time that of ^{238}U (Faul, 1954). However, since the specific activity of ^{232}Th is 4.07 Bq kg^{-1} compared to 12.21 Bq kg^{-1} of ^{238}U, the radioactivity of the two nuclides is more or less the same. Wrenn et al. (1985) have reported the body content of ^{232}Th of about 80 mBq, of which 60 per cent is in the skeleton.

^{228}Ra which heads the subseries of ^{232}Th series is similar to ^{226}Ra in toxic behavior. ^{228}Ra has a half-life of 5.76 years and decays to ^{228}Ac by beta emission. ^{228}Ra is much more available to plants and animals than ^{232}Th. Therefore the activity concentrations of ^{228}Ra in humans are mostly due to the dietary intake of ^{228}Ra itself and not due to the decay of ^{232}Th. The annual activity intake arising from inhalation is estimated to be 0.01 Bq, while that from ingestion of food is considerably larger, about 15Bq in areas of normal radiation background. The estimated average activity mass concentrations in bone and tissues in humans is 50 mBq and 4 mBq respectively in areas of normal background radiation. Another nuclide of some interest in this series is ^{220}Rn which is also known as thoron. It has a half-life of 55 sec.

2.3 Technologically enhanced natural radiation

The modern scientific and technological practices contribute, though slightly, to the prevailing natural background radiation level in the environment, but the phosphate industries and coal-fired electric power stations are the major contributors. The coal, like most mineral materials found in nature, contains trace quantities of ^{40}K, ^{238}U, ^{232}Th and their decay products. By burning coal the activities of these naturally occurring radionuclides are dispersed to various parts of biosphere. The world annual production of coal was about 3.1×10^{12} kg in 1985 (UNSCEAR 1987). Coal is most commonly used for industrial purposes, power generation and space heating. The average activity concentrations in coal are 50 Bq kg^{-1} of ^{40}K and 20 Bq kg^{-1} each of ^{238}U and ^{232}Th respectively (UNSCEAR 1988). The average concentrations in escaping fly-ash are ^{265}Bq Ky^{-1} for ^{40}K, 200 Bq kg^{-1} for ^{238}U; 240 Bq kg^{-1} for ^{226}Ra; 930Bq kg^{-1} for ^{210}Pb; 1700 Bq kg^{-1} for ^{210}Po; 70 Bq kg^{-1} for ^{232}Th; 110 Bq kg^{-1} for ^{228}Th and 130 Bq kg^{-1} for ^{228}Ra (UNSCEAR, 1982).

Mining and processing of phosphate ores distribute ^{238}U and its decay products among various products, and wastes of the phosphate industry. Industrial effluents, the use of phosphate fertilisers in agriculture, the use of its by-products in the building industry are the possible sources of enhanced levels of natural radiation. In 1982 the estimated world production of phosphate rock was about 130 million tonnes. The consumption of phosphate fertilisers was about 30 million tonnes. The average consumption of phosphate fertiliser per unit area of agricultural land varied in 1982 from 3.6 kg per hectare in the developing countries to 10.9 kg per hectare in the developed countries, the world average being 6.7 kg per hectare (FAO, 1984). The typical concentration of ^{238}U in sedimentary phosphate is 1,500 Bq kg^{-1} (UNSCEAR, 1988).

Building materials containing higher concentration of ^{226}Ra may enhance the indoor exposures to radon and its decay products. The additional exposure due to cosmic rays incurred during the flights, vary according to the altitude and, to a smaller extent, to the latitude and to the solar activity. Consumer products such as radioluminous products, electronic and electrical devices, smoke detectors and ceramics, glasswares, alloys, etc. containing uranium or thorium also contribute to the technologically enhanced natural radiation levels (UNSCEAR, 1988).

2.4 Sources of artificial radionuclides

The use of radioisotopes in medicine, nuclear wepon tests and nuclear power reactors are the major sources of artificial radionuclides. Among the more than 200 different radionuclides used in these artificial sources, the long lived radionuclides ^{90}Sr (28.8 years) and ^{137}Cs (30.2 years) contribute significantly to the background radiation level in the environment.

The ^{90}Sr is chemically similar to calcium and therefore enters the human body following a path similar to that of calcium. It accmulates in the skeleton causing internal exposure. On the other hand, ^{137}Cs is chemically similar to potassium and enters the human body following the path of potassium. However, ^{137}Cs is tightly bound by soil and thus the uptake by plants from the soil is relatively less compared to that of ^{90}Sr. However the direct contamination is possible. In fact 60 per cent of the effective dose equivalent to the external radiation associated with past atmospheric nuclear weapon testing has been attributed to ^{137}Cs (Kevin et al., 1990).

A large number of nuclear tests have been conducted since 1945. Totally 423 atmospheric tests have been conducted in different parts of the world which have resulted in an estimated fission yield of 217 Mt. Intensive nuclear test programmes in the atmosphere took place during 1954-58 (128) and 1961-62 (128) resulting in a fission yield of 44 Mt and 102 Mt respectively. Underground nuclear explosions have been, and still are, being conducted but the resulting environmental contamination from these is relatively minor. The large-yield nuclear explosions carry radioactive debris into the stratosphere from where it is dispersed and deposited around the world through radioactive fallout. Although several hundred radionuclides are produced by nuclear explosions, only the radionuclides ^{131}I, ^{90}Sr, ^{137}Cs and ^{14}C contribute significantly to human exposure. The doses from ^{131}I are delivered in a matter of weeks, those from ^{90}Sr and ^{137}Cs are delivered for few decades, while from ^{14}C will be delivered over thousands of years. The collective dose due to all atmospheric nuclear explosions was estimated to be 3×10^{-7} man Sv. This is found to be equivalent to about three years of exposure to natural sources for the present population of the world on the basis of an annual per capita exposure to natural sources of 2.4 mSv (UNSCEAR, 1988).

At the end of 1992 the 423 nuclear reactors of 31 countries had 330 GW of installed generating capacity. This was responsible for some 17.7% of the world's electricity generated in 1992. Ten countries are generating one third or more of their total electricity using nuclear power while some 15 countries rely upon unclear power plants to supply at least one fifth of their total electricity needs (IAEA, 1993). Projections for world nuclear generating capacity for the year 2000 are in the range of 400-500 GW. The per capita dose calculated is about 3 nSv (GW a)$^{-1}$ (UNSCEAR, 1988). These figures show that nuclear power reactors in their normal operation contribute very little to the global radiation level. However, nuclear accidents release a spectra of long and short lived radionuclides into the environment.

In the operation of a nuclear reactor, utmost care is taken to prevent the release of ^{90}Sr, ^{137}Cs and other radionuclides into the environment. In spite of these precautions, however, a few reactor accidents took place (Sellafield, UK in October 1957, Oak Ridge Radiochemical Processing Plant, USA in November 1959, Three Mile Island, USA in March 1979 and Chernobyl, USSR, in April 1986). In some of these accidents ^{90}Sr and ^{137}Cs have been released into the environment. The reported effective dose equivalents in the first year for Chernobyl accident in different parts of the world range from 1.8 to 760 µSv for European countries, 260 µSv for USSR, 2.1 to 190 µSv for Asian countries, 1.4 for µSv for Canada and 1.5 µSv for USA.

In nuclear weapons testing and in nuclear reactors, a considerable amount of radioactivity is produced and, if adequate precautions are not taken, would be released into environment. The use of large numbers of radioisotopes in medicene, agriculture, and industry also contribute to environmental radioactivity. We are also aware that a number of redionuclides are produced, and ionising radiations emitted, during the production of electrical power in fossil-fuel based power plants. In view of these studies on background radiation levels and distribution of radionuclides in the environment have acquired a greater significance today than ever before. The preoperational surveillance on the radiation level of the region would help to assess the impact of major industries and technological activities in general and nuclear power reactors in particular on the environment and to optimise the human exposure to technologically enhanced as well as man-made radiations.

3. Kaiga nuclear power plant

Kaiga (Fig. 14.1) is a new addition to the map of nuclear power producing centres of India. Other centres are Tarapur, Rawatbhata, Kalpakkam, Narora and Kakrapar. It is planned to have 6 units of Pressurised Heavy Water Reactor (PHWR) type with a capacity of 235 MWe each. Out of six planned units the Government of India has given a sanction for two units of 235 MWe and are under construction. The environmental clearance was also obtained from the Department of Environment in July, 1985. Subsequently the financial sanction was also obtained on 30.7.1987 for Rs. 731 crores for the sanctioned two units of 235 MWe each.

Kaiga (14°51' 08 "N, 74° 26'40"E) is situated on the left bank of river Kali about 13 km upstream of Kadra Dam and about 56 km east of Karwar. The site is surrounded by the hill ranges of Western Ghats with varying elevations from 300 m to 500 m. The ground level at the site varies between +35 m to +50 m with average ground level of +40 m. The Nagjari hydro-power station of 810 MWe (6 X 135 MWe) capacity is situated at a radial distance of 45 km and a paper mill at Dandeli situated about 47 km from the site. Both are on the upstream side of river Kali.

The population densities around the plant based on 1981 census within 10 km and 5 km around the site are 9 persons/Sq. km. and 8 persons/Sq. Km. respectively. The total population within 10 km radius is about 2800. The population within 2.3 km radius is 395 and within 5 km radius is 681. Out of the population of 395 residing inside the exclusion zone of 2.3 km radius about 120 persons are affected by the submergence of Kadra reservoir, constructed for hydroelectric power generation. About 85 per cent of area within 10 km comprises forest land, 9 per cent falls within water spread of

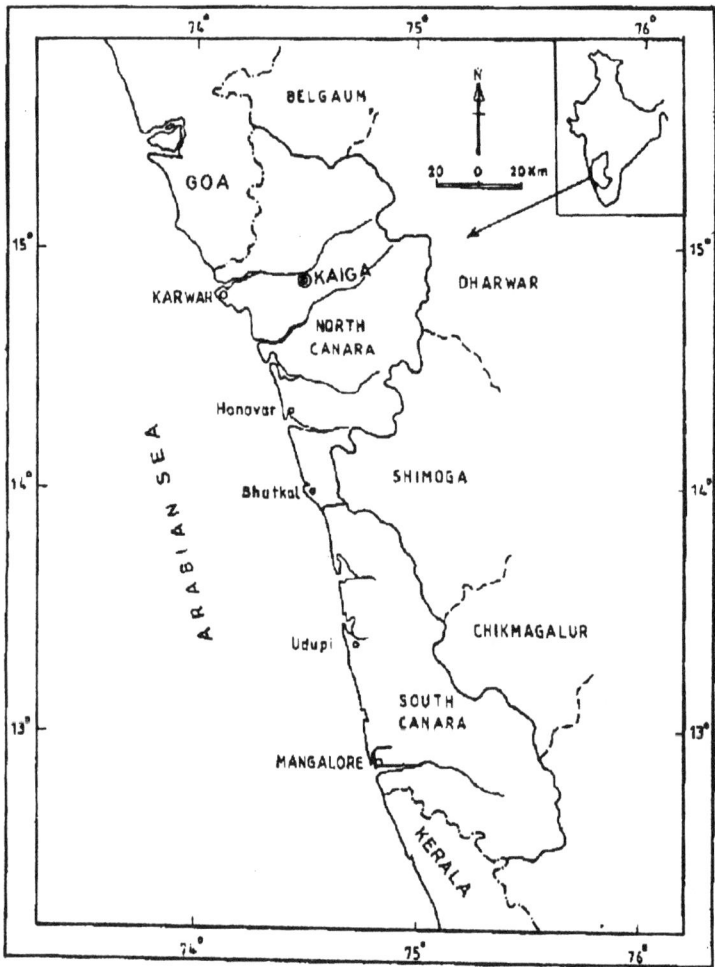

Fig. 14.1 : Location map of Kaiga area

the Kadra reservoir, 4 per cent is agricultural and 2 per cent is barren. A plain land measuring 2 km × 1 km is available for locating plant upto 2000 MWe potential. The plant site area required 2 × 235 MWe units with once through system of cooling is about 680 m × 400 m. The site area with spread area of about 25 per cent and rest is agricultural/barren land. Details of land required for the project is as follows:

Area required for plants (forest)	:	120 Ha.
Forest land in exclusion zone	:	545 Ha.
Government land in excusion zone	:	5Ha.
Private land in exclusion zone	:	166 Ha.
Total land under exclusion zone excluding submergence	:	836 Ha.
Forest land used as corridor for routing 400 KV and 220 KV transmission lines from Kaiga to Sirsi	:	612 Ha.
Total forest land to be deforested	:	732 Ha.
Total forest land which will be inside the exclusion zone but remain with the Department of Forest	:	545 Ha.

There are 85 families affected among the private land holders due to the acquisition of land for the plant site. These families were paid compensation as per the State government's order. Nuclear Power Corporation (NPC) provided employment to one member each to the affected families. The NPC also paid rehabilitation grant of Rs. 38 lakhs as per State Government's order and agreed to develop infrastructural facilities on the piece of land identified by the State Government for rehabilitating the 85 families. 48 families have lost their land for the Township. NPC paid them compensation and also provided job for one member in each family.

The total forest area to be cleared for the Kaiga project for the plant capacity of 2000 MWe is 120 Ha. This amounts to 0.06 Ha. per MWe of installed capacity (for the hydal project, on the average, the same has been estimated to be 18.6 Ha per MWe installed capacity). The forest area cleared till now is 27.5 Ha as against the estimated area of 40 Ha for the first 2 units.

Both the reactors are pressurised heavy water type. In this type of reactors both the moderator and coolant is the heavy water. Reactor-boiler comprises two heavy water circuits, reactor shields with cooling circuits, control mechanism and an enclosure with a recirculated conditioned atmosphere. One of the heavy water circuits is the high pressure, high temperature primary coolant circuit, which is used to extract heat from the reactor fuel bundles loaded in the calendria and to generate steam in the boiler. The other heavy water circuit is the low pressure, low temperature moderator circuit. The arrangement of the primary circuit is such that there is a circulation of coolant by thermal convation. This is sufficient to cool the reactor in shut down conditions except for the initial 50 seconds following the shut down which is taken care by the fly wheels provided on the pumps through generating the sufficient power to circulate the coolant. The calendria is surrounded by water in a vault lined with carbon steel. This water acts as a shield and virtually eliminates the argon activity. The power of the reactor is regulated by a regulating system consisting of 4 absorbers and 4 shim elements in 4 locations. These absorbers are used to maintain criticality at the required power level.

The fuel which will be used in these reactors is natural uranium dioxide. There are 8 fuel bundles in each coolant channel and totally there are 306 channels in each reactor. The average residence time is 2 years and maximum residence time of fuel is 4 years. The fuelling machines are operated by remote control from the plant control room. The removal and insertion of fuel bundle is done when reactor is in operation. Two different types of fast acting shut down systems have been provided which are entirely independent of each other to shut down the reactor from any operating condition. Primary shut down system consists of mechanical shut off rods. Secondary shut down system consists of liquid poison tubes which are normally maintained empty and are filled with borated heavy water when shut down is called for. This system also has a similar capability as the primary shut down system. The entire system is placed inside the double containment primary and secondary containment building. The primary containment is made up of prestressed concrete cylindrical wall of 33.64 m radius and of 610 mm thick with prestressed segmental dome at top. The secondary containment is made up of reinforced cement concrete cylindrical wall of radius 39.60 m and of 1610 mm thick with a segmental dome at the top.

The source of water is the Kadra reservoir of Kali river. State Government has assured a minimum out-flows of 185 cusecs from the Kadra reservoir. The water taken from Kali river is primarily used for power generation. No link up with irrigation or any other major use exists. The normal out-flows is 100 cusecs. The temperature difference of water between inlet and outlet of the condenser will be around 8 to 10° C (Rao, 1990).

4. Objectives

This study was undertaken with the objectives: (i) to establish a reliable baseline data of background radiation level and concentration of prominent natural and artificial radionuclides in the environment of Kaiga for future impact assessment for the operation of nuclear power plants, (ii) to study the origin, distribution and transportation of radionuclides in the region, (iii) to study the biological up-take of radionuclides by plants in the vicinity and (iv) to study the external and internal exposure to human beings from natural and artificial radionuclides through various means.

Well established nuclear techniques were employed to carry out these investigations. Radiochemical methods were followed to separate the individual radionuclides wherever necessary. A sensitive plastic scintillometer was used to measure the ambient gamma radiation level in the environment. The low background alpha, beta and gamma counting systems were used for counting the samples. The details of radiochemical methods and detectors/counting systems used are given in the following paragraphs.

The ambient gamma radiation level survey was conducted using a high sensitive plastic scintillometer. The activity concentration of some important radionuclides of both natural and artificial origin such as Ra-226, Pb-210 and Po-210 of U-238 series, and U-238 itself, Ra-228 of Th-232 series, and Th-232 itself, K-40, Cs-137 and Sr-90 were measured in natural samples of soil, vegetation, vegetables, fish, fruits and diet samples of Kaiga environment by carefully processing the samples and counting the activity.

5. Materials and Methods

The measurement of low levels of radionuclides in environmental and biological materials often depends on separation of the nuclide of interest from a bulky matrix containing interfering radio-nuclides. It is of course possible to determine some radionuclides directly or by methods requiring the minimum of sample preparation. Examples are the determination of gamma emitting nuclides by high resolution gamma ray spectrometry and the determination of alpha emitting nuclides by direct alpha spectrometry of large area sources. However, where the radionuclides of interest is present at a very low concentration or emits no penetrating radiation, radiochemical analysis is usually required. The majority of radiochemical methods can be divided into five basic stages: (i) Sample preparation, (ii) Preconcentration, (iii) Chemical separation, (iv) Source preparation and (v) Counting.

Sample preparation, as the name implies, includes all treatments or additions to the sample before the preconcentration stage begins. It may consist of drying, grinding and homogenising a solid sample, the filtration and acidification of an aqueous sample, the addition of radiotracers and/or stable element carriers or other similar steps.

Preconcentration is a common preliminary step of chemical separation. Its object is to reduce the bulk of the sample before separative chemistry begins, although it may also achieve a certain amount of chemical separation. Perhaps the most common preconcentration steps are the evaporation of aqueous samples to reduce their volume and the ashing of solid samples to remove organic material. However, other procedures frequently used are the acid leaching of a solid sample followed by evaporation of the leachate, and bulk co-precipitation steps.

Chemical separation involves the removal of the radionuclide of interest from both the stable elements and interfering natural or artificial radioelements in the matrix. At this stage the sample is usually in solution and in a suitable volume for processing by a wide variety of chemical procedures. Those most commonly used are cation and anion exchange, solvent extraction, extraction chromatography, coprecipitation and distillation. These steps may be used singly or in various combinations to obtain the required degree of purification.

The counting stage in a procedure may also be part of the separation process. Because of background and counting efficiency considerations, it is generally true of the various techniques that the lower limits of detection are obtained (Lakins, 1984). High resolution gamma spectrometry, alpha spectrometry, low background alpha counting system and low background gas flow beta counting systems are some of the counting systems used for counting and estimating the activity/concentration of radionuclides.

5.1 Sample collection and processing

The environmental samples such as soil, sand, sediments, land and sea food items and vegetation were collected from selected stations up to 32 km of radius from Kaiga plant site (Fig 14.2). The vegetable, fish, vegetation and grass samples were charred under flame and converted into white ash in a muffle furnance at 450°C temperature. Soil, sand and sediment samples were ground, mixed uniformally seived and dried in an oven at 110° C temperature. These samples were then taken for further processing and analysis of individual radionuclide activity.

5.2 Chemical separation and counting methods

The activities of some important natural and artificial radionuclides were estimated by following standard radiochemical methods and employing well established nuclear techniques (Iyengar et al., 1990). The method adopted are explained below.

* Ra-226 by emanometry method, and followed by alpha counting employing scintillation cell
* Ra-228 by allowing equilibrium growth of its daughter Ac-228 ($T_{1/2}$ 6.1 hr) and separation of the latter on LaF_3 and counting in a low beta counting system
* Pb-210 by electro-chemical deposition on a silver planchet and subsequent alpha counting
* Pb-210 by allowing equilibrium growth of Bi-210 ($T_{1/2}$ 5.01 hr) and separation on $BiPO_4$ and counting in a low beta counter
* Cs-134+137 and Sr-90 by fusion mixture method followed by beta counting

5.3 Alpha and beta counting systems

An alpha counting system with ZnS (Ag) detector which is having a background of 0.3 cpm and an efficiency of 30 per cent was used for alpha particle counting of chemically separated samples.

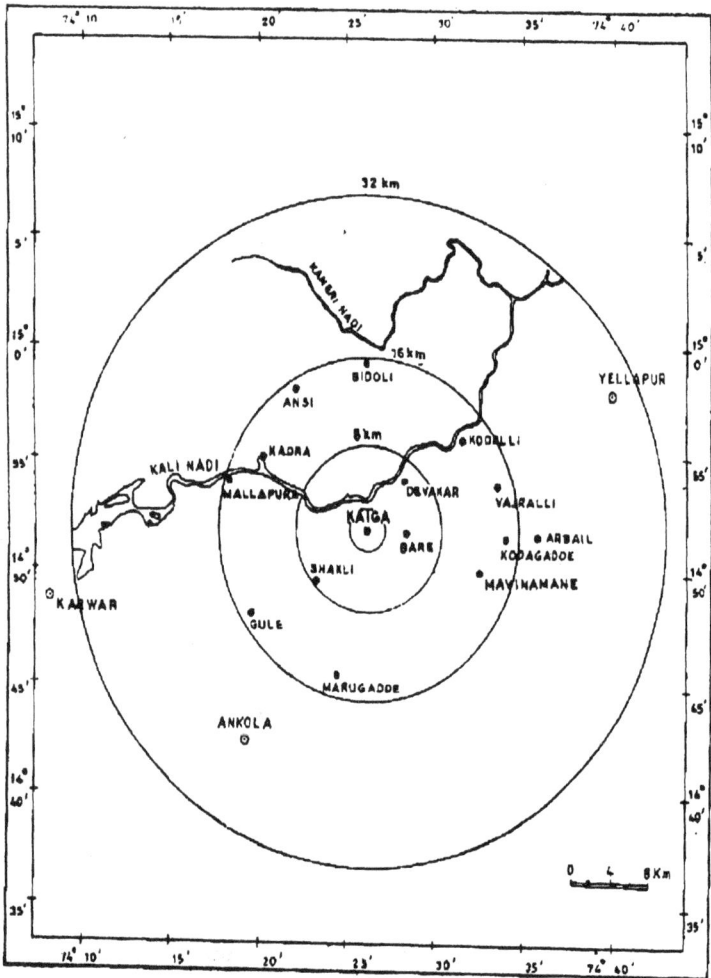

Fig. 14.2 : Area covered under investigation and sampling stations at the Kaiga environment

The beta counting system with an argon gas flow type beta detector which is having a background of 1-2 cpm with an efficiency 40 per cent was used for beta particle counting of chemically separated samples.

5.4 Gamma spectrometer

Gamma spectrometry offers a convenient, direct and non destructive method for the measurement of the activity of different radionuclides in environmental samples from their characteristic gamma line. The P C based gamma spectrometer which consists of a HPGe detector of volume 90 cc with a relative efficiency of 18 per cent and a 4K Multi-Channel Analyser was employed in this study. The gamma spectrometer with a high efficiency well type NaI (TI) detector of 10 cm X 10 cm with a well dimension 2.5 cm in diameter and 7.5 cm in depth was also employed for counting the samples of very low activity.

6.   Results and Discussion

6.1 Gamma absorbed dose in air

The results of ambient gamma radiation level survey were used to estimate the gamma absorbed dose rates in air in the environment of Kaiga. The estimated absorbed dose rates in air were ranged from 35 to 87 nGy h^{-1} with a mean value of 59 nGy h^{-1}. The results of gamma absorbed dose in air reported for 23 countries representing about one half of the world population were reviewed (UNSCEAR, 1988) and the values range from 24-85 nGy h^{-1} with a mean value of 55 nGy h^{-1}. The results of the present work for Kaiga environment compare quite well with these world average values.

6.2 Radioactivity in Soil

6.2.1 Uranium-238, Thorium-232 and Potassium-40

The activities of ^{238}U, ^{232}Th and ^{40}K in soils samples collected from different sampling stations around Kaiga (see Fig. 14.2) were estimated from gamma spectrometry method employing a HPGe spectrometer. The results obtained show that the ^{238}U activity varies from BDL (Below Detectable Level) to 30.8 Bq kg^{-1} with a mean of

9.3 Bq kg^{-1}, ^{232}Th activity from BDL to 50.4 Bq kg^{-1} with a mean of 17.4 Bq kg^{-1} and ^{40}K activity from BDL to 591 bBq kg^{-1} with a mean of 150.6 Bq kg^{-1}. Mishra and Sadasivan (1971) have carried out the similar study throughout the country, they reported all India average values for ^{238}U and ^{232}Th as 14.82 Bq kg^{-1} and 18.31 Bq kg^{-1} respectively. And the reported world range and average values are 10-50 Bq kg^{-1} and 25 Bq kg^{-1} for ^{238}U, 7-50 Bq kg^{-1} and 25 Bq kg^{-1} for ^{232}Th and 100-700 Bq kg^{-1} and 370 Bq kg^{-1} for ^{40}K respectively (UNSCEAR, 1982).

6.2.2 Other important radionuclides in soil

The results of activities of ^{226}Ra, ^{210}Pb, ^{210}Po, are important radionuclides of ^{238}U series and ^{228}Ra of ^{232}Th series estimated by employing suitable chemical methods and nuclear techniques. The results obtained show that the activity of ^{226}Ra varies from 3.18-21. 58 Bq kg^{-1} with a mean of 13.26 Bq kg^{-1}, ^{228}Ra varies from 5.16-12.86 Bq kg^{-1} with a mean of 12.04 Bq kg^{-1}, ^{210}Po varies from 38.15-158.95 Bq kg^{-1} with a mean of 81.46 Bq kg^{-1} and ^{210}Po from 3.69-112.75. Bq kg^{-1} with a mean of 29.22 Bq kg^{-1}.

The reported ^{226}Ra activity for different parts of India by Mishra and Sadasivan (1971) ranged from 2.59-26.3 Bq kg^{-1} and Russel and Smith (1996) have reported for different countries which ranged from 2.96-140.6 Bq kg^{-1}. Radhakrishna (1993) has reported the ^{210}Pb activity which ranges from 3.6 to 45.2 Bq kg^{-1} for the soils of Mangalore region, Jasorowski and Grzybowska (1977, 1986) have reported for different parts of Poland which ranges from 14.06 to 45.88 Bq kg^{-1}. Similar studies conducted by Schuttelkopf and Kiefer (1981) in an uranium area Black Forest of Germany ranges from 22.2 to 122.1 Bq kg^{-1}. The reported ^{210}Po activity in soils are ranging from 1.3 to 13.7 Bq kg^{-1}, for Mangalore region (Radhakrishna, 1993), 8.14 to 128.4 Bq kg^{-1} for different parts of US (AEC, US Spl. Rep., 1980), 27-47 Bq kg^{-1} for Brazil (Santos et al., 1990), 33.3-207.2 Bq kg^{-1} for Black Forest, uranium mining area, Germany (Schuttelkopf and Kiefer, 1981) and the world range varies from 8.14-219 Bq kg^{-1} Parfenov (1974). The reported activity of ^{228}Ra for Kalpakkam, Tamil Nadu is 28.86 Bq kg^{-1} (Iyengar et al., 1978), 33.33 Bq kg^{-1} for New York (Linsalata et al., 1989) and it ranges from 3.8 to 16.9 Bq kg^{-1} in Mangalore region (Radhakrishna, 1993).

Even though the activity concentration of ^{226}Ra in the soils of

Kaiga environ is normal and comparable with the other environments of normal background regions but the concentrations of ^{210}Pb and ^{210}Po were relatively higher and not in equilibrium with their parent nuclide ^{226}Ra. Relatively higher concentrations of ^{210}Pb and ^{210}Po observed for Kaiga environ may be traced to the tropical weather and humic soil of the region. The climate of the Kaiga region is tropical with a high precipitation of about 3000 mm per annum during the southwest monsoon (June - mid Nov.). Winters (Nov. - Feb.) and summers (March-May) are mild, and humidity is high (70%) for 9 to 10 months of the year. The natural vegetation of the region is of a moist deciduous and tropical evergreen type and soils are pedalfer in nature consequently, the soil of the region is humic, clay and organic in nature. The ^{210}Pb is very effectively sequestered by organic matter in the top layer of the soil (Durrance 1986) and Lewis (1977) reported the strong positive correlation of ^{210}Pb and organic matter in soil profiles. Berger et al . (1965) showed that in humic soil the ^{210}Po content is approximately three times higher than in mineral soils. Durrance (1986) have also pointed out that the concentrations of ^{210}Pb and ^{210}Po will be greater in clayey soils than sandy soils. The results of the present work support these earlier findings.

6.2.3 Cesium-137 and Strontium-90 in soil

The activities of important, artificially produced radionuclides such as ^{137}Cs and ^{90}Sr were estimated in the soils of Kaiga environment. The results obtained show that the activity of^{137}Cs varies from BDL to 12.04 Bq kg^{-1}with a mean of 2.51 Bq kg^{-1} and ^{90}Sr varies from BDL to 3.12 Bq kg^{-1} with a mean value of 0.94 Bq kg^{-1}. It is clear from the results obtained that the activity of ^{137}Cs is comparatively higher than the activity of^{90}Sr and this can be attributed to high capacity of adsorption of ^{137}Cs to the clay particles and to organic matter of the soil (Durrance 1986). When compared to the activity of ^{137}Cs the activity of^{90}Sr is lower. This must be due to the fact that the mobility of ^{90}Sr is comparatively higher than the mobility of ^{137}Cs (Menzel, 1965).

The activity of ^{137}Cs reported in literature ranged from BDL to 14.43 Bq kg^{-1} for Kalpakkam (Iyengar et al., 1978), 5.2 to 9.5 Bq kg^{-1} for Bombay (Shukla et al. 1987), 1.74 to 8.29 Bq kg^{-1} for Tarapur (Bhat et al., 1970), and it was < 1.30 Bq kg^{-1} for Kakrapar (Vashi et al., 1992). The reported values of ^{90}Sr ranges from BDL to 6.67 Bq kg^{-1} for

Kalpakkam (Iyengar *et al.* 1979), and from 0.89 to 5.12 Bq kg⁻¹ for
Tarapur (Bhat *et al.*, 1970).

6.2.4 Radioactivity in food items, grass and in vegetation

The locally grown food items to be consumed by the people and
grasse samples of Kaiga area were analysed to understand the biological uptake of radionuclides from the soil and their contribution
to radiation dose due to ingestion to the human population of the
region. Grass, vegetation, rice, jack fruit, banana, fish, composite
meal and vegetables such as brinjal, cucumber, spinach and ladies
finger were collected from different sampling stations (see Fig.
14.2) and analysed for different radionuclides. The mean values
calculated from the results obtained are presented in Table 65. The
data revealed that relatively higher activity of ^{210}Pb and ^{210}Po in the
soils of the Kaiga resulted in relatively higher biological uptake
except in fruits. Consequently relatively higher intake of ^{210}Pb
through the composite meal can be observed (Table 65). The
effective dose equivalents for different radionuclides are found to
be 10.36 μSv y⁻¹ for ^{226}Ra, 165.26 μSv y⁻¹ for ^{210}Pb, 20.23 μSv y⁻¹ for
^{210}Po, 30.83 μSv y⁻¹ for ^{228}Ra, 1.12 μSv y⁻¹ for ^{137}Cs and 1.61 μSv y⁻¹ for
^{90}Sr respectively. Total effective dose equivalent due to dietary
intake was 230.41 μSv y⁻¹. It is evident from the above results that
the major contribution to the total dose is from natural radionuclide
^{210}Pb (72.0%) which was followed by ^{228}Ra (13.4%), ^{210}Po (8.8%), ^{226}Ra
(4.5%), ^{90}Sr (0.7%) and ^{137}Cs (0.5%).

7.   Conclusions

Based on these studies in the environment of Kaiga, the
following observations are made.

1.   The absorbed dose rates of Kaiga environment are compa-
     rable to the reported dose rates of other Indian environ-
     ments of normal background areas. The dose rates also
     compare quite well with world average values.

2.   The concentrations of ^{210}Pb and ^{210}Po in soils of Kaiga envi-
     ronment were higher than the concentrations of ^{226}Ra and
     ^{228}Ra. Their concentration was also higher as compared to
     the values reported for other normal background areas.

3. The ^{226}Ra activity in the soils of Kaiga environment was comparable with values reported for Bombay and all India average and lower compared to all other values reported for other environments. The ^{228}Ra concentration is comparable with the value reported for Mangalore environment and it was low as compared to other environments. However, the distribution of ^{226}Ra and ^{228}Ra are almost uniform in the soils of Kaiga environment.

4. The intake of radionuclides through food by the people of Kaiga region was comparable with their counterparts in other places of normal background areas of the country and abroad.

5. Among the radionuclides investigated, the dietary intake of ^{210}Pb and the corresponding annual effective dose is found to be maximum (excluding ^{40}K) and it was followed by ^{228}Ra. The intake of ^{137}Cs and ^{90}Sr and corresponding effective doses were quite low.

From the interpretation of the above findings the following inferences are drawn:

1. The absorbed gamma dose rates in air in the environment of a Kaiga are normal and comparable with world average values.

2. The higher concentration of ^{210}Pb and ^{210}Po in the soils of Kaiga environment may be attributed to the presence of tropical forest, heavy precipitation and higher emanation rate and atmospheric concentrations of ^{222}Rn.

3. The concentrations of important radionuclides in biological matrices of Kaiga environment follow a pattern similar to that of other environments.

4. The relatively higher intake of ^{210}Pb and the higher annual effective dose due to it compared to other radionuclides of Kaiga environment can be attributed to its higher concentration in soils and vegetable samples.

Table 61 : Singly occurring natural radionuclides produced by cosmic rays

Radionuclide	Half Life	Principal Mode of decay and Energy (MeV)
^3H	12 years	β-0.0186
^7Be	53 days	γ0.477
^{10}Be	1.6×10^6 years	β-0.555
^{14}C	5730 years	β-0.156
^{22}Na	2.6 years	β+0.545, γ1.28
^{24}Na	15.0 hr	β-1.4, γ1.37, 2.75
^{32}P	14 days	B-1.71
^{33}p	24 days	β-0.246
^{35}S	88 days	β-0.167
^{36}Cl	3.1×10^5 years	β-0.714
^{38}S	2.87 hr	β-1.1, γ1.88
^{38}Cl	37 min	β-4.91, γ1.60, 2.17
^{39}Cl	55 min	β-1.91, γ0.25, 1.27, 1.52

Eisenbud, 1987

Table 62 : Singly occurring natural radionuclides of terrestrial origin

Radio-nuclide	Abundance (%)	Half life (years)	Principal mode of decay, Energy (MeV) and yield (%)	Specific Activity (elemental) (Bq/kg)
^{40}K	0.012	1.26×10^9	β 1.33, 89% γ with EC 1.46, 11%	31635
^{50}V	0.25	6×10^{15}	γ with β- 0.78, 30% γ with EC 1.55, 70%	0.111
^{87}Rb	27.9	4.8×10^{10}	β-0.28, 100%	8.9×10^5
^{115}In	95.8	6.0×10^{14}	β-0.48, 100%	184.3
^{123}Te	0.87	1.2×10^{13}	EC	78.1
^{138}La	0.089	1.12×10^{11}	β-0.21, 80% γ with EC (0.81, 1.43) 70%	766
^{142}Ce	11.07	$>5 \times 10^{16}$	α	0.207
^{144}Nd	23.9	2.4×10^{15}	α 1.83	9.25
^{147}Sm	15.1	1.05×10^{11}	α 2.23	1.3×10^5
^{148}Sm	11.27	$>2 \times 10^{14}$	—	50.7
^{146}Sm	13.82	$>1 \times 10^{15}$	—	12.2
^{152}Gd	0.20	1.1×10^{14}	α 2.1	1.59
^{156}Dy	0.052	$>1 \times 10^{18}$	—	4.4×10^{-6}
^{174}Hf	0.163	2×10^{15}	α 2.5	0.06
^{176}Lu	2.6	2.2×10^{10}	β-0.43 γ0.089, 0.203, 0.306	8.9×10^4
^{180}Ta	0.012	$>1 \times 10^{12}$	—	8.84
^{187}Re	62.9	4.3×10^{10}	β-0.003	1.04×10^5
^{190}Pt	0.013	6.9×10^{11}	α 3.18	13.3

Eisenbud, 1987

Table 63 : Uranium series

Isotope	Half life	Type of radiation	Energy of radiation (MeV)
^{238}U	4.5×10^9 years	α	4.20
			4.15
		γ	0.048
^{234}Th	24 days	β	0.192
			0.100
		γ	0.092
^{234m}Pa	1.2 min	β	2.29
			1.53
			1.25
		γ (IT)	0.39
		γ	0.817
^{234}U	2.5×10^5 years	α	4.77
			4.72
			0.093
^{230}Th	8.0×10^4 years	α	4.68
			4.61
			4.51
		γ	0.068
			0.253
^{226}Ra	1622 years	α	4.78
			4.59
		γ	0.186
			0.26
^{222}Rn	3.8 days	α	5.48
		γ	0.510
^{218}Po	3.05 min	α	6.0
^{214}Pb	26.8 min	β	0.72
		γ	0.053
			0.242
			0.295
			0.352
^{218}At	1.5–2 sec	α	6.70
			6.65
^{214}Bi	19.7 min	β	3.26
			1.51
			1.00

contd....

Isotope	Half life	Type of radiation	Energy of radiation (MeV)
		α	1.88
			5.52
			5.45
^{214}Po	1.64×10^{-4} Sec	α	5.27
		γ	7.68
			0.799
^{210}Tl	1.3 min	β	1.9
			1.3
		γ	2.3
			0.296
			0.795
^{210}Pb	22 years	β	1.31
			0.015
		γ	0.061
			0.0465
^{210}Bi	5.0 days	β	1.17
		α	5.0
^{210}Po	138 days	α	5.3
		γ	0.80
^{206}Tl	4.2 min	β	1.51
^{206}Pb	Stable	—	—

Eisenbud, 1987

Table 64 : Thorium series

Type of radio Isotope	Half life	Type of radiation	Energy (MeV)
^{232}Th	1.4×10^{10} years	α	4.01
			3.95
		γ	0.055
^{228}Ra	6.7 years	β	0.055
^{228}Ac	6.13 hr	β	2.18
			1.85
			1.72
			1.11
			0.64
			0.46
		γ	0.058
			0.129
			0.184
^{228}Th	1.9 years	α	5.42
			5.34
		γ	0.083
^{224}Ra	3.64 days	α	5.68
			5.45
			5.19
		γ	0.241
^{220}Rn	55 sec	α	6.28
		γ	0.50
^{216}Po	0.16 sec	α	6.77
^{212}Pb	10.6 hr	β	0.33
			0.57
—		γ	0.176
			0.238
			0.300
^{212}Bi	60.5 min	β	2.25
		α	6.086
			6.047
		γs with β	1.81
			1.61
			1.03
			0.83
			0.72
		γs with α	0.040
			0.288
			0.46
^{212}Po	3.04×10^{-7} Sec	α	10.55
			8.785
^{208}Tl	3.1 min	β	1.80
		γ	2.61
			0.86
			0.58
			0.51
^{208}Pb	Stable	—	—

Eisenbud, 1987

Table 65 : Geometric mean activities in different samples of Kaiga environment (Bq kg⁻¹)

Radio nuclide	Soil	Grass	Vegetation	Food Items									
				Rice	Brinjal	Cucumber	Spinach	Ladies finger	Jack fruit	Banana	Fish	Composite meal
^{226}Ra	13.26	0.19	0.32	0.05	0.04	0.02	0.10	0.01	0.04	0.05	0.42	0.04
^{210}Pb	81.46	1.96	4.22	0.60	0.09	0.27	0.15	0.34	0.05	0.12	0.59	0.10
^{210}Po	29.22	1.98	20.7	0.03	0.06	0.07	0.28	0.30	0.06	0.04	9.23	0.04
^{228}Ra	12.04	0.19	0.30	0.30	0.28	0.37	0.15	0.19	0.33	0.21	0.25	0.09
^{137}Cs	2.66	0.23	2.31	0.10	0.08	0.09	0.08	0.07	BDL	0.05	0.28	0.07
^{90}Sr	0.94	0.03	0.16	BDL	BDL	BDL	0.05	0.03	BDL	0.28	0.09	0.04

BDL—Below Detection Level

References

Berger, K.C.; Erhardth, W.H. and Francis, C.W. 1965. Po-210 analysis of vegetables, cured and uncured tobacco and associated soil. *Science*. 150 : 1738-1739.

Bhat, I.S.; Khan, A. A. and Kamath, P.R. 1964-1968. *Radiation Environment; Preoperational Measurements*. BARC (TAPS); India.

Durrance Eric Michael. 1986. *Radioactivity in Geology; Principles and Application*. Ellis Horwood Ltd.

Eakins, J. D. 1984. The application of radiochemical separation procedures to environmental and biological materials, *Nucl. Inst. and meth. in Phy. Res.* 223; 194-199.

Faul, H. 1954. *Nuclear Geology*, Wiley. New York.

Food and Agriculture Organisation of the United Nations. 1962. Report of an expert committee. Organisation of surveys for radionuclides in foods and agriculture. FAO atomic energy series No 4. Rome.

IAEA Bulletin. 1993. Quarterly journal of the International Atomic Energy Agency. Vienna, Austria. 35 (1).

Iyengar, M.A. R.; Bhat, I.S. and Kamath, P.R. 1979. Progress Report of Environmental Survey Laboratory, Kalpakkam 1974-78. Barc/I-536.

Iyengar, M.A.R.; Ganapathy, S.; Kannan, V. ; Rajan, M.P. and Rajaram, S. 1990. Procedure Mannual. Workshop on Environmental Radioactivity, Kaiga, India. April 16-18.

Jaworowski, Z. and Grzybowska. 1980. Natural radionuclides and heavy metals in soils. *Proc. symp. Management of Environment* (Patel, B. ed.). 275-286.

Jaworowski, AZ. and Grazybowska. 1977. Natural radionuclides in industrial and rural soils. *The Sci. Total Environ.* 7: 45-52.

Kevin, M. Miller; John Kuiper; Irene K. and Helfer. 1990. Cs-137 fallout depth distributions in forest verses field sites; Implications for external gamma dose rates. J. *J. environ. radioactivity*, 12 ; 23-47.

Menzel, R. G. 1965. Soil plant relationships of radioactive elements. *Health Phys.* 11; 1325-1332.

Mishra, U.C. and Sadasivan, S. 1971. Natural radioactivity levels in Indian soils,. *J. Sc. Industrial Res.* 30; 59-62.

Mishra, U.C. and Sadasivan, S. 1972. Fallout radioactivity in Indian soils. *Health Phys.* 11;1-8.

Morgan, K.Z.; Snyder, W.S. and Ford, M. R. 1964. Relative hazards of the various radioactive materials. *Health Phys.* 10 : 151-69.

National Council on Radiation Protection and Measurements. 1975. Natural background radiation in the United States. *NCRP Rep. No 45.* Bethesda, Maryland.

Parfenov, Y.D. 1974. Po-210 in the environment and in the human organism. *Atomic Energy Rev.* 12;75-143.

Radhakrishna, A.P. 1993. *Studies on the baseline radiation background in the environment of Mangalore*. Thesis submitted to Mangalore Univ. for the award of Ph D Degree. Mangalore, India.

Rao, M. V. 1990. *Kaiga project - A bird's eye view.* Workshop on Environmental Radio-activity, Kaiga, India. April 16-18.

Russel, S.R. and Smith, K.A. 1966. Naturally occurring radioactive, substances; The uranium and thorium series in Radioactivity and Human diet (Edited by schott Russel, Pergamon Press). 365-379.

Santos, P. L. ; Goouvea, R. C.; Dutta, I. R. and Gouvea, V. A. 1990. Accumulation of Po-210 in foodstuffs cultivatd in farms around the Brazilian mining and milling facilities on Pocos de Caldas Plateau. *J Environ. Radioactivity.* 1: 141-149.

Schuttelkopf, H. and Kiefer, H. 1982. The radium-226 and polonium-210 concentration of the Black Forest. In; *Natural Radiation Environment* (Proc. 2nd Special Symp. Bombay, Vohra, K. G. ; Mishra, U.C.; Pillai, K. C. ; Sadasivan, S. Eds.) Wiley Eastern Ltd. New Delhi, India. 194-200.

Shukla, V. K. ; Menon, M. R. ; Lalit, B. Y. 1987. Environmental contamination from Chernobyl fall out. *Bulletin of Radiation Protection.* 10 (1&2) : 93-96.

United Nations, Yearbook of world energy statistics 1985, 1987.United Nations, New York.

United Nations Environment Programme. Radiation Doses, effects, risks. 1985. Sales No ESL II D 4 ISBN 92-807-1104-0 (Geoffrey Lean. ed).

United Nations Scientific Committee on the Effects of Atomic Radiation 1977. 32nd annual session, United Nations, New York.

United Nations Scientific Committee on the Effects of Atomic Radiation. 1982. 32nd session, suppl. No 45 (A /37/45). United Nations, New York.

United Nations Scientific Committee on the effects of atomic radiation. 1988. Sources, effects and risks of ionising radiation. Report to the general assembly. United Nations, New York.

U S Atomic Energy Commission Special Report. 1980. Po-210 in soils and plants. A.T. (11-1) -1733.

Vashi, V.D.; Krishnamony, K. and Pillai, K. C. 1992. A report on preoperational environmental monitoring for the Kakrapar. Atomic power project (KAPP). BARC, India.

Wrenn, M. E. ; Durbin, P.W.; Howard, B. ; Lipszstein, J. ; Still, E. T. and Willis. D. 1985. L. Metabolism of ingested U and Ra. *Health Phys.* 48:601-603.

CHAPTER 15

FOOD PRESERVATION BY IONIZING RADIATION AND ITS IMPACT ON ENVIRONMENTAL MANAGEMENT

M.D. Alur and S.N. Doke

Since ages, man wanted to conserve food for his future needs in times of natural adversity. Centuries ago, some conventional methods such as sun-drying, salting, pickling etc. were developed to conserve perishable and non-perishable food items. As science advanced, freezing, refrigeration and pasteurization techniques were developed. In the early twentieth century, use of antibiotics and some chemicals were advocated for preservation of foods. However, antibiotics were discarded as microorganisms developed resistance to them and several food preservatives after scrutinizing their safety were also rejected. Even though at the turn of the twentieth century, refrigeration, freezing, drying, controlled atmosphere storage, salting etc. are still being continued, a new technology of radiation preservation has been developed in the middle of the century. The technology was thoroughly studied and it is a flexible process, i.e., a single process which by varying the radiation doses multipurpose objectives in food conservation could be achieved.

In this chapter, an attempt is made to highlight the salient features of food irradiation, particularly, with respect to its mechanism of action on food components, its use to inhibit sprouting in tubers, decontaminate spices and preserve highly perishable items such as meat and fish. The natural contamination from human and industrial activities which pose threat not only to aquatic life, but also to the health of human beings is discussed lucidly.

By the year, 2020, world population is expected to reach 10 billion while arable land is expected to increase from 1.5 billion hectares in 1980 to only 1.6 billion hectares during the same period (FAO, 1981). This land will not be sufficient to produce for the entire population of the world. At present, 30 per cent of the total food production is lost due to spoilage. Major losses occur in developing countries of the world due to lack of proper storage and preservation facilities. Developing countries face energy crisis, cannot afford cold storage for food storage. Food irradiation is emerging as a potential method of food preservation. It is used to enhance the shelf life or improve the hygienic quality of raw and processed food materials. At present, 37 countries allow the use of this technology on food items (Loharanu, 1994). Food irradiation extends the shelf life of foods by delaying ripening, inhibiting sprouting and killing insects and microorganisms.

Mechanism of Action of Ionizing Radiation

The preservative effects of ionizing radiation are due to the primary and secondary effects resulting from reactions with atoms and molecules of food and organisms present in the food.

Primary Effects

When ionizing radiations penetrate into foods all or part of their energy is absorbed by the food. The absorbed energy leads to the ionization or excitation of the atoms and molecules of the medium. Depending upon the energy of the electron, a molecule can be ionized by losing an electron:

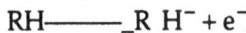

$$RH \longrightarrow \_R \ H^- + e^-$$

or be dissociated by splitting:

$$RH \longrightarrow \_R + \_H$$

If the energy associated with the electron is low, it may cause excitation of the molecule:

$$RH \longrightarrow RH^*$$

Water, the main constituent of foods, absorbs the radiation energy and undergoes radiolysis forming the following species:

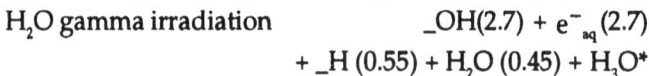

$$H_2O \ gamma \ irradiation \qquad \_OH(2.7) + e^-_{aq}(2.7)$$
$$+ \_H \ (0.55) + H_2O \ (0.45) + H_3O^*$$

Figures in parentheses indicate G values of each species (Thakur and Singh, 1994).

In large molecules, the absorbed radiation energy causes breakage of bonds in food constituents resulting in the degradation products (Nawar, 1983).

Secondary effects

Free radicals, electrons and cations produced by primary effects of ionizing radiations are very reactive and these reactions are very fast of the order of 10^{10} to 10^{11} m^{-1}s^{-1} (Simic, 1983).

The reactive species formed may undergo the following reactions:

_R + _H ——— R-H (combination)

_R + _R ———R-R (dimerization)

_RH*e ⁻——RH (electron capture)

Among primary radiolytic products of water, the hydroxyl radical is a powerful oxidizing agent and reacts rapidly with unsaturated compounds. It can add to these compounds or abstract hydrogen from C-H and S-H bonds.

R-H + _ OH ——— _R + H$_2$O

RS-H + _OH ——— RS_ + H$_2$O (Nawar, 1983).

The hydrated electrons, are very reactive and react with aromatic compounds, carboxylic acids, ketones, aldehydes and thiols. Hydrogen atoms react by abstracting hydrogen from C-H bonds or they can add to olefinic compounds (WHO, 1981).

Effect on Food Components

Besides water, carbohydrates, proteins and lipids are the major components of food systems. Effects of ionizing radiations on these components are given below:

Carbohydrates

During irradiation in the presence of water, carbohydrates are attacked by OH radicals. They abstract hydrogen of C-H bonds forming alpha-hydroxyl and alpha, beta-dihydroxy radicals (Simic et al., 1969). Both alpha-hydroxy and alpha, beta-dihydroxy radicals undergo disproportionation, dimerization and dehydration.

Polysaccharides degrade upon irradiation by cleavage of the glycosidic bonds. This leads to the formation of smaller carbohydrates, resulting in the softening and losses of texture in some fruits and low viscosity of polysaccharide solutions.

Irradiation of corn starch (12-13% moisture) in the presence of oxygen, yielded malonaldehyde, acetaldehyde, acetone, glyoxal, glucose and ribose as products (Nawar, 1983). Firmness of tomato fruits decreases when irradiated at 2.5 kGy dose. Molecular weight of pectins decreased upon irradiation with 2 kGy dose. Dry state irradiation of carbohydrates results in changes in their melting point, optical rotation etc. and gases like H_2, CO_2 CH_4 and CO are released.

Carbohydrates as a component of food system are less sensitive to radiolytic degradation. Thus, radiolytic products formed from pure starch at a dose of 5 kGy were equal to from wheat flour exposed to 50 kGy. No difference in amylose content of irradiated (0.3 kGy) and non-irradiated brown rice was reported (Sabularse *et al.*, 1992).

Proteins

Amino acids are the building blocks of proteins. They are joined together via peptide bonds to form protein molecules. Since free amino acids and peptides are present in foods, in aqueous solution and in absence of oxygen, simple amino acids undergo hydrogen abstraction, reductive deamination and decarboxylation.

Sulfur and aromatic amino acids react more readily with free radicals than aliphatic amino acids. Sulfur containing amino acids are important in food irradiation since these radiolytic products impart off odour e.g., H_2S.

Due to rigid spatial structure of proteins, radicals formed as a result of irradiation are held together in position and have high chance of recombination. Thus, owing to their configuration, proteins are more resistant to irradiation compared to isolated amino acids (Diehl, 1990).

A heat denatured protein yields more free radicals than its

native form (Urbain, 1989). In the absence of water, the action of irradiation is restricted to direct action, while in the presence of water, indirect action predominates. Irradiation can cause denaturation of proteins by breaking hydrogen bonds involved in secondary and tertiary structures. These changes may expose other embedded groups, viz., disulphide bonds, for further reactions, with radiolytic products of water. Irradiation may cause splitting of protein molecules into smaller units or at higher doses, aggregation due to cross-linking may take place. Changes produced by irradiation up to 10 kGy in food proteins are negligible due to the protective effects of other food constituents (Simic et al., 1985). Available lysine is not affected by irradiation at 1 to 5 kGy. Legumes exposed to gamma irradiation doses of 1, 10 or 50 kGy did not show any changes in protein contents (Delincee and Bognar, 1993).

Lipids

Triglycerides are the main components of lipids. On irradiation of lipids, cleavage occurs in the region of oxygen atoms or at the double bonds resulting in the formation of free radicals. Major degradation products of free fatty acids are CO_2, CO, H_2, hydrocarbons and aldehydes. In the presence of oxygen, the free radicals may yield hydroperoxides. Since lipids exist in non-aqueous phase in food systems, radiolytic products of water do not play a major role in their degradation. Primary effects of irradiation on lipids lead to the formation of cation radicals and excited molecules.

Irradiation of milk fats in sealed cans produced H_2, CO, CO_2, C_{15-17} n-alkanes and 1-alkene, C_{10-18} and C_{16-17} alka-dienes (Khatri et al., 1966).

Radiolytic products from beef, pork, soybean, corn, olive, safflower and cottonseed oil exposed to radiation doses ranging from 5 to 60 kGy include hydrocarbons, aldehydes, methyl ester and free fatty acids as the major volatile compounds (Champagne and Nawar, 1969; Dubravcis and Nawar, 1969).

Vitamins

Vitamins are essential nutrients present in food systems. Radio-sensitivity of vitamins differs depending upon whether they are in pure solutions or in foods. Radiolytic degradation of vitamins occurs from their reaction with free radicals or carbonyl compounds. Therefore, their destruction is dependent on the content of water and oxygen (Rosenthal, 1992). Fat soluble vitamins react with radicals produced by direct effects of radiation on lipids. Thus, radiolytic degradation of Vitamin A has been reported to be higher in iso-octane and vegetable oil compared to aqueous emulsions (Bhushan and Kumta, 1977). In aqueous medium, vitamins are exposed to radiolytic products of water. Due to small amounts of vitamins present in foods, the _OH radicals react predominantly with carbohydrates and proteins, their reactions with vitamins being of less significance (Simic, 1983). However, 30 to 60% loss of ascorbic acid has been reported by irradiating orange juice at a radiation dose of 1 kGy, while no loss was found in onion powder exposed to a high dose of 270 kGy (Galetto et al., 1979).

Loss of vitamins in foods is lesser in foods compared to pure solutions. Thus, vitamin B_1 was destroyed to the extent of 50 per cent with a radiation dose of 0.5 kGy in aqueous solution, while only 5 per cent loss occurred in dried whole egg after irradiation. Similarly, corn flour exhibited no significant loss of thiamine content when irradiated to 1 to 10 kGy of gamma radiation even after storage for a year (Tobback, 1977). No effect of irradiation on thiamine and riboflavin contents or irradiated wheat flour immediately or during storage after irradiation has been reported (Chappel and MacQueen, 1970).

Irradiation of Foods

Food irradiation is treatment of foods with energy from gamma rays, x-rays or electrons for specific purpose. It offers advantages to the producers, processors, retailers and consumers by prevention of food losses after harvesting and assuring quality and safety of food consumed. At very low doses, it inhibits regrowth or sprouting in stored potatoes, onion and garlic; delays ripening and over-ripening of fruits; substitutes chemicals used to control insects and pests in stored rice, wheat flour, sooji, pulses, gram flour, dry fruits, nuts, spices, dry fish etc. At higher doses of irradiation, pasteurizes or retards spoilage of meat, poultry and seafood by inactivating bacteria causing spoilage of these foods and also ensures food safety by destroying food-borne pathogenic bacteria such as Salmonella (responsible for typhoid fever), vibrio (causing cholera), E. coli, Shigella, Listeria, Yersinla, Staphylococcus, campylobacter and parasitic organisms such as protozoa *(Entamoeba, Toxoplasma)*, roundworms *(Trichinella)*, tapeworms and flukes. At still higher doses, irradiation can improve quality and microbial safety of spices and dried herbs. In 1964, an International Group of Microbiologists suggested the following terminology for irradiation of foods (Goresline *et al.*, 1964) and is also depicted in Table 66.

Table 66 : Microbiological effects of gama rays (dose dependent)

Disinfestation	Insects, larvae and eggs
Decontamination	Non-spore forming pathogenic microbes
Pasteurization	Reduction of microorganisms, extension of shelf life
Sterilization	Destruction of all microorganisms

Radappertization : Equivalent to radiation sterilization or commercial sterilization. Typical levels of irradiation doses are 30 to 40 kgy.

Radicidation: Equivalent to pasteurization of milk, for example, it refers to the reduction of number of viable specific non-spore forming pathogens, other than viruses, so that none is detectable by any standard method. Typical doses to achieve this process are 2.5 to 10 kGy.

Radurization: Equivalent to pasteurization. It refers to the enhancement of the keeping quality of a food by causing substantial reduction in the numbers of viable specific spoilage microbes by radiation. Common does levels are 0.75 to 2.5 kGy for fresh meats, poultry, seafood, fruits, vegetables and cereal grains.

A joint Expert Committee of the Food and Agricultural Organization, International Atomic Energy Agency and the World Health Organization has allowed only the following types of ionizing radiations for food irradiation: (i) gamma rays from ^{60}Co and ^{137}Cs at an energy level of 5 MeV. The gamma radiations emitted by ^{60}Co and ^{137}Cs is limited to 1.33 and 0.66 MeV levels, respectively (ii) X-rays generated from machine sources below an energy level of 5 MeV; (iii) electrons generated from machine sources below an energy level of 10 MeV (Diehl, 1990).

Some conventional methods of preservation of foods and their purposes are given in Table 67 for comparative purposes.

Table 67 : Some processes for preservation of food

Method	Purpose
Refrigeration	To reduce spoilage by slowing down the growth of microorganisms, to prevent the growth of most pathogenic bacteria.
Heat treatment	To pasteurize (reducing the level of microorganisms) and destroying all relevant pathogens without significantly affecting the taste of the food e.g. milk
Canning	To prolong shelf life by eliminating the microoganisms including pathogens that grow in the food
Chemical treatment	(a) to prevent sprouting (b) to destroy insects and thus prevent insect attack

Use of Radiation in Inhibition of Sprouting in Tubers and Bulbs

Potatoes and onions are widely grown and exported vegetable crops. Storage losses of potatoes and other root crops occur mainly due to sprouting, moisture losses leading to shriveling and storage losses. Sprouting during storage can be prevented by post-harvest applications of chemical inhibitors, such as, isopropyl n-phenyl carbamate (IPC) or methyl ester of naphthalene acetic acid (MENA) on potatoes or pre-harvest spraying of onion with maleic hydrazide or ethepon or storing at refrigerated temperature (Rosenthal, 1992). Use of chemical inhibitors has limitation regarding their effectiveness, regulation in different countries and their toxicity (Diehl, 1990). Refrigeration is an expensive method and it also causes an increase in reducing sugars resulting in browning, if these potatoes are used for chips. Irradiation of potatoes (0.06-0.15 kGy) and onion (0.02 to 0.15 KGy) inhibited sprouting in these crops for eight months at room temperature. Ionizing radiation caused morphological and histological changes in dormant buds which induced necrosis at growing points during storage as a result, these buds are unable to form roots and stems. Ionizing radiations have also been used to inhibit sprouting in garlic, sweet potato, yams, carrot and brinjal (Thomas, 1984; Kwon et al., 1985; Adesuyi and Mackenzie, 1973).

Irradiation of tubers viz. potatoes stored at 15° C showed in creased accumulation of free sugars. Initially, loss of ascorbic acid was observed. Irradiation had no apparent effect on the characteristics of the starch. There was significant increase in the activity of phenyl alanine ammonia lyase (PAL) (about 30 fold) in the bud tissue within three hours after irradiation. PAL in gamma irradiated potatoes has shown that the faster component being radiation-induced activation in the rate of synthesis of the enzyme (Pendharkar and Nair, 1975).

Delay of senescence, maturation and ripening control in fruits and vegetables

Senescence is the phase of plant growth from full maturation to death especially of the fruits and leaves. This phase is characterized by an accumulation of metabolic products, an increase in respiration rate and loss of dry weight. Therefore, a delay in senescence would enhance the shelf life of fresh fruits and a reduction in spoilage. Irradiation could be used to extend the shelf life of fruits and vegetables (Thomas, 1986). Irradiation of strawberries and storage at $2°$-$4°C$ extended the shelf life by 9 days, but treatment at higher dose (4 kGy) resulted in tissue softening due to degradation of cellulose and pectins in cell walls.

Irradiation (2 kGy) of mushrooms reduced microbial contents resulting in increase in shelf life by 4 days. Radiation caused delay in senescence of carrot tissue, while reducing mircobial spoilage (Champagne and Nawar, 1969).

Non-seasonal fruits such as citrus ripened on the tree itself. Therefore, irradiation is directed towards extending their shelf life by preventing spoilage due to fungal growth in order to prevent spreading of insects and pests. Several countries required quarantine treatment of imported fruits and vegetables. Several countries have banned use of chemicals for quarantine treatment due to their toxicity. Irradiation proved to be an effective alternative for quarantine treatment of fruits and vegetables (Thomas, 1986; Akamine and Moy, 1983).

Spoilage of fruits and vegetables

Bacterial soft rot

This type of spoilage caused by *Erwinia carotovora* and *Pseudomonas marginalis*. The causative organism breakdown pectin giving rise to a soft and bad odour. Some of the vegetables affected by this disease are carrot, onion, garlic, tomato etc.

Gray mould rot

This condition is caused in vegetables viz. asparagus, onions, garlic, beans, carrots, celery, tomatoes, cabbage, cauliflower, raddish etc. by *Butrytis cinerea* which produces a gray mycelium. In this disease, the causal fungus grows on decayed area in the form of a prominent gray mould.

Sour rot (watery soft rot)

This condition of vegetables is caused by *Geotrichum candidum* in garlic, carrots, asparagus, onions, beans etc. The causal fungus is widely distributed in soil and on decaying fruits and vegetables. *Drosophila melanogaster* (fruit fly) carries spores and mycelial fragments on its body from decaying fruits and vegetables to wounds in healthy fruits and vegetables (McColloch *et at.,* 1968).

Rhizopus soft rot

This condition is caused by *Rhizopus stolonifer* that makes vegetables soft and mushy. Cottony growth of the mould with small black dots of sporangia often covers the vegetables. Normally, beans, carrots, sweet potatoes, cabbage etc. are affected by this disease. In order to control fungal spoilage of fresh fruits the use of some chemical agents is being practised, which are listed in Table 68.

Table 68 : Some chemical agents employed to control fungal soilage of fresh foods

Chemical agents	Fruits
Thiabendazole	Fruits, apples, pears, citrus fruits, pineapples
Benomyl	Apples, pears, bananas, mangoes, papayas, pineaples, cherries
Biphenyls	Citrus fruits
Sulphur dioxide fumigation	Grapes

Eckert, J.W. 1979

Control of Sprouting and Germination

Low dose irradiation treatment inhibits sprouting of potatoes, yams, onions, garlics, ginger and chestnuts. In the case of potatoes, irradiation has marked advantage over the current methods of sprout control. In order to last through a season without spoilage due to sprouting, potatoes are treated with chemicals or left in cold storage at 3.3-4.4°C. The process is expensive. Chemicals such as maleic hydride, isopropyl-n-chloro-phenyl carbamate (chlorophan) are currently being used in potatoes. These chemical sprout inhibitors can leave residues on the products. Food irradiation is a safe and remarkably effective sprout inhibitor. It leaves no residues and allows storage of the treated commodity at higher temperature.

Low dose irradiation in conjunction with other post-harvest treatments such as skin coating or storage at modified atmosphere was found to delay ripening and senescence in bananas and mangoes. At the dose levels employed for delaying ripening, the oriental and melon fruit flies at different developmental stages could be eliminated (Thomas *et al.*, 1975).

Irradiation at sprout inhibiting dose eliminates the eggs and early larval instars of potato tuber moth which causes damages in stored tubers. Long term storage of irradiated potatoes is feasible if stored at 15°C. The after-cooking darkening observed in irradiated potatoes could be avoided by removing skin prior to cooking (Thomas and Joshi, 1977). To achieve maximum sprout inhibition in onions, the bulbs should be irradiated at 0.06-0.10 KGy soon after harvest when they are in the dormant state (Thomas *et al.*, 1975). Some GRAS chemical food preservatives affecting the growth of microorganisms in spices, bread, fruits etc. are detailed in Table 69.

Table 69 : Some GRAS chemical food preservatives

Preservatives	Organism affected	Foods
Propionic acid	Moulds	Bread, cakes and cheese
Sorbic acid	Moulds	Syrups, cheeses, jelly, cakes, salad dressings
Benzoic acid	Yeasts and moulds	Soft drinks, tomato ketchup
Parabens	Yeast and moulds	Pickles, tomato ketchup
Sulphites	Insects	Dried fruits
Ethylene/propylene oxide	Yeasts and moulds	Spices
Sodium nitrite	Clostridia	Meat

James M. Jay, 1992.

Preservation of Foods of Animal Origin by Ioning Radiation

Microorganisms exhibit enormous differences in their responses to ionizing radiations. Thus gram negative organisms such as *Pseudomonas fluorescens, Aeromonas hydrophila, Proteus vulgaris* etc. are extremely sensitive to radiation, while gram positive organisms such as *Microccocus radiodurans, Bacillus spp. exhibit* extreme resistance (Lewis *et al.,* 1971; 1974). The resistance offered by microorganisms to radiation is expressed in terms of D_{10} value, i.e., dose required to inactivate 90 per cent of initial cell population. Table 70 incorporates the D_{10} values of different types of bacteria.

Table 70 : D10 values (dose required for 90% destruction) of some selected bacteria

Bacteria	Medium	Irridation Temperature	D10 (kGY)
Vibrio parachaemolyticus	fish homogenate	ambient	0.03-0.06
Pseudomonas fluorescens	ground beef	ambient	0.12
Campylobacter jejuni	ground beef	ambient	0.14-0.16
Aeromonas hydrophila	ground beef	2°C	0.12-0.14
Proteus vulgaris	Oyster homgenate	5°C	0.2
Yersinia enterocolitica	ground beef	ambient	0.1-0.2
Shigella dysenteria	Shrimp homogenate	forzen	0.22
Salmonella typhimurium	ground beef	ambient	0.55
Staphylococcus aureus	minced chicken meat	4°C	0.42
Micrococcus radiodurans	nutrient broth	ambient	3.4
Escherichia coli	surface of prawn	−10°C	0.35

Source : Diehl, J.F., 1990.

Microorganisms also exhibit considerable differences in their spoilage profiles in flesh foods. Thus gram negative spoilage bacteria are inactivated by low doses of radiation and radiation survivors such as *Micrococcus spp.* and *Bacillus spp.* being biochemically inert do not bring about spoilage of irradiated flesh foods (Alur *et al.,* 1971; Alur and Lewis, 1980; Alur *et al.,* 1989).

Thus, spoilage in foods of animal origin occurs mainly through microbial spoilage. Refrigeration, freezing, curing and canning are meant to arrest microbial spoilage. Lipids in fish and meats can undergo oxidation in the presence of oxygen causing rancid odours. Irradiation contributes to their preservation by preventing microbial spoilage. Further, gamma irradiation exerts no deleterious effects on the nutritional quality of fish (Armstrong *et al.,* 1994).

Foods of animal origin may contain a variety of pathogenic organisms relevant to public health, *e.g. Salmonella, Escherichia coli, Vibrio cholerae, Vibrio parahaemolyticus* and *Yersinia enterocolitica.* There are other parasites - e.g. *Toxoplasma gondii, Trichinella spiralis, Cysticercus bovis* and *Cysticercus cellulosae* - present in several meats and fishes (Engel *et al.,* 1988). These organisms pose a serious health hazard to consumers and cause loss of billions of dollars in terms of health care and lost work hours. In USA alone, 6.5 - 33 million cases of food-borne diseases occur each year, of these 2 million are caused by *Salmonella* (Loharanu, 1994). Elimination of nonsporing pathogens from these products requires doses of 3-10 KGy (El-Zawahry and Rowley, 1979; Lambert and Maxy, 1984). The maximum dose permitted by the US Food and Drug Admininstration (FDA) for irradiating poultry products is 3 kGy.

Fish in tropical countries is preserved by sun drying. Infestation by flies during sun drying results in heavy losses during storage and marketing. The losses of unprotected dried products can be as high as 50-70 per cent in the world. Irradiation has been used to prevent losses in dried, smoked and fresh fishes.

Irradiation lengthens the storage life in the case of products of animal origin, such as fish, meat and poultry, which would otherwise suffer from bacterial spoilage. With these products, however, the prevention of spoilage and decontamination from pathogenic organisms go hand in hand.

It is well known that high percentages of foods of animal origin are contaminated with Salmonellae, Campylobacter, Staphylococci, *Clostridium perfringens* - to mention only the most important. This is the result of the increase in food infections and food poisoning in all countries. Apart from bacteria, parasites along with cysticercae, echinococci, trichinae and toxoplasmas, still threaten the health of man. These dangers can be avoided either by breeding animals that are pathogen-free or by treating the food to decontaminate it or by letting the consumer know that all potentially dangerous foods should be held under refrigeration and heated before being eaten. However, the first method is not attainable on large scale while the last method makes the consumer alone responsible. The ideal situation would be to treat food in such a way that any danger to the health of the consumer would be eliminated. The best example of this is the pasteurization of milk, into a totally harmless food. But this method of treatment is limited to liquid components and other heat treatments, e.g. canning, can not be used in the case of raw meat and poultry, i.e. for foods that are the most contaminated. In these cases irradiation is a possible method of decontamination, although it will not provide a 100 per cent solution to the problem.

Another food that could be irradiated is raw poultry which

contains a high percentage of pathogenic organism and there is no alternative method of decontaminating it. A reduction in Salmonella of 3 decimal points and still greater reduction in Campylobacter can be achieved with a dose of 3 KGy. Other organism e.g. staphylococci and the vegetative form of *Clostridium perfringens* are destroyed by the dose mentioned above. As poultry and poultry products are incriminated in food infections, decontamination, if widely used, a significant reduction in the number of cases of food infections would occur. In the Netherlands, Canada and Russia irradiated poultry has been released for test marketing (Thayer *et al.*, 1986).

Fish, crustaceans and molluscs are groups of foods that have been found to respond well to irradiation with T-rays. The treatment with T-rays not only extends the shelf-life of sea-foods by several folds but also destroys salmonellae, staphylococci, pathogenic *E.coli* strains, Vibrio and pathogenic parasites. No chemical changes associated with flavour of irradiated meat have been reported (Merritt *et al.*, 1975).

Grains

Irradiation technology is also used for disinfestation of cereal grain (Tilton and Burditt, 1983). A dose of 1 KGy completely kills the adult beetle in legumes within a week. This dose is sufficient to control infestation of beans over six months of storage [Roy and Prasad, 1993] . A minimum dose of 0.3 KGy is recommended for control of stored products insects. No significant loss of nutrients is reported in grains and cereals after irradiation.

The compositional changes in irradiated (0.2 to 2.0 KGy) wheat with special reference to physico-chemical properties of starch and proteins and their susceptibility to amylase and protease, respectively, have been studied. Sensitivity of starch and protein molecules in wheat endosperm to their respective hydrolyses increase following irradiation. This can be attributed to their fragmentation to low molecular entities which are easily attacked by the hydrolytic enzymes. However, beta-amylase present in resting seeds is radio-resistant (Ananthaswamy *et al.*, 1970; Srinivas *et al.*, 1972).

Studies have been carried out on compositional aspects of red gram *(Cajanus cajan)*. Reduction in the cooking time of irradiated red gram and pulse may be attributed to physico-chemical changes of food components. Depolymerization of starch by radiation treatment (10 kGy) is revealed by its increased susceptibility to amylase action. Thus, irradiation of red gram improves its cooking and textural qualities as well as digestibility (Nene *et al.*, 1975a, 1975b).

Irradiation of Spices

India has been considered the spice centre of the world. Today India leads the world in spice production and markets about two million tons of different spices valued approximately at Rs. 3500 crores. At present, India exports pepper, ginger, chillies, cardamom, turmeric etc. to USA, Canada, Europe, Russia, Australia, Japan and Gulf countries. Quality concepts all over the world are becoming stringent and the importing countries insist on standards stipulated by them. Gamma radiation is the answer to the problem faced by the producers and exporters of spices. Optimum doses of irradiation for decontamination of spices are:

Doses (kGy)	Effects
1	Ramoval of all insects at different stages of growth.
5	Elimination of fungal growth.
10	Total decontamination of microbes and insects.

Spices are often contaminated causing difficulties during further processing e.g. in meat products. In addition to spoilage organisms, pathogenic organisms are also encountered in spices leading to food infection in man. For this reason, spices are regularly decontaminated with gases such as ethylene oxide. This treatment is not harmless from the toxicological point of view. An alternative method is to irradiate these with 5-10 kGy which gives good results having the advantage that spices can be pre-packed, thus avoiding renewed contamination. The irradiation of spices has been carried out for a number of years in Netherlands and Hungary with success. In USA, it has become mandatory to irradiate spices up to 10 kGy. Various radiation processing for food preservation with respective dose ranges are given in Table 71.

Table 71 : Various radiation processes for food preservation

Application	Dose range (kGy)
Sterilization of spicies, herbs etc.	7-10
Inactivation of pathogens, e.g. *Salmonella*	2.5-5.0
Control of moulds	2-5
Prolongation of chilled shelf life, e.g. fresh fish	2-5
Inactivation of parasites e.g.	0.3-1
Trichinells spiralis	
Disinfestation e.g insect control	0.1-1
Inhibition of sprouting	0.1-0.2

Spicies are contaminated with molds and heat - resistant bacterial spores during harvesting and processing [Kiss and Frakas, 1981; Govindaranjan, 1985). Fumigation with ethylene oxide or propylene oxide is widely used for reducing the microbial contamination of spices. But fumigation does not only kills molds and it also poses health hazards due to chlorohydrin formation. Ethylene oxide has been banned in some countries due to health problem. Irradiation has proved an effective alternative in reducing the microbial load of spices without causing any sensorial changes in them (Saint-Lebe *et al.*, 1985]. Irradiation doses of 10 kGy or less are capable of killing food poisoning microorganism (Singh *et al.*, 1988).

Environmental Concerns

Irradiation of food will reduce the food losses and the amount of energy used in the food industry [Brynjolfsson, 1978]. It will reduce the use of insecticides, such as halogenated hydrocarbons and it will reduce the use of some other undesirable chemicals, such as ethylene oxide. It will also reduce the amount of food borne diseases of microbial or parasitic origin. If Caesium-137 sources are used, a waste product from the nuclear industry would be put into the under tight control. Cobalt-60, on the other hand, would be intentionally produced for the purpose of irradiation and would increase the total load of radioactive isotopes in the environment. Although the technical problems of operating safely in the hands of skilled people are small, the administrative problems of assuring safe operation may be difficult. As long as controls are strictly enforced, the environmental impact of the isotope sources is small. When electron accelerators are used, the technical and administrative problems of assuring operation are small and undesirable effects on the environment are small.

Food-borne Diseases

Primary sources of microorganism found in foods are incorporated in Table 72. Among the microorganisms, food-borne pathogens can cause serious health problems in developed and developing countries alike. Interest in some countries in the use of radiation for Salmonella control has been enocouraged by clearance of some specific processes by national health authorities. Frozen shrimps, prawns and froglegs are irradiated in commercial quantities in the Netherlands using doses upto 4 kGy with excellent results [Kampelmacher, 1984]. This dose would also deal effectively with *Shigella* which was implicated with a serious food poisoning incident in 1983 and also with *Campylobacter* more recently identified and associated with enteritis in Australia (Tsuji, 1983).

Table 72 : Primary sources of microorganisms found in foods

Source	Predominant microorganisms
Soil and water	These two environments share many bacteria in common. Soil organisms may enter by wind and rain. *Alteromonas* are aquatic forms requiring seawater salinity for growth.
Plants and plant products	Lactic acid bacteria and some yeasts adhere to plant surfaces. Among other viz. *Corynebacterium, psuedomonas, xanthomonas* and several fungal pathogens.
Animal feeds	Important sources of *salmonella* in poultry, *Listeria monocytogenes* are found in dairy and meat animals.

James M. Jay, 1992.

A dose of 6-8 kGy results in a hundred fold reduction in total count and reduction of *Staphylococcus aureus, Salmonella* sp. and *Escherichia coli* to below levels of detection. A dry product like tea was effectively freed of salmonella by irradiation in commercial scale operations in South Africa.

Microorganisms from Sewage

Microorganisms that are present on fish can be pathogenic or nonpathogenic. It is non-pathogenic bacteria that are responsible for main routes of spoilage. The pathogenic bacteria that contaminate fish in a variety of ways (i) untreated sewage being discharged into water from which fish and shellfish are harvested; (ii) from water of unsatisfactory microbiological quality that is used in the processing of fish; (iii) by transfer from unwashed hands or improperly sanitized equipments.

Sewage contains bacteria and viruses. Bacteria include *Salmonella* causing typhoid and parathyphoid, *vibrio cholerae* which causes Cholera and *Shigella*, which causes dysentery. Among viruses infectious hepatitis are important (Connell, 1995).

Contamination with Pollutants

A large number of harmful matter and elements are known as pollutants, but mercury has been implicated in disease to man caused by eating fish. Other elements that cause injury to health include cadmium, lead, selenium and arsenic. A disease affecting the central nervous system occurred in persons living in the region of Minammata Bay and Nigata in Japan during 1953 to 1964. Several hundred persons were affected and fifty died. In both cases, the cause was industrial effluents containing mercury that entered the sea. High concentrations of mercury in the environment occur more often in rivers and lakes. Consequently, a large number of freshwater fish species have been found to contain high levels of mercury. A large quantities of mercury compounds has been discharged by industrial concerns into rivers and lakes in Sweden, Finland and Canada. That is why, these countries have placed an indefinite ban on fishing in certain rivers and lakes known to be contaminated with mercury.

In shellfish taken in some inshore area, high concentrations of cadmium, zinc and lead have been found. If eaten in high amounts, such fish could cause risks to health in some individuals. Therefore, many countries are taking mandatory action to reduce pollution of the aquatic environment with heavy metals (Ruivo, 1972; Muller and Lloyd, 1994).

Organic Chemicals

An enormous variety of chemicals from industrial processes finds their way into the aquatic environment. Eventually, minute quantities of some of them end up in fish. Notable among a group of chlorinated hydrocarbons including insecticides, DDT and its breakdown products, aldrin, benzene hexachloride (BHC) and polychlorinated biphenyls (PCBs). Massive increase in the use of these chemicals has led to increasing concentrations in animal and plant tissues. Decrease in bird, sea trout and salmon populations have been attributed to their use, but no effect on the health of man has ever been demonstrated. However, wide use of chlorinated hydrocarbons is severely restricted in many countries. In U.K. these chemicals have been continuously monitored resulting in considerable reduction in the concentration of DDT and dieldrin in foods.

Occasionally catches become grossly contaminated with chemicals, especially, mineral oils, resulting from accidental or other kinds of large scale release. Evidence of such contamination is apparent in tainted odours or flavours of the fish. However, there is no evidence that hydrocarbons in fish present a hazard to health (Tidmars, 1985). Some of the chemicals intentionally added as food preservatives with the their limitations are described in Table 73.

Specific toxins

Aflatoxins

Aflatoxins are produced by fungi which can grow on certain foods. Commonly *Aspergillus flavus* and *A, parasiticus* produce aflatoxin. Aflatoxins at levels of 3 ppm are extremely poisonous and are involved in liver cancer. They are mutagenic and cause deformities in foetuses. They also suppress immune system. Toxin is produced in grain and ground nuts in warm and moist conditions by the growing fungus. Several studies have indicated that an increase in the content of toxin with the process of irradiation [Anon, 1987].

Botulism Toxin

Irradiation at doses of 10 kGy or less inactivates a wide range of bacteria and fungi in food. However, some bacteria in the form of resistant spores are not killed at these doses of irradiation. When conditions are favourable, *Clostridium botulinum* spores can germinate which produce a lethal type of toxin. Usually when *C.botulinum* has reproduced and the food displays some indications of inedibility viz. smells or discoloration or slimy feel is produced by other

Table 73 : Chemicals used as food preservatives and their limitations

Compound	Use	Product in which it is used	Limitations and status
Diethyl pyrocarbonate	Control of microorganisms	Beverges	Reacts with ammonia to form urathane, a known carcinogen; it is irritant to the eyes and skin. Banned.
Ethylene and prophylene oxides	Destroys yeasts, moulds, bacteria and viruses	Fumigant for dried fruits, ground spices, edible gums and dried mushrooms	Toxic reaction products such as ethylene glycol and ethylene chlorohydrin are formed. Both epoxides are highly combustible.
Nitrites and nitrates	Curing agents, impart characteristic flavour, colour and microbial stability associated with curing; retards the growth of C. botulinum and toxin production	Meat, fish and cheese manufacture	Sodium nitrite combines with secondary and tertiary amines forming nitrosamines which are carcinogenic. Nitroso-pyrrolidine found in fried bacon is a known carcinogen.
Sorbic acid	Effective against moulds and yeasts, inhibits growth and toxin production by C. botulinum	Meat	Irritant to eyes, GRAS
Propionic acid	Mould inhibitor, sodium salt effective against rope causing bacteria, B. mesentericus.	Cakes, leavened bread	Acid is corrosive in nature, GRAS
Parabens	Effective against moulds and yeast.	Animal products	Ineffective against bactria. Methyl and propyl parabens are GRAS
Sulphites and SO_2	Spray or fumes; sanitizing agent	Beverges, dried fruits, vegetables, syrups and fruit juices	SO_2 is corrosive sulphites products H_2S causing blackening of cans, GRAS, but not for meat

Adapted from : Dziezak, 1986

microbes which can act as a warning. *C. botulinum* Type E can grow and produce toxins at temperatures 3 ºC and above this is of greatest concern is fish. It is therefore, recommended that if it were permitted to irradiate fish, the label should clearly state "Irradiated fish - store below 2 ºC and do not eat raw".

Irradiation is a powerful technique for the destruction of parasites and microorganism in meats. A dose of 1 kGy can be used to destroy parasites such as *Tirichinella spiralis* in pork. Doses of 2-5 kGy destroy vegetative bacteria and effectively reduce the population of pathogens and spoilage organism. To destroy spores and viruses or to sterilize meat, much higher doses are required. If meat is to be stored at ambient temperature, *Clostridium botulinum* spores must be eliminated which requires a dose of 45 kGy [Shay *et al.*, 1988; Anellis *et al.*, 1979).

Food poisoning caused by *Vibrio parahaemolyticus*

This has been recorded in Japan where 70 per cent of food poisoning out breaks are caused by eating fish. The disease is mild and the organism is easily destroyed by heat, radiation and susceptible to chill temperature. Further, *Vibrio cholerae* is often associated with faecal contamination.

Food Safety Assessment

Food poisoning usually refers to food contaminated by microorganisms or by toxins from microorganisms. The effect of irradiation vary with the type of food and microoganisms. Some microorganisms are very susceptible to irradiation, whereas others are resistant. With any food processing method, one must ensure that elimination of one organism does not promote the growth of other organisms of greater health concerns. Food must be heated sufficiently to eliminate that most heat resistant of pathogens, *Clostridium botulinum*. Pasteurised food required continued refrigeration.

Irradiation and thermal processing are common in causing mutations in organisms, making them pathogenic or more virulent.

Table 74 : Cost of radiation applications in foods

Purpose of treatment	Dose (kGy)	Cost per pound (cents)
Radurization of fish	2.0	1–2
Radurization of meat	2.0	1–2
Sprout inhibition of potatoes	0.1	1.0
Disinfestation of grains	0.3	0.01
Rot control in strawberries	2.0	4.0
Salmonella control	5.0	1.0

Source : Sutton, 1969.

Table 75 : Energy needed for various methods of preservation

Method of preservation	Energy required
Storage for 5.5 days in refrigerator	318 KJ/Kg
Heat sterilization	918 KJ/Kg
Deep freezing	7552 KJ/Kg
Irradiation at 3 kGy	17.8–95 KJ/Kg

Kampelmacher, E.H., 1984

Although mutations can be caused by irradiation or heat, such mutations are not beneficial to the organisms and hence there is no evidence of radiation-induced mutations (Pauli, 1995; WHO, 1992).

Tables 74 and 75 describe the cost of irradiation application in foods and the energy needed for various methods of preservation. Cost of irradiation depends upon the purpose as well as the food commodity to be irradiated. For bulk items such as food grain, tubers like potato and onions, it is extremely low while irradiation of spices costs quite high which could be absorbed by charging to the importers. Irradiation process is less energy-intensive as evidenced by the values given in Table 75. Compared to heat sterilization and deep freezing, the energy required for the radiation process is 10 to 100 - fold less. Thus, during the energy crunch throughout the world, food irradiation is a boon to the mankind by providing durable, hygienic, cost-effective food items to the hungry people of the under-developed countries of the world. This will also ameliorate the malnutrition prevalent among the children in the developing countries by providing nutritious and wholesome foods.

References

Adesuyi, S.A. and Mackenzie, J.A. 1973. The inhibition of sprouting in stored yams (*Dioscorea rotundata*) POIR by gamma radiation and chemicals. In: *Proceedings of the Radiation Preservation of Food;* International Atomic Energy Agency, Vienna, pp. 127-136.

Akamine, E.K. and Moy, J.H. 1983. Delay in post-harvest ripening and senescence of fruits. In : *Preservation of Food by Ionizing Radiation,* Vol. III, pp. 129-158. E.S. Josephson and M.S. Peterson (Eds.). CRC Press, Boca Raton, Florida.

Alur, M.D., Lewis, N.F. and Kumta U. S. 1971. Spoilage potential of predominant organisms and radiation survivors in fishery products. *Ind. J. Exp. Biol. 9,* 48-52.

Alur, M.D. and Lewis N. F. 1980. Influence of storage temperature on microflora of Indian mackerel (*Rastrelliger kanagurta*). *Die Fleischwirstschaft, 60,* 453-455.

Alur, M.D., Venugopal, V. and Nerkar, D. P. 1989. Spoilage potential of some contaminant bacteria isolated from Indian mackerel (*Rastrelliger kanagurta*). *J. Food Sci. 54,* 1111-1115.

Ananthaswamy, H.N., Vakil, U.K. and Sreenivasan, A. 1970. Susceptibility to amylosysis of gamma-irradiated wheat *J. Food Sci. 35,* 792-794.

Anellis, A. Rowly, D. B. and Ross, E. W. 1979, Microbiological safety of radappertized beef. *J. Food Prof. 42,* 927-794.

Anonymous, 1987. How safe is food irradiation? *Science Age, 4 (9),* 51-55.

Armstrong, S.G., Grant Wyllie, S. and Leach, D.N. 1994. Effects of preservation by gamma-irradiation on the nutritional quality of Australian fish. *Food Chem., 50, 351-357.*

Bhushan, B. and Kumta, U.S. 1977. Radiation response of Vitamin A in aqueous dispersions. *J. Agric. Food Chem. 25,* 131-135.

Brynjolfsson, A. 1978. Energy and food irradiation. In : *"Food Preservation by irradiation"* Proc. of a symposium in Wageningen 21-25 Nov. 1977 IAEA/FAO/WHO Vol. II pp. 285-299 IAEA, Vienna.

Champagne, J.R. and Nawar, W.W. 1969. The volatile components of irradiated beef and pork fats. *J. Food Sci., 34,* 335-339.

Chappel, C.I. and McQueen, K.f. 1970. Effect of gamma irradiation on vitamin content of enriched flavour. *Food Irradiation, 10,* 8-10.

Connell, J.J. 1995. *Control of Fish Quality. Fishing News Books* Blackwell Science Ltd. Oxford.

Delincee, H. and Bognar, A. 1993. *Bioavailability-93* Federation of European Chemical Society, p. 367.

Diehl, J.F. 1990. *Safety of Irradiated Foods,* Marcel Dekker Inc., New York.

Dubravcis, M.F. and Nawar, W.W. 1969. Effects of high-energy radiation on the lipids of fish *J. Agric. Food Chm., 17,* 639-644.

Dziezak, J.D. 1986. Preservatives : Antimicrobial agents A means towards product stability. *Food Technol., 40,* 104-111.

Eckert, J. W. 1979. Fungicidal and fungistatic agents. Control of pathogenic microorganisms on fresh fruits and vegetables after harvest, In : *Food Mycology,* ed: Rhodes, M.E., pp 164-199 Boston.

Ei-Zawahru, Y. A. and Rowley, D. B. 1979. Radiation resistance and injury of *Yersinia Entercolitica Applied Environ. Microbiol. 37* 50-54.

Engel, R.E., Post, A. R. and post R. C. 1988. Implementation of, irradiation of pork for Trichina control, *Food Technol. 42*, 71-75.

Farkas, J. 1973. *Aspects of the introduction of Food irradiation in developing countries.* International Atomic Energy Agency, Vienna,PL/518/8, pp. 43-59.

FAO, 1981. Food and Agriculture Organization of the United Nations. In *"Agriculture Towards 2000"* FAO, Rome.

Galetto, W., J. Kahan, M. Eiss, J. Welbourn, A. Bednarczyk and O. Siberstin. 1979. Irradiation treatment of onion, *J. Food Sci. 44*, 591-595.

Goresline, H.E., Ingram, M., Macuch, P., Mocquot, G., Mossel, D.A.A. Niven, C.F. and Thatcher, F. S. 1964. Tentative classification of food irradiation processes with microbiological objectives. *Nature , 204*, 237-238.

Govindarajan, V.S. 1985. Capsicum-production, technology, chemistry and quality part 1. History Botany cultivation and primary processing *Food Sci. Crit. Rev. Nutr., 22*, 109-176.

Jay, J. 1992. Food Preservation with Chemicals. In: *Modern Food Microbiology*, 4th edition. Chapman and Hall, New York.

Kampelmacher, E.H. 1984. Irradiation of food : A new technology for preserving and ensuring the hygiene of foods. *Fleischwirtschaft, 64*, 322-327.

Khatri, L., Libbey, L. and Day, E. 1966. Gas chromatoraphic and mass spectral identification of some volatile components of gamma irradiated milk fat. *J. Agric. Food Chem., 14*, 465-469.

Kiss, I. and Farkas,J. 1981. Combined effects of gamma radiation and heat treatment on microflora of spices. In : *Proceedings of Food Irradiation*, pp. 107-115, International Atomic Energy Ageney, Vienna.

Kwon, J.H., Byun, M.W. and Cho, H.O. 1985. Effects of gamma irradiation dose and timings of treatment after harvest on the storageability of garlic bulbs. *J. Food Sci., 50*, 379-381.

Lambert, L.D. and Maxcy, R.B. 1984. Effect of gamma radiation on *Campylobacter jejuni J. Food Sci., 49*, 665-667.

Lewis, N.F., Alur, M. D. and Kumta, U. S. 1971. Radiation sensitivity of fish microflora. *Ind. J. Exp. Biol. 9*, 45-47.

Lewis, N. F., Alur, M.D. and Kumta, U.S. 1974. Role of carotenoid pigments in radio-resistant *Micrococcus. Can. J. Microbiol. 20*, 455-459.

Loharanu, P. 1994. Cost/benefit aspects of Food Irradiation. *Food Technol., 48*, 104-108.

Mr Colloch, L.P., Cook, H.T. and Wright, W.R. 1968 . *Market Diseases of tomatoes, peppers and egg plants. Agricultural Handbook No. 28*, Washington D. C. Agricultural Research Service, USDA.

Merritt, N.P., Angelini, E., Wierbicks, G.W. and Shults, G. W. 1975. Chemical Changes associated with flavour in irradiated meat. *J. Agric. Food Chem. 23*, 1037-1041.

Muller, R. and Lloyd, R. 1994. *Sub-lethal and chronic effects of pollutants on fresh water fish Fishing* News Books, Blackie Sci. Publ. Oxford.

Nawar, W.W. 1983. Comparison of chemical consequences of heat and irradiation treatment of lipids. In : *Recent Advances in Food Irradiation*. pp. 115-127. P.S. Elias and A. J. Cohen, eds. Elsevier, Amsterdam.

Nene, S.P., Vakil, U.K. and Sreenivasan, A. 1975a. Effects of gamma irradiation on red gran (*Cajanus cajan*) proteins. *J. Food Sci. 40*, 815-819.

Nene, S.P., Vakil, U.K. and Sreenivasan, A. 1975b. Effect of gamma-irradiation on some physico-chemical characteristics of red gram starch. *J. Food Sci. 40*, 943-947.

Pauli, G.M. 1995. *Evaluating the safety of irradiated foods* chapter 8, pp. 89-100. Published by American Chemical Society.

Pendharkar, M.B. and Nair, P.M. 1975. Induction of phenylalanine ammonia - lyase in gamma-irradiated potatoes. *Radiation Botany, 15*, 191-197.

Rosenthal, I. 1992. *Electro-megnetic Radiations in Foods Science* Springer - Verlag, New York.

Roy, M.K. and Prasad, H. H. 1993. Gamma radiation in the control of important storage pests of three grain legumes. *J. Food Sci. Technol., 30*, 275-278.

Ruivo M. (ed) 1972. *Marine Pollution and Sealife Fishing News Books*, Oxford, England.

Sabularse, V.C., Liuzzo, J.A., Rao, R.M. and Grodner, R.M. 1992. Physico-chemical characteristics of brown rice as influenced by gamma radiation. J. Food Sci. 57, 143-145.

Saint-Lebe, L, Henon, Y. and Thery, V. 1985. Le Traitment Ionisant Des Produits seis et Deshydrates Cas des plants medicinales a infusion. In : *Proceedings of food irradiation processing*. pp. 9-16, International Atomic Energy Agency, Vienna.

Shay, B. J. , Egan, A. F. and Wills, P. A. 1988. The use of irradiation for extending the storage life of fresh processed meats. *Food Technology in Australia, 40*, 310-313.

Simic, M. G., Neta, P. and Heyon, E. 1969. Pulse radiolysis study of alcohols in aqueous solution. *J. Physical Chem. 73*, 3794-3800.

Simic, M. G. 1983. Radiation chemistry of water-soluble food components. In : *Preservation of Food by Ionizing Radiation*. Vol. 2, eds. Josephson, E. S. and Peterson, M. S., pp. 1-73, CRC Press Boca Raton FL.

Simic, M. G., Gajewski, E. and Dizdanglu, M. 1985. Kinetics and mechanisms of hydroxyl radical-induced crosslinks between phenylalanine peptides. Radiat. Phys. Chem. 24, 465-473.

Singh, L., Mohan, M. S., Desai, S.B.P., Sankaran. R. and Sharma, T. R. 1988. The use of gamma irradiation for improving microbiological qualities of spices. *J. Food Sci. Technol. 25*, 357-360

Srinivas, M., Ananthaswami, H. N., Vakil, U.K. and Sreenivasan, A. 1972. Effects of gamma-radiation on wheat proteins. *J. Food Sci, 37*, 715-718.

Sutton, H.C. 1969. Radiation and food preservation Food Technology in New Zealand 4 (3) March, 1969.

Thakur, B.R. and Singh, R.K. 1994. Food Irradiation-Chemistry and applications. *Food Reviews International*, 10, 437-473

Thayer, D.W., Lachica, R.V., Huhtanen C.N. and Wierbicki, E. 1086. Use of irradiation to ensure the microbiological safety of processed meats. *Food technol. 40*, 159-162.

Thomas, P. and Joshi, M.R. 1977. Prevention of after-cooking darkening of irradiated potatoes. *Potato Res. 20*, 77-84.

Thomas, P. Srirangarajan, A. N. and Limaye, S. P. 1975. Studies on sprout inhibition of onions by gamma irradiation 1. Influence of time interval between harvesting and irradiation, radiation dose, and environmental conditions on sprouting. *Rad. Bot., 15*, 215-222.

Thomas, P. 1984. Radiation preservation of foods of plant origin. Part 2. Onion and other bulk crops. *Crit. Rev. Food Sci. Nutr. 21*, 95-136.

Thomas, P. 1986. Radiation preservation of foods of plant origin. Part 3. Tropical Fruits, bananas, mangoes and papays. *Crit. Rev. Food Sci. and Nutr. 23*, 147-205.

Tidmars W. G. 1985. *Taintings in Fishery Resources*, Department of Fisheries and Oceans, Ottawa

Tilton, E. W. and Burditt, A. K. 1983. Insect disinfestation of grains and fruits. In: *Preservation of Foods by Ionizing Radiation*. pp. 215-229, Eds. Josephson, E. S. and Peterson, M. S., CRC Press, Boca Raton, Fl.

Tobback, P. P. 1977. Radiation chemistry of vitamins. In : *Radiation Chemistry of Major Food Components*. Eds. Elias, P. S. and Cohen, A. J. pp. 187-220, Elsevier, Amsterdam.

Tsuji, K. 1983. Low-dose cobalt-60 irradiation for reduction of microbial contamination in raw materials for animal health products. *Food Technol., 37*, 48-52.

Urbain, W. M. 1989. *Food Irradiation*. Acad. Press, Orlando, Fl.

WHO, 1981. *Wholesomeness of irradiated foods*. WHO Tech. Rep. Ser. 659, Geneva.

WHO, 1992. *Review of the Safety and Nutritional Adequacy of irradiated Foods*. WHO/ HPP/FOS/922, World Health Organization, Geneva.

INDEX